Aptamer Generation and Bioapplication

Aptamer Generation and Bioapplication

Editors

Sai Wang
Wei Wu
Long Wu

MDPI

Basel • Beijing • Wuhan • Barcelona • Belgrade • Novi Sad • Cluj • Manchester

Editors

Sai Wang
College of Food Science
and Engineering
Ocean University of China
Qingdao
China

Wei Wu
College of Food Science
and Engineering
Qingdao Agricultural University
Qingdao
China

Long Wu
College of Food Science
and Engineering
Hainan University
Haikou
China

Editorial Office
MDPI
St. Alban-Anlage 66
4052 Basel, Switzerland

This is a reprint of articles from the Special Issue published online in the open access journal *Molecules* (ISSN 1420-3049) (available at: www.mdpi.com/journal/molecules/special_issues/Aptamer_Generation).

For citation purposes, cite each article independently as indicated on the article page online and as indicated below:

Lastname, A.A.; Lastname, B.B. Article Title. *Journal Name* **Year**, *Volume Number*, Page Range.

ISBN 978-3-7258-1376-6 (Hbk)
ISBN 978-3-7258-1375-9 (PDF)
doi.org/10.3390/books978-3-7258-1375-9

Contents

molecules

Article

The Ability of Nuclease-Resistant RNA Aptamer against *Streptococcus suis* Serotype 2, Strain P1/7 to Reduce Biofilm Formation In Vitro

Apinyapat Matchawong [1], Chatchawan Srisawat [2], Sirikwan Sangboonruang [1,3] and Chayada Sitthidet Tharinjaroen [1,3,*]

1 Division of Clinical Microbiology, Faculty of Associated Medical Sciences, Chiang Mai University, Chiang Mai 50200, Thailand; apinyapat_m@hotmail.com (A.M.); sirikwan.sang@cmu.ac.th (S.S.)
2 Department of Biochemistry, Faculty of Medicine, Siriraj Hospital, Mahidol University, Bangkok 10700, Thailand; chatchawan.sri@mahidol.ac.th
3 Infectious Diseases Research Unit (IDRU), Faculty of Associated Medical Sciences, Chiang Mai University, Chiang Mai 50200, Thailand
* Correspondence: chayada.si@cmu.ac.th; Tel.: +66-539-35068 (ext. 19)

Abstract: *Streptococcus suis*, a Gram-positive bacterium, is an important swine and human pathogen, with serotype 2 being the most prevalent strain found worldwide. Deafness, meningitis, and death (in severe cases) are observed in *S. suis*-infected cases. Development of the ligands that can bind to *S. suis* with high affinity and specificity could be beneficial for the diagnosis and treatment of *S. suis* infection. Herein, the nuclease-resistant RNA aptamers based on 2′-fluoropyrimidine modification against *S. suis* serotype 2, strain P1/7, were established using the cell- Systematic Evolution of Ligands by Exponential enrichment (SELEX) technique. One of the aptamers, R8-su12, could bind to the *S. suis* target strain as well as other *S. suis* serotypes, i.e., 1, 1/2, 9, and 14, but not to other bacteria tested, i.e., *S. pneumoniae* ATCC 49619, *Staphylococcus aureus* ATCC 25923, *Escherichia coli* ATCC 25922, and *Pseudomonas aeruginosa* ATCC 27853. Moreover, the R8-su12 RNA aptamer was also capable of inhibiting the biofilm formation of the *S. suis* target strain, making it potentially useful for the study of biofilm formation and the treatment of *S. suis* infection in humans and pigs in the future.

Keywords: RNA aptamer; nuclease-resistant aptamer; *Streptococcus suis*; biofilm formation; *S. suis* infection

Citation: Matchawong, A.; Srisawat, C.; Sangboonruang, S.; Tharinjaroen, C.S. The Ability of Nuclease-Resistant RNA Aptamer against *Streptococcus suis* Serotype 2, Strain P1/7 to Reduce Biofilm Formation In Vitro. *Molecules* **2022**, *27*, 3894. https://doi.org/10.3390/molecules27123894

Academic Editors: Sai Wang, Wei Wu and Long Wu

Received: 9 May 2022
Accepted: 14 June 2022
Published: 17 June 2022

1. Introduction

Streptococcus suis is a zoonotic pathogen causing infectious diseases in pigs and humans such as meningitis, pericarditis, and septic shock [1,2]. To date, *S. suis* has been classified into 29 serotypes (SS) based on the specific antigens that appear on the polysaccharide capsule [3,4]. *S. suis* SS 2 infection was the most prevalent human case in Thailand, especially in the northern region [5,6]. Conventional methods for the identification of *S. suis* are culture and biochemical tests [7], which are time-consuming and sometimes yield controversial results as other bacteria such as *S. pneumoniae*, *S. bovis*, enterococci, viridans streptococci or even *Listeria monocytogenes* [8]. This can lead to a delayed or incorrect diagnosis. One of the important virulence characteristics of *S. suis* is biofilm formation [9,10], which allows for the bacteria to colonize permanently, increases resistance to the host immune system and antibiotics, and promotes their survival and proliferation [11]. An understanding of the biofilm formation involved in *S. suis* pathogenesis is needed for the effective control of *S. suis* infection in humans and animals.

Different types of ligands could be generated to bind *S. suis* with high affinity and specificity for research, diagnostic, and therapeutic purposes. One example is aptamers, which are single-stranded DNA or RNA [12]. The aptamers are generated using the technique called "Systematic Evolution of Ligands by EXponential enrichment" or SELEX, originally established in 1990 by Ellington and Tuerk [13,14]. In this study, the *S. suis*-specific RNA aptamers were developed because RNA has a greater ability to fold into more diverse three-dimensional conformations, which are capable of binding to target molecules, compared with DNA [15,16]. In addition, the RNA aptamers can be modified to become more resistant to the nucleases that could interfere with their functions. One of the modification methods is the substitution of the 2′-hydroxyl group of ribose sugar with the fluoro (-F), amino (-NH_2), or methoxy (-OMe) groups [16,17].

In this study, nuclease-resistant RNA aptamers against *S. suis* SS 2, type strain P1/7 were successfully generated using the whole-cell SELEX approach. One of them, the R8-su12 RNA aptamer, could specifically recognize *S. suis* and inhibit biofilm formation, making it potentially useful for the study of biofilm formation and the treatment of *S. suis* infections in humans and pigs in the future.

2. Results

2.1. Nuclease-Resistant RNA Aptamer Selection Based on the Cell-SELEX Technique

To select the nuclease-resistant RNA aptamers specific to *S. suis* SS 2, P1/7, approximately 10^{14} random RNA library pool was produced. A total of 10 mg/mL baker's yeast tRNA was also added to the library pool as a binding competitor to reduce non-specific RNA binding. Then, the random RNA library mixture was incubated with 10^7 live *S. suis* SS 2, P1/7 cells. The cell-SELEX was performed with a total of eight rounds. The enrichment of aptamer pool during selection rounds 4 (R4), 6 (R6), and 8 (R8) were measured using real-time reverse transcription (RT) qPCR compared with the random RNA library (Lib). The fold change in the relative binding affinity of RNA aptamer was increased in the later round of cell-SELEX (data not shown). The latest R8 RNA aptamer pool was used to determine the relative binding affinity compared to the starting N40 Lib. After calculation, the % binding of R8 RNA aptamer pool was higher than % binding of Lib with statistical significance ($p = 0.023$), implying the enrichment of the R8 RNA aptamer pool against *S. suis* SS 2, P1/7 (Figure 1A). The 8th round cell-SELEX pool was also screened for specificity using target and non-target cells, which were *S. pneumoniae* and *S. pyogenes*. The R8 RNA aptamer pool tested with *S. suis* SS 2, P1/7 showed the highest % binding when compared to the testing against *S. pneumoniae* and *S. pyogenes*, with statistical significance ($p = 0.034$ and $p = 0.034$, respectively) (Figure 1B). These results suggested that the R8 RNA aptamer pool was enriched and able to bind to *S. suis* SS 2, P1/7.

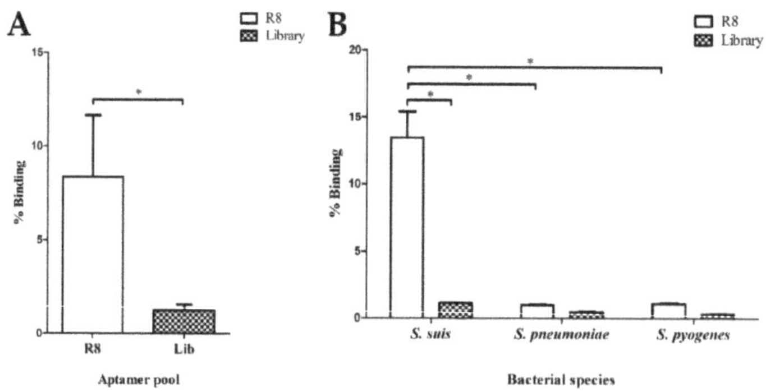

Figure 1. Characterization of RNA aptamer pool. (**A**) Enrichment of RNA aptamer pool derived from round eighth (R8) of cell-SELEX assay. The % binding of R8 RNA aptamer pool compared to the library (Lib) was presented. (**B**) Specificity of R8 RNA aptamer pool. The R8 RNA aptamer pool and random RNA Lib were tested with *S. suis*, *S. pneumoniae*, and *S. pyogenes*. * indicates the statistically significant difference using a Mann–Whitney U test ($p < 0.05$).

Therefore, the aptamer pool from R8 was cloned and sequenced. According to the sequencing analysis of 26 clones, they can be classified into four distinct groups as shown in Table 1. In the first group contained seven clones of short 40-nt long, calculated as 26.9%. The second group was also 40-nt long, with two corresponding sequences (7.7%). The third group consisted of four clones (15.4%), which were 39-nt long. The last group, named "ungrouped", comprised of 13 different clones with various sequences, counting as 50% of the total number of all sequences (Table S1, Supplementary Materials).

Table 1. Summarization of consensus randomized sequences of the RNA R8 aptamer pool.

Group	Representative Clone	Consensus Randomized Sequences	Nucleotides	Frequency (%)
1	R8-su12	CAUACUGAGUAAGAUCGGAAAUUUCGGUGUAAGGCCACGG	40	7/26 (26.9%)
2	R8-su057	UGGAUGUAUGGAACUUGCAGAUCUUAACUGCACGAAGCGU	40	2/26 (7.7%)
3	R8-su15	ACACGUUGCUGAAACAUACCGAGUAACAUAAAGCGGGUG	39	4/26 (15.4%)
4	Ungrouped	Different clones with various sequences	40	13/26 (50%)

Thus, group 1 showed the highest consensus sequences frequency among the 26 clones, suggesting that they might have a high affinity toward *S. suis* SS2, P 1/7. The representative R8-su12 RNA aptamer from group 1 was then selected for the prediction of secondary structures; the folding temperature was set at 37 °C. The R8-su12 RNA aptamer had one predicted secondary structure with a folding energy (initial ΔG) equal to −24.30 kcal/mol. The predicted secondary structure consists of an external loop with four bases and one closing helices. Moreover, loops and stems were also found, as shown in Figure 2. Hence, the R8-su12 RNA aptamer was chosen for further characterization.

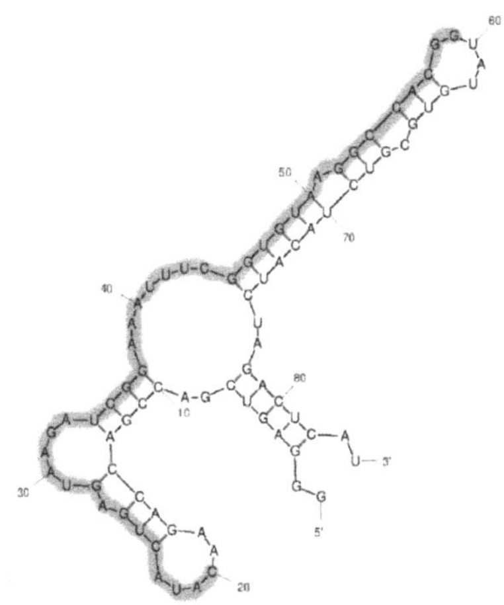

Figure 2. Predicted secondary structure of the R8-su12 RNA aptamer. The randomized region is 40-nt long and shown in the shaded area.

2.2. Specificity of R8-su12 RNA Aptamer

To test the specificity of the R8-su12 RNA aptamer, the target *S. suis* SS 2, P1/7 and non-target cells including the American Type Culture Collection (ATCC) strains of *S. pneumoniae*, *S. aureus*, *E. coli*, and *P. aeruginosa*, were evaluated. The negative control—L2 RNA aptamer (5′-GGGAGUCGACCGACCAGAAGAAGCUGCUUCAAAAUAGAUCUACACUCACAU AAGGCGAAUAUGUGCGUCUACAUCUAGACUCAU-3′) was used in this experiment, instead of a random RNA Lib. The R8-su12 and L2 RNA aptamer were incubated with target and non-target cells. As shown in Figure 3A, the R8-su12 RNA aptamer show statistically significant higher binding affinities to *S. suis* SS 2, P1/7 when compared with *S. pneumoniae*, *S. aureus*, *E. coli*, and *P. aeruginosa*. (p = 0.021, 0.020, 0.020, and 0.021, respectively).

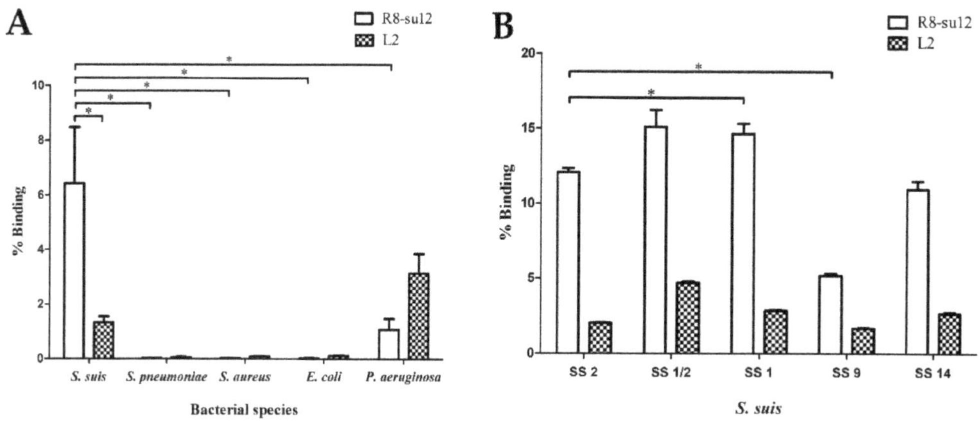

Figure 3. Specificity of the R8-su12 RNA aptamer. (**A**) The specificity of R8-su12 RNA aptamer was tested using target *S. suis* SS 2, P1/7 and non-target cells. (**B**) The specificity of R8-su12 RNA aptamer was tested using target *S. suis* SS 2, P1/7 and other *S. suis* serotypes. * indicates a statistically significant difference using a Mann–Whitney U test ($p < 0.05$).

Moreover, the R8-su12 RNA aptamer was also tested with *S. suis* SS 1/2, 1, 9, and 14. These strains are closely related to the target cells and showed a high prevalence of *S. suis* infection strains in Thailand [18]. The R8-su12 RNA aptamer showed an ability to bind to these *S. suis* strains. Interestingly, the R8-su12 RNA aptamer showed the highest % binding when tested with *S. suis* SS 1/2, followed by *S. suis* SS 1, 2, 14, and 9, respectively (Figure 3B). In addition, the R8-su12 RNA aptamer showed a significant higher binding affinity against *S. suis* SS 1 (p = 0.021) compared to the target cells. On the other hand, a lower binding affinity against *S. suis* SS 9 (p = 0.021) was shown. No significant binding was found in *S. suis* SS 1/2 (p = 0.275) and *S. suis* SS 14 (p = 0.127) (Figure 3B).

2.3. The Effect of R8-su12 RNA Aptamer on S. suis SS 2, P1/7 Biofilm Formation

Biofilm formation is associated with the virulence of *S. suis* [11] and one biofilm-production-related molecule—polysaccharide intercellular adhesin (PIA) is located on the cell surface [19]. Hence, we hypothesized that the derived aptamer that targeted the surface molecules on *S. suis* cells might affect the biofilm formation. As shown in the result, the biofilm formation of *S. suis* SS 2, P1/7 cultured with R8-su12 RNA aptamer, L2 RNA aptamer and baker's yeast tRNA was significantly reduced by 61.2%, 12.8% and 12.9 %, respectively, when compared to control ($p < 0.05$). Noticeably, the influential activity that decreased biofilm formation was also observed in R8-su12 RNA aptamer treated *S. suis* SS 2, P1/7, showing 55.5% and 55.4% reductions when compared to L2 RNA aptamer ($p < 0.001$) and baker's yeast tRNA ($p < 0.001$), respectively. While the presence of L2 RNA aptamer and baker's yeast tRNA in the culture medium, biofilm formation showed no statistically significant difference (p = 0.988) (Figure 4). These results implied that the R8-su12 RNA aptamer can reduce the production of *S. suis* SS 2, P1/7 biofilm formation.

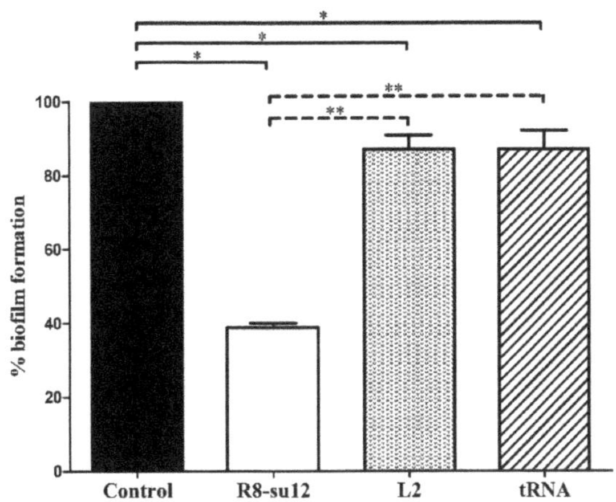

Figure 4. Reduction in biofilm formation by R8-su12 RNA aptamer. The *S. suis* SS 2, P1/7 was cultured in BHI broth with supplements (control) and in the presence of R8-su12 RNA aptamer (R8-su12), L2 RNA aptamer (L2), and baker's yeast tRNA (tRNA). After 3 days of incubation without shaking, production of biofilm was determined using crystal violet biofilm assay. Percentage of *S. suis* SS 2, P1/7 biofilm formation was calculated. * indicates $p < 0.05$ and ** indicates $p < 0.001$ using the one-way ANOVA test.

3. Discussion

S. suis is reported as a zoonotic pathogen causing infectious diseases in pigs and humans, such as meningitis, pericarditis, and septic shock [1,2]. *S. suis* serotype 2 infection is the most prevalent human case in the world, including Thailand, especially in the northern region [5,6]. Nowadays, no vaccine is available and rapid disease progression is of great concern. Thus, the prevention and control of *S. suis* infection is needed. The development of ligands that can bind to *S. suis* with high affinity and specificity could be beneficial for the diagnosis and treatment of *S. suis* infection.

In this study, the nuclease-resistant RNA aptamer, R8-su12, against *S. suis* SS 2, P1/7 was first established. RNA aptamers were chosen over DNA aptamers for their ability to fold into a more diverse, three-dimensional structures [15,16]. Moreover, they can be generated with 2′-fluoropyrimidines (2′-fluoro-dCTP and 2′-fluoro-dUTP) substitution to improve their stability and resistance to nucleases [20–22]. The whole cell-SELEX approach was used because the identification and production of unique proteins as selection targets are difficult, as the molecules on the *S. suis* cell surface show high levels of similarity [23]. This approach would also help to generate the aptamers that target multiple surface antigens, and would not require the isolation and purification of target proteins, thus avoiding the misfolding or denaturation of the target molecules [24]. In addition, we developed the electrophoretic separation technique to partition non-binding RNAs from the aptamers that bound to bacterial cells, and this was used in the last two rounds of selection. Using this technique, non- or weak-binding RNAs were actively separated into the agarose gel under an electric field due to their negative charges, while the aptamers bound with high affinity to *S. suis* remained in the sample well with bacterial cells. Theoretically, the active separation using electrophoresis would remove non-binding RNAs from the target cells more efficiently than the washing separation, which relies on the passive dissociation of RNAs. Therefore, the stringency of selection was increased, resulting in the successful recovery of the high-affinity *S. suis* aptamers.

For the specificity determination, the data showed that the R8-su12 RNA aptamer bound to target cells, *S. suis* SS 2, P1/7, as well as the others *S. suis* serotypes, 1, 1/2, 2, and 14. Unfortunately, the target molecule of the aptamer was not determined in this study. We speculated that the potential targets of the R8-su12 RNA aptamer could be some components of capsular polysaccharides (CPSs) that are shared between these serotypes. There are several studies revealed that *S. suis* serotypes 1, 1/2, 2, and 14 contain sialic acid in their CPSs [9,25,26]. Therefore, this might be one of the candidate molecules to which R8-su12 RNA aptamer can bind. However, it would be helpful to identify the potential targets of this aptamer in further studies.

One of the important virulence characteristics of *S. suis* is biofilm formation. This leads to an increased resistance to host defenses and antimicrobial agents, making pathogens difficult or impossible to eliminate [10]. Therefore, the R8-su12 RNA aptamer was further tested for its effects on *S. suis* biofilm formation. We found that the derived aptamer can reduce the biofilm formation of *S. suis* SS 2, P1/7. The studies on the bacteria biofilm revealed that the regulatory mechanism can be divided into eight categories, depending on the different communication signaling molecules produced by bacteria [20]. These consist of: (1) bacterial death and dissolution mechanism, which plays an important role in the formation and development of bacterial biofilms; (2) extracellular polymeric substance matrix (EPS) that consists of some extracellular polysaccharides, DNA, proteins, and some other macromolecules; (3) two-component systems (TCS), which mainly involve receptor histidine protein kinases (HPK) and response regulators (RR); (4) extracytoplasmic function (ECF) signaling pathway that regulate the production of alginate, which is important for the biofilm structure; (5) intracellular second messenger cyclic diguanylate (c-di-GMP), which can induce the production of extracellular polymeric substances (EPS) and surface adhesins, leading to the formation of bacterial biofilms; (6) small RNAs (sRNAs), which are involved in the regulation of expression or activity of important transcriptional regulators and components required for cell attachment and biofilm formation; (7) bacterial biofilm intracellular signal transduction system called bacterial quorum sensing (QS) system; (8) the polysaccharide intercellular adhesin (PIA) which plays an important role in the bacterial aggregation stage of bacterial biofilm formation. Moreover, the study by Zhou [27] revealed that the reduction in O-acetylserine (thiol)-lyase B (CysM), a key enzymatic regulator of cysteine synthesis, can inhibit *S. suis* biofilm formation. The study also showed that the ability of the complementary mutant (CΔcysM) strain to form a biofilm may indirectly related to the QS system. It is possible that the R8-su12 RNA aptamer might bind to those molecules, indirectly or directly playing a role in biofilm formation. However, additional studies are required to investigate how the aptamer affects biofilm formation.

4. Materials and Methods

4.1. Bacteria Used and Culture Condition

The *Streptococcus suis* serotype 2, type strain P1/7 (*S. suis* SS 2, P1/7) (kindly provided by Division of Bacteriology, Department of Microbiology, Faculty of Medicine, Chiang Mai University) was used as the target cell in this study. For specificity testing, other bacterial species, including *Escherichia coli* ATCC 25922, *Pseudomonas aeruginosa* ATCC 27853, *Staphylococcus aureus* ATCC 25923, *Streptococcus pneumoniae* ATCC 49619, *S. pyogenes* ATCC 19615, *S. suis* serotype 1 DMST, *S. suis* serotype 1/2 DMST, *S. suis* serotype 14 DMST, and *S. suis* serotype 9 clinical isolates were used. *S. suis* SS 2, P1/7 was cultured onto 5% blood agar (BA) (Becton Dickinson, NJ, USA) and incubated at 37 °C, 5% CO_2 for 18–24 h. For other bacterial species were incubated overnight for 18–24 h at 37 °C, 5% CO_2 on 5% BA (Becton Dickinson, NJ, USA) or trypticase soy agar (Becton Dickinson, NJ, USA) depending on their optimal medium. To prepare the target cells for the SELEX process, fresh colonies from BA (Becton Dickinson, NJ, USA) were harvested and washed thrice with 1 mL of Phosphate Buffer Saline (PBS). Then, the bacterial cells were adjusted to reach at the desired OD_{600} with a total of 1 mL of binding buffer before being loaded into a centrifugal filter (Ultrafree®-MC with Microporous Durapore® PVDF Membrane, pore size 0.45 μm, Merck, NJ, USA).

4.2. Cell-SELEX Technique

The starting 2′-F pyrimidine modified RNA Lib was prepared by in vitro transcription of a N40 DNA Lib containing a 40-nt random region and primer binding sites (5′-AGTAATACGACTCACTATAGGGAGTCGACCGACCAGAA-N40-TATGTGCGTCTACAT CTAGACTCAT-3′) using T7 R&D polymerase (50 U/μL) (Lucigen Corporation, WI, USA).

To obtain the practical diversity of the Lib about 10^{14} molecules, 100 pmol of random RNA Lib was used in the first round of cell-SELEX. The 2′-F random RNA Lib was thermally equilibrated for proper folding in a total volume of 100 μL binding buffer (0.05 M HEPES pH 7.4, 0.1 M NaCl, and 0.01 M $MgCl_2$) with 10 mg/mL baker's yeast tRNA (Merck, NJ, USA) at 65 °C, followed by a cool-down at room temperature for 5 min. The freshly prepared 10^7 cells of *S. suis* SS 2, P1/7 were incubated with the equilibrated random RNA Lib at room temperature in the centrifugal filter for 60 min in rounds 1–7 and 50 min in round 8, with gentle rotation. The unbound or weakly bound aptamers were partitioned by washing in binding buffer for various specific volumes, numbers of washes, and incubation times on the rotator according to the selection protocol (Table 2). For rounds 7 and 8, the separation was performed using an electrophoresis technique to increase the selection stringency. In brief, after the binding step, the bacterial pellet was washed once with 100 μL of binding buffer for 1 min, and then resuspended with 50 μL of binding buffer. The suspension was transferred into a well of agarose gel (3% in 0.5× TAE). After that, 0.5× TAE was carefully added into an electrophoresis chamber. Gel electrophoresis was run using 100 V for 10 min. The target cells with bound aptamers were collected and washed thrice with 100 μL of binding buffer for 1 min each time before the elution step.

Table 2. The aptamer selection protocol for cell-SELEX using *S. suis* SS 2, P1/7 as target cells.

SELEX Round	Input RNA (pmol)	*S. suis* SS 2, P1/7 (Cells)	Washing
1	100	1.03×10^7	100 μL × 5 times, 1 min each
2	50	1.02×10^7	100 μL × 5 times, 1 min each
3	50	1.00×10^7	100 μL × 5 times, 3 min each
4	25	1.11×10^7	100 μL × 5 times, 3 min each
5	10	1.21×10^7	100 μL × 5 times, 3 min each
6	10	1.29×10^7	100 μL × 5 times, 5 min each
7	10	1.01×10^7	100 μL for 1 min, followed by electrophoretic separation
8	10	1.07×10^7	100 μL for 1 min, followed by electrophoretic separation

The bound RNAs were eluted using 100 μL of elution buffer (8 M urea, 5 mM EDTA pH 8.0), followed by ethanol precipitation. The RNA pool was reverse-transcribed using Superscript® III Reverse Transcriptase (Thermo Fisher Scientific, Waltham, MA, USA) and 3′-N40 primer (5′-ATGAGTCTA GATGTAGACGCACATA-3′). Then, the cDNA was amplified with *Taq* DNA polymerase (Thermo Fisher Scientific, Waltham, MA, USA) using 3′-N40 primer and 5′-primer containing T7 promoter (underlined) (5′-AG<u>TAATACGACTCACTATAG</u> <u>GG</u>AGTCGACCGACCAGAA-3′). The PCR products were transcribed in vitro, purified using the Qiaquick PCR Purification Kit (Qiagen, Hilden, Germany), and used for the next round of cell-SELEX. The cell-SELEX was performed until the 8th round (R8).

4.3. Evaluation of Relative Binding Affinity and Specificity of the RNA Aptamer

The enrichment of the aptamer pool in selected rounds (Rx), which were R4, R6, and R8, was quantified by relative binding affinity using the binding assay. One hundred picomoles of either the control RNA Lib or selected aptamer pool in the presence of baker's yeast tRNA was incubated with approximately 10^7 cells of *S. suis* SS 2, P1/7. Both Lib and Rx were saved for approximately 10 μL as input RNA. After binding, washing, and elution steps, as described above, the amount of input and bound RNAs were quantified using real-time RT qPCR. The real-time master mix was generated in a total 10 μL per reaction comprising the final concentration of 1× SensiFAST™ SYBR® No-ROX Kit (Meridian Bioscience, TN, USA),

0.5 μM of each 3′-N40 primer and 5′-primer (5′-GGGAGTCGACCGACCAGAA-3′), and 0.5 μL of cDNA from each RT reaction (Lib input, Lib elute, Rx input, Rx elute, NTC). Each sample was performed in triplicate. For non-template control (NTC), real-time PCR master mix and cDNA from NTC-RT reaction were used. The real-time RT qPCR was performed under the following conditions: 94 °C for 30 s, 60 °C for 30 s, 72 °C for 30 s, with a total of 45 cycles. The fluorescence signal was read at 60 °C and the association curve was set at 65 °C to 95 °C. The binding percentage was calculated using Formula (1):

$$\%\ \text{binding} = 2 - (\text{Ct}_{\text{elute}} - \text{Ct}_{\text{input}}), \tag{1}$$

where Ct_{elute} refers to the cycle threshold of elute sample and Ct_{input} refers to the cycle threshold of input sample. The % binding of random RNA library (% $\text{binding}_{\text{Lib}}$) and the % binding of Rx (% $\text{binding}_{\text{Rx}}$) were then calculated into the fold change using Formula (2):

$$\text{Fold change} = \%\ \text{binding}_{\text{Rx}} / \%\ \text{binding}_{\text{Lib}}, \tag{2}$$

For specificity testing, the R8 aptamer pool and representative R8-su12 RNA aptamers were screened toward *S. suis* SS 2, P1/7 and non-target cells, including other species of *Streptococcus* (*S. pyogenes, S. pneumoniae*), Gram-positive bacteria (*S. aureus*), Gram-negative bacteria (*E. coli, P. aeruginosa*), and other *S. suis* serotype (*S. suis* SS1/2, 1, 9, and 14). One hundred picomole of the control (random RNA Lib pool or L2 aptamer) and the derived nuclease-resistant RNA aptamers were incubated with 10^7 of target cells, *S. suis* SS 2, P1/7 and non-target cells. Then, the cell-SELEX was processed until the eluate was obtained. After that, it was used as a template in the binding assay using real-time RT qPCR. The binding percentage and fold change were calculated as described above.

4.4. Aptamer Characterization

The aptamer pool from cell-SELEX round 8 was cloned using pTG19-T PCR cloning vector B (Vivantis, Selangor, Malaysia). Twenty-six clones were sequenced and analyzed for similarity using the Clustal Omega Multiple Sequence Alignment (https://www.ebi.ac.uk/Tools/msa/clustalo/ accessed on 11 November 2020). The selected RNA aptamer was also predicted for secondary structure using the mfold Web Server (http://www.unafold.org/mfold/applications/rna-folding-form.php accessed on 7 February 2021) to determine the binding sites between the aptamer and target cells.

4.5. Biofilm Formation Assay

To determine the effect of the RNA aptamer on the biofilm formation of *S. suis* SS 2, P1/7, the selected R8-su12 RNA aptamer was tested. The determination of *S. suis* SS 2, P1/7 biofilm formation was preliminarily tested based on the crystal violet biofilm assay [28].

To examine the RNA aptamer's ability to inhibit biofilm formation, the detection of biofilm formation was carried out according to Meng et al. [28] with some modifications. In brief, 200 μL of *S. suis* SS 2, P1/7 from the overnight culture at 37 °C was mixed with 100 μL of 4.76 ng/μL (176 nM) of R8-su12 RNA aptamer with 10 mg/mL baker's yeast tRNA, negative control L2 RNA aptamer with 10 mg/mL baker's yeast tRNA, and 10 mg/mL baker's yeast tRNA alone. Then, the mixture was transferred into a 24-well polystyrene plate containing 1.8 mL of BHI broth. The plates were incubated at 37 °C for 3 days without shaking. After staining with 0.1% crystal violet for 30 min, the OD_{600} was measured. The BHI broth with the supplements was performed as a negative control, and its OD_{600} was set as blank. These blank absorbance values and cut off values (ODc) were used to calculate and interpret the biofilm formation result as described [28]. Each sample in the biofilm formation assay was performed in quadruplicate in three independent experiments.

4.6. Statistically Analysis

All samples were tested in triplicate with at least three independent experiments. The one-way analysis of variance and Mann–Whitney U test were used to determine the statistically significant when $p < 0.05$ or 0.001. All the calculations were performed using the SPSS Statistics 6.0 software (IBM Corp., Armonk, NY, USA).

5. Conclusions

Herein, the nuclease resistant RNA aptamer specific to *S. suis* SS 2, strain P1/7, R8-su12, was successfully constructed. Moreover, the derived RNA aptamer revealed the ability to bind to other pathogenic *S. suis* (serotype 1/2, 1, 9, and 14). Remarkably, the R8-su12 RNA aptamer can significantly reduce the biofilm formation, one of the *S. suis* virulence factors. However, the target ligand of the aptamer on the *S. suis* SS 2, strain P1/7 cell surface would be further identified. This might be useful for understanding the pathogenesis, leading to an early diagnostic, treatment, and effective control of *S. suis* infection.

Supplementary Materials: The following supporting information can be downloaded at: https://www.mdpi.com/article/10.3390/molecules27123894/s1, Table S1: The sequencing analysis of the RNA R8 aptamer pool.

Author Contributions: Conceptualization, C.S.T.; methodology, A.M., C.S. and S.S.; formal analysis, C.S.; investigation, A.M.; resources, C.S.T, C.S. and S.S.; data curation, A.M. and C.S.T.; writing—original draft preparation, A.M.; writing—review and editing, C.S.T. and C.S.; funding acquisition, C.S.T. All authors have read and agreed to the published version of the manuscript.

Funding: This study was supported by scholarships from the Department of Medical Technology, Faculty of Associated Medical Sciences, Chiang Mai University, and the National Research Council of Thailand (NRCT), Ministry of Higher Education, Science, Research and Innovation, Thailand with grant no. 517803.

Institutional Review Board Statement: Not applicable.

Informed Consent Statement: Not applicable.

Data Availability Statement: Not applicable.

Acknowledgments: We thank Prasit Tharavichitkul, Division of Bacteriology, Department of Microbiology, Faculty of Medicine, Chiang Mai University, for providing the *S. suis* serotype 2, type strain P1/7. We thank Theeraphong Pho-iam Siriraj, Genomics, Office of the Dean, Faculty of Medicine Siriraj Hospital, Mahidol University for providing some reagents.

Conflicts of Interest: The authors declare no conflict of interest.

Sample Availability: Samples of the compounds are not available from the authors.

References

1. Staats, J.J.; Feder, I.; Okwumabua, O.; Chengappa, M.M. *Streptococcus Suis*: Past and present. *Vet. Res. Commun.* **1997**, *21*, 381–407. [CrossRef] [PubMed]
2. Tang, J.; Wang, C.; Feng, Y.; Yang, W.; Song, H.; Chen, Z.; Yu, H.; Pan, X.; Zhou, X.; Wang, H.; et al. Streptococcal toxic shock syndrome caused by *Streptococcus suis* serotype 2. *PLoS Med.* **2006**, *3*, e151. [CrossRef] [PubMed]
3. Okura, M.; Osaki, M.; Nomoto, R.; Arai, S.; Osawa, R.; Sekizaki, T.; Takamatsu, D. Current taxonomical situation of *Streptococcus suis*. *Pathogens* **2016**, *5*, 45. [CrossRef]
4. Dutkiewicz, J.; Sroka, J.; Zając, V.; Wasiński, B.; Cisak, E.; Sawczyn, A.; Kloc, A.; Wójcik-Fatla, A. *Streptococcus suis*: A re-emerging pathogen associated with occupational exposure to pigs or pork products. part I—epidemiology. *Ann. Agric. Environ. Med.* **2017**, *24*, 683–695. [CrossRef] [PubMed]
5. Wangkaew, S.; Chaiwarith, R.; Tharavichitkul, P.; Supparatpinyo, K. *Streptococcus suis* infection: A series of 41 cases from Chiang Mai University Hospital. *J. Infect.* **2006**, *52*, 455–460. [CrossRef]
6. Kerdsin, A.; Dejsirilert, S.; Puangpatra, P.; Sripakdee, S.; Chumla, K.; Boonkerd, N.; Polwichai, P.; Tanimura, S.; Takeuchi, D.; Nakayama, T.; et al. Genotypic profile of *Streptococcus suis* serotype 2 and clinical features of infection in humans, Thailand. *Emerg. Infect. Dis.* **2011**, *17*, 835–842. [CrossRef]
7. Hommez, J.; Devriese, L.A.; Henrichsen, J.; Castryck, F. Identification and characterization of *Streptococcus suis*. *Vet. Microbiol.* **1986**, *11*, 349–355. [CrossRef]
8. Goyette-Desjardins, G.; Auger, J.P.; Xu, J.; Segura, M.; Gottschalk, M. *Streptococcus suis*, an important pig pathogen and emerging zoonotic agent—An update on the worldwide distribution based on serotyping and sequence typing. *Emerg. Microbes. Infect.* **2014**, *3*, e45. [CrossRef]

9. Van Calsteren, M.R.; Gagnon, F.; Lacouture, S.; Fittipaldi, N.; Gottschalk, M. Structure determination of *Streptococcus suis* serotype 2 capsular polysaccharide. *Biochem. Cell. Biol.* **2010**, *88*, 513–525. [CrossRef]
10. Donlan, R.M.; Costerton, J.W. Biofilms: Survival mechanisms of clinically relevant microorganisms. *Clin. Microbiol. Rev.* **2002**, *15*, 167–193. [CrossRef]
11. Grenier, D.; Grignon, L.; Gottschalk, M. Characterisation of biofilm formation by a *Streptococcus suis* meningitis isolate. *Vet. J.* **2009**, *179*, 292–295. [CrossRef] [PubMed]
12. Stoltenburg, R.; Reinemann, C.; Strehlitz, B. SELEX—A (r)evolutionary method to generate high-affinity nucleic acid ligands. *Biomol. Eng.* **2007**, *24*, 381–403. [CrossRef] [PubMed]
13. Ellington, A.D.; Szostak, J.W. In vitro selection of RNA molecules that bind specific ligands. *Nature* **1990**, *346*, 818–822. [CrossRef] [PubMed]
14. Tuerk, C.; Gold, L. Systematic evolution of ligands by exponential enrichment: RNA ligands to bacteriophage T4 DNA polymerase. *Science* **1990**, *249*, 505–510. [CrossRef] [PubMed]
15. Darmostuk, M.; Rimpelova, S.; Gbelcova, H.; Ruml, T. Current approaches in SELEX: An update to aptamer selection technology. *Biotechnol. Adv.* **2015**, *33*, 1141–1161. [CrossRef]
16. Shigdar, S.; Macdonald, J.; O'Connor, M.; Wang, T.; Xiang, D.; Al Shamaileh, H.; Qiao, L.; Wei, M.; Zhou, S.F.; Zhu, Y. Aptamers as theranostic agents: Modifications, serum stability and functionalisation. *Sensors* **2013**, *13*, 13624–13637. [CrossRef]
17. Sampson, T. Aptamers and SELEX: The technology. *World Pat. Inf.* **2003**, *25*, 123–129. [CrossRef]
18. Thongkamkoon, P.; Kiatyingangsulee, T.; Gottschalk, M. Serotypes of *Streptococcus suis* isolated from healthy pigs in Phayao Province, Thailand. *BMC Res. Notes* **2017**, *10*, 53. [CrossRef]
19. Wang, Y.; Wang, Y.; Sun, L.; Grenier, D.; Yi, L. *Streptococcus suis* biofilm: Regulation, drug-resistance mechanisms, and disinfection strategies. *Appl. Microbiol. Biotechnol.* **2018**, *102*, 9121–9129. [CrossRef]
20. Eaton, B.E.; Pieken, W.A. Ribonucleosides and RNA. *Annu. Rev. Biochem.* **1995**, *64*, 837–863. [CrossRef]
21. Pieken, W.; Tasset, D.; Janjic, N.; Gold, L.; Kirschenheuter, G.P.; (Inventors); Gilead Sciences Inc.; (Assignee). High Affinity Nucleic Acid Ligand Containing Modified Nucleotides. U.S. Patent US5660985A, 26 August 1997. Available online: https://patents.google.com/patent/US6184364B1/en (accessed on 9 December 2021).
22. Polisky, B.; Jenison, R.D.; Gold, L.; (Inventors); Gilead Sciences Inc.; (Assignee). High-Affinity Nucleic Acid Ligands That Discriminate between Theophylline and Caffeine. U.S. Patent US5580737A, 3 December 1996. Available online: https://patents.google.com/patent/US5580737A/en (accessed on 9 December 2021).
23. Baums, C.G.; Valentin-Weigand, P. Surface-associated and secreted factors of *Streptococcus suis* in epidemiology, pathogenesis and vaccine development. *Anim. Health Res. Rev.* **2009**, *10*, 65–83. [CrossRef] [PubMed]
24. Nosaz, Z.; Rasoulinejad, S.; Mousavi Gargari, S.L. Development of a DNA aptamer to detect *Brucella abortus* and *Brucella melitensis* through cell SELEX. *Iran. J. Vet. Res.* **2020**, *21*, 294–300. [PubMed]
25. Van Calsteren, M.R.; Gagnon, F.; Calzas, C.; Goyette-Desjardins, G.; Okura, M.; Takamatsu, D.; Gottschalk, M.; Segura, M. Structure determination of *Streptococcus suis* serotype 14 capsular polysaccharide. *Biochem. Cell. Biol.* **2013**, *91*, 49–58. [CrossRef] [PubMed]
26. Van Calsteren, M.R.; Goyette-Desjardins, G.; Gagnon, F.; Okura, M.; Takamatsu, D.; Roy, R.; Gottschalk, M.; Segura, M. Explaining the Serological Characteristics of *Streptococcus suis* Serotypes 1 and 1/2 from Their Capsular Polysaccharide Structure and Biosynthesis. *J. Biol. Chem.* **2016**, *291*, 8387–8398. [CrossRef]
27. Zhou, Y.; Yu, F.; Chen, M.; Zhang, Y.; Qu, Q.; Wei, Y.; Xie, C.; Wu, T.; Liu, Y.; Zhang, Z.; et al. Tylosin Inhibits *Streptococcus suis* Biofilm Formation by Interacting With the O-acetylserine (thiol)-lyase B CysM. *Front. Vet. Sci.* **2022**, *8*, 829899. [CrossRef]
28. Meng, X.; Shi, Y.; Ji, W.; Meng, X.; Zhang, J.; Wang, H.; Lu, C.; Sun, J.; Yan, Y. Application of a bacteriophage lysin to disrupt biofilms formed by the animal pathogen *Streptococcus suis*. *Appl. Environ. Microbiol.* **2011**, *77*, 8272–8279. [CrossRef]

molecules

MDPI

Article

A Simple Structure-Switch Aptasensor Using Label-Free Aptamer for Fluorescence Detection of Aflatoxin B1

Chao Wang [1,2], Hao Yu [1,2] and Qiang Zhao [1,2,3,*]

[1] State Key Laboratory of Environmental Chemistry and Ecotoxicology, Research Center for Eco-Environmental Sciences, Chinese Academy of Sciences, Beijing 100085, China; wangchao@lyu.edu.cn (C.W.); tina_3127@163.com (H.Y.)
[2] University of Chinese Academy of Sciences, Beijing 100049, China
[3] School of Environment, Hangzhou Institute for Advanced Study, UCAS, Hangzhou 310000, China
* Correspondence: qiangzhao@rcees.ac.cn

Abstract: Aflatoxin B1 (AFB1) is one of the mycotoxins produced by *Aspergillus flavus* and *Aspergillus parasiticus*, and it causes contamination in foods and great risk to human health. Simple sensitive detection of AFB1 is important and demanded for food safety and quality control. Aptamers can specifically bind to targets with high affinity, showing advantages in affinity assays and biosensors. We reported an aptamer structure-switch for fluorescent detection of aflatoxin B1 (AFB1), using a label-free aptamer, a fluorescein (FAM)-labeled complementary strand (FDNA), and a quencher (BHQ1)-labeled complementary strand (QDNA). When AFB1 is absent, these three strands assemble into a duplex DNA structure through DNA hybridization, making FAM close to BHQ1, and fluorescence quenching occurs. In the presence of AFB1, the aptamer binds with AFB1, instead of hybridizing with QDNA. Thus, FAM is apart from BHQ1, and fluorescence increases with the addition of AFB1. This assay allowed detection of AFB1 with a detection limit of 61 pM AFB1 and a dynamic concentration range of 61 pM to 4 μM. This aptamer-based method enabled detection of AFB1 in complex sample matrix (e.g., beer and corn flour samples).

Keywords: aptamer; aflatoxin B1; fluorophore; quencher; structure-switch

Citation: Wang, C.; Yu, H.; Zhao, Q. A Simple Structure-Switch Aptasensor Using Label-Free Aptamer for Fluorescence Detection of Aflatoxin B1. *Molecules* **2022**, *27*, 4257. https://doi.org/10.3390/molecules27134257

Academic Editors: Sai Wang, Wei Wu and Long Wu

Received: 28 May 2022
Accepted: 29 June 2022
Published: 1 July 2022

1. Introduction

Many agricultural products (e.g., maize, corn, wheat, oilseed, peanut, and several other cereals) and fruits suffer contamination of aflatoxins [1], toxic metabolites produced by *Aspergillus flavus* and *Aspergillus parasiticus*. Aflatoxins may also be found in other contaminated food products (e.g., beer, liquor, and grape wine) [2]. The ingestion of aflatoxins threatens human and animal health [3]. Aflatoxin B1 (AFB1) is the most toxic component of aflatoxins fractions [4,5]. The International Agency for Research in Cancer (IARC) has classified AFB1 as a Group 1 carcinogen [6]. Detection of AFB1 is important in food safety and quality control. The conventional AFB1 detection methods (e.g., HPLC and LC-MS) are time consuming and require sophisticated instruments, having limitations in rapid, high-throughput and on-site test [7]. Antibody based assays allow rapid detection of AFB1 [8], but antibodies are costly and vulnerable to denature [9]. Hence, it is still desirable to develop simpler methods for rapid, high-throughput, and cost-effective detection of AFB1 on site.

Aptamer is single stranded DNA or RNA that can specifically bind to a target molecule with high affinity [10]. The aptamer is a promising rival of the antibody, and shows some advantages (e.g., in vitro selection, easy chemical synthesis, small size, high thermal stability, and low cost) [11]. It has wide applications in assays and biosensors [12]. Some aptamer-based assays for AFB1 have been developed, including fluorescence, colorimetry, electrochemistry, surface plasmon resonance, etc. [13–18]. Among these assays, fluorescence methods have attracted more attention due to their simplicity and sensitivity [19]. Fluorescence aptasensors usually rely on the target-binding induced fluorescence change, and labeling of an aptamer with fluorophore/quencher is often needed, which may change the initial affinity of the aptamer and increase cost [20]. Some nanomaterials have been used for fluorescence aptasensors, benefiting from their unique quenching ability or fluorescence property [21]. However, preparation of nanomaterials is complicated, and functionalization of nanomaterials is laborious. Nutiu and Li reported a structure-switch signaling aptamer strategy [22] in which intact label-free aptamers were used as recognition units for fluorescence detection of target molecules. This strategy is general and has few restrictions on the size and secondary structure of aptamers. To date, this structure-switch signaling aptamer strategy has been successfully applied to detections of some targets, such as adenosine triphosphate (ATP), thrombin [22], and chloramphenicol [23].

In this study, we describe a fluorescent switch aptasensor for detection of AFB1 using a label-free aptamer, a fluorescein (FAM)-labeled complementary DNA (denoted as FDNA), and a quencher (BHQ1)-labeled complementary DNA (denoted as QDNA). In the absence of AFB1, a DNA duplex of the aptamer, FDNA, and QDNA is formed through DNA hybridization. This DNA assembly brings FAM and BHQ1 into proximity, causing fluorescence quenching. When AFB1 is present, the aptamer prefers to bind with AFB1, as the aptamer–AFB1 complex is more stable than the DNA duplex assembly. Aptamer–AFB1 binding causes release of QDNA from DNA duplex, and fluorescence intensity increases. Thus, detection of AFB1 is achieved by measuring the fluorescence change. Under optimized assay conditions, we achieved quantitative detection of AFB1 in a concentration range of 61 pM to 4 μM, with a detection limit of 61 pM AFB1. This structure-switch aptasensor allowed the detection of AFB1 in a complex sample matrix (e.g., beer and corn flour). It shows the advantages of the use of label-free aptamers and a larger enhancement of the fluorescence signal. This analytical method has potential in rapid, high-throughput, on-site and cost-effective detection of AFB1.

2. Results and Discussions

2.1. Working Principle of Fluorescent Switch Aptasensor for AFB1 Detection

Figure 1 illustrates the working principle of the fluorescent switch aptasensor. FDNA is labeled with a fluorescein (FAM) at the 5′end, and QDNA is labeled with a quencher (BHQ1) at the 3′end. The unlabeled aptamer sequence contains a FDNA-hybridizing sequence and an AFB1 binding motif. In the absence of AFB1, an FDNA, QDNA, and aptamer sequence together construct a tripartite DNA duplex, which makes FAM close to BHQ1, causing fluorescence quenching. In the presence of AFB1, aptamer binds with AFB1, and QDNA is released from the DNA duplex, resulting in a fluorescence increase. AFB1 can be detected by measuring the increase in fluorescence intensity of this aptasensor.

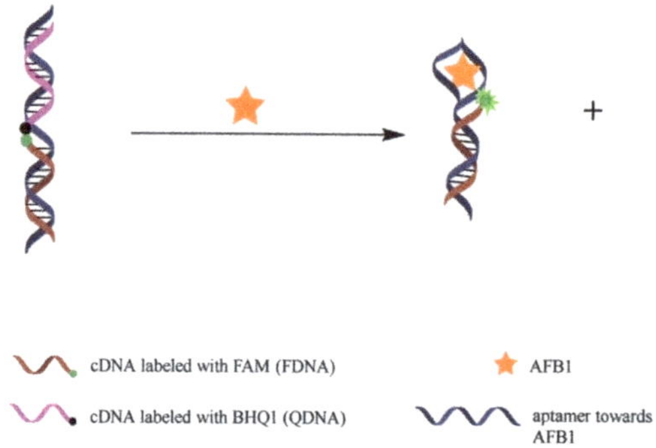

Figure 1. Working principle of fluorescent switch aptasensor for AFB1 detection. In the absence of AFB1, aptamer hybridizes with both FDNA and QDNA to form a tripartite DNA duplex structure in which FAM and BHQ1 are close to each other, causing fluorescence quenching. In the presence of AFB1, aptamer prefers to bind with AFB1 rather than QDNA, resulting in disassembly of the tripartite DNA duplex and fluorescence recovery. AFB1 can be detected by measuring the fluorescence change.

2.2. Feasibility of Fluorescent Structure-Switch Aptasensor

We tested the feasibility of the fluorescent switch aptasensor for AFB1 detection. As Figure 2 shows, the sample containing FDNA and QDNA displayed high fluorescence intensity similar to that of the sample containing only FDNA. This result means FDNA fluorescence cannot be quenched by QDNA in the absence of aptamer. After the addition of aptamer, a significant decrease in fluorescence intensity was observed. This result shows the tripartite DNA duplex was successfully constructed, and FDNA fluorescence was quenched. Fluorescence obviously increased upon addition of AFB1 (200 nM) into the solution containing FDNA, QDNA, and the aptamer. Thus, the fluorescent switch aptasensor is feasible for AFB1 detection.

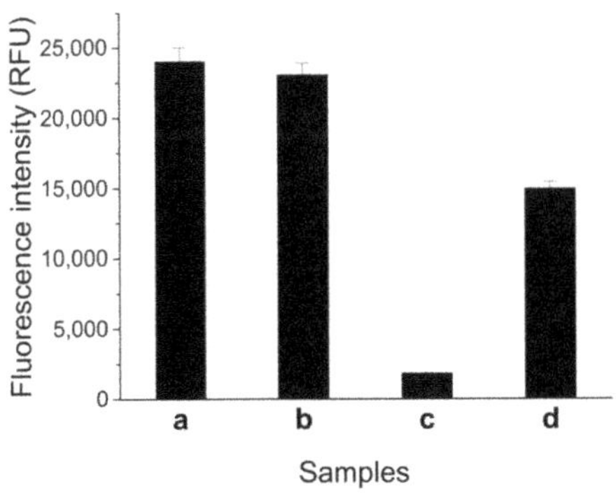

Figure 2. Feasibility test of the fluorescent switch aptasensor for AFB1 detection by measuring different samples. (**a**) FDNA; (**b**) FDNA+QDNA; (**c**) FDNA+QDNA+aptamer; and (**d**) FDNA + QDNA + aptamer + AFB1. Assay buffer was 10 mM Tris-HCl (pH 7.5) solution containing 50 mM $MgCl_2$, 50 mM NaCl, and 0.1% Tween 20. FDNA concentration was 50 nM. Aptamer (Af27) concentration was 100 nM. QDNA (Af27-C13Q) concentration was 100 nM. AFB1 concentration was 200 nM. Incubation at 4 °C for 60 min was applied.

2.3. Optimization of Experimental Conditions

To achieve better assay performances, we optimized some important experimental conditions including sequences of aptamer and QDNA, concentration of FDNA, concentration ratio between FDNA, aptamer and QDNA (C_{FDNA}:$C_{Aptamer}$:C_{QDNA}), $MgCl_2$ concentration, NaCl concentration, incubation temperature, and incubation time. In the following discussions, F_0, F_{blank}, and F_{AFB1} represent the fluorescence intensity of sample containing only FDNA; the fluorescence intensity of sample containing FDNA, QDNA, and aptamer; and the fluorescence intensity of sample containing FDNA, QDNA, aptamer, and AFB1, respectively. Quenching efficiency was calculated by $(1 - (F_{blank}/F_0)) \times 100\%$.

A suitable DNA duplex structure is essential for the structure-switch aptasensor to generate signal response. An unstable structure gives rise to a high background, while a too stable structure might lead to weak signal response to AFB1, causing a reduction in sensitivity. We explored different combinations of the aptamer and QDNA. As shown in Figure 3, when Af27 and Af27-C13Q were employed, a larger ratio of F_{AFB1} to F_{blank} (F_{AFB1}/F_{blank}) was obtained. Therefore, aptamer Af27 and QDNA Af27-C13Q were applied in the further experiments.

To investigate the effect of the concentration ratio between FDNA, aptamer, and QDNA (C_{FDNA}:$C_{Aptamer}$:C_{QDNA}), we fixed FDNA concentration at 50 nM and changed the aptamer concentration and QDNA concentration. As displayed in Figure 4A, F_{blank} decreased with the increase in concentrations of aptamer and QDNA, meaning a lower background. However, F_{AFB1} also decreased with the increase in aptamer and QDNA. Finally, C_{FDNA}:$C_{Aptamer}$:C_{QDNA} was chosen to be 1:2:3, respectively, because the quenching efficiency and the ratio of F_{AFB1}/F_{blank} approached to larger values at this condition (Figure 4B,C).

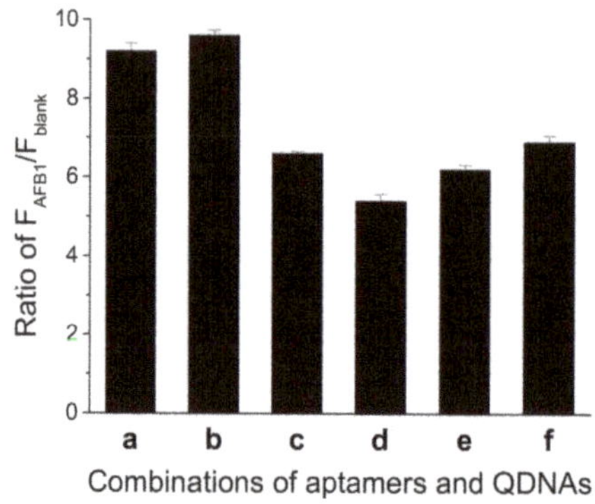

Figure 3. Effects of different combinations of aptamer and QDNA on the ratio of F_{AFB1}/F_{blank}. (**a**) Af27 + Af27 − C12Q; (**b**) Af27 + Af27 − C13Q; (**c**) Af27 + Af27 − C14Q; (**d**) Af29 + Af29 − C13Q; (**e**) Af29 + Af29 − C14Q; and (**f**) Af29 + Af29 − C15Q. Assay buffer was 10 mM Tris-HCl (pH 7.5) solution containing 50 mM $MgCl_2$, 50 mM NaCl, and 0.1% Tween 20. FDNA concentration was 50 nM. Aptamers (Af27 or Af29) concentrations were 100 nM. QDNAs concentrations were 100 nM. AFB1 concentration was 200 nM. Incubation at 4 °C for 60 min was applied.

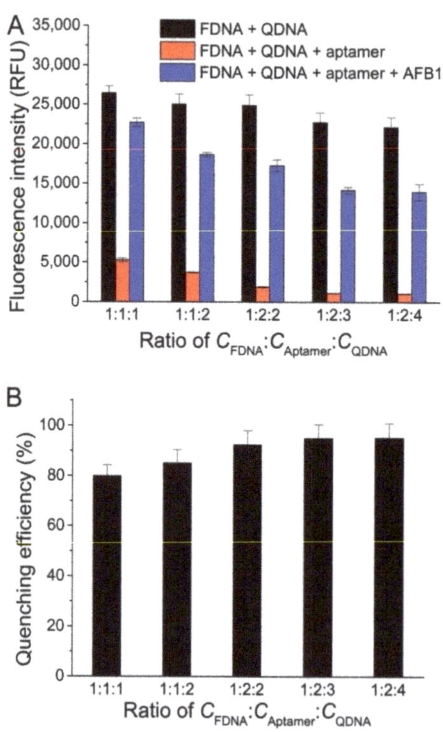

Figure 4. *Cont.*

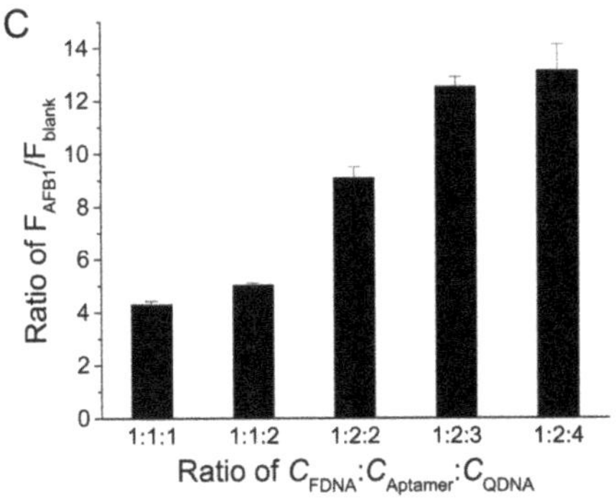

Figure 4. Effects of different concentration ratios between FDNA, aptamer, and QDNA (C_{FDNA}:$C_{Aptamer}$:C_{QDNA}) on (**A**) fluorescence intensity of samples, (**B**) quenching efficiency, and (**C**) ratio of F_{AFB1}/F_{blank}. Assay buffer was 10 mM Tris-HCl (pH 7.5) solution containing 50 mM $MgCl_2$, 50 mM NaCl, and 0.1% Tween 20. FDNA concentration was fixed at 50 nM. AFB1 concentration was 200 nM. Af27 and Af27-C13Q were used as aptamer and QDNA, respectively. Incubation at 4 °C for 60 min was applied.

Then, the FDNA concentration was optimized under the condition that C_{FDNA}:$C_{Aptamer}$:C_{QDNA} was fixed at 1:2:3, respectively. With an increase in FDNA concentration, fluorescence intensities F_0 and F_{AFB1} increased (Figure S1A), while the quenching efficiency had slight change (Figure S1B). The ratio of F_{AFB1}/F_{blank} reached a maximum value when 50 nM FDNA was used (Figure S1C). Therefore, 50 nM FDNA was chosen in the further experiments.

We investigated the effect of $MgCl_2$ concentration on detection of AFB1 (Figure S2). Upon addition of $MgCl_2$, the quenching efficiency and the ratio of F_{AFB1}/F_{blank} all increased, which can be explained due to the fact that Mg^{2+} promotes both DNA hybridization and aptamer–AFB1 binding [16,17]. Finally, 50 mM $MgCl_2$ was chosen because F_{AFB1}/F_{blank} reached the maximum value at this condition. Only a slight effect of NaCl concentration on fluorescence intensity was observed (Figure S3) in the presence of 50 mM $MgCl_2$, consistent with previous study [16,17]. Finally, we chose to add 50 mM NaCl in assay buffer.

Incubation temperature has important influence on performance of this aptasensor. As Figure S4 shows, when the incubation temperature was set at 25 °C or 37 °C, the presence of AFB1 did not cause a significant change in fluorescence. This result can be attributed to the fact that, at these temperatures, the aptamer has low affinity [15], and AFB1 binding cannot effectively replace QDNA. In contrast, a large signal change caused by AFB1 was achieved by employing incubation at 4 °C, because the aptamer has enhanced affinity at 4 °C [15]. Therefore, incubation at 4 °C was chosen. Furthermore, incubation time was also optimized (Figure S5), and incubation for 60 min was applied to find a larger value of the ratio of F_{AFB1}/F_{blank}.

2.4. Detection of AFB1

We detected different concentrations of AFB1 under optimized conditions. As Figure 5 shows, the fluorescence intensity gradually increased with the increasing concentration of AFB1 in a dynamic concentration range from 0 to 4 μM. A linear relationship between fluorescence intensity and AFB1 concentration ranging from 0.2 to 31.2 nM was obtained ($Y = 186X + 1143$ ($R^2 = 0.99613$), where Y is the fluorescence intensity and X is the concentration of AFB1). The lowest concentration of AFB1 that could be detected was 61 pM, based on three times the standard deviation of the blank sample, which is lower than or comparable to the detection limits of some previously reported methods (listed in Table S1 in ESM) [13,21,24–33].

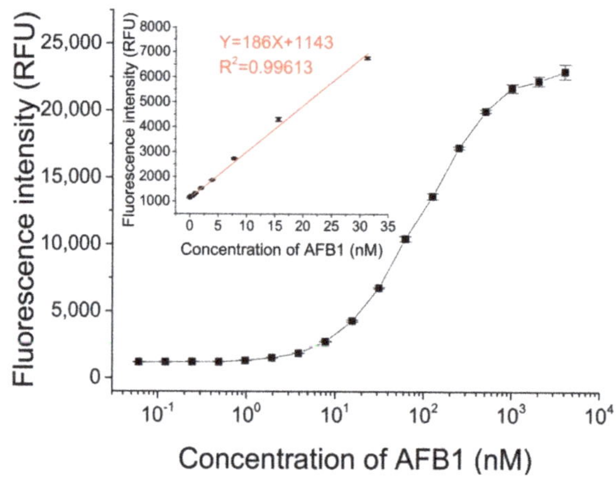

Figure 5. Detection of different concentrations of AFB1 using this fluorescent switch aptasensor. Assay buffer contained 10 mM Tris-HCl (pH 7.5) solution, 50 mM $MgCl_2$, 50 mM NaCl, and 0.1% Tween20. Af27 and Af27-C13Q were used as aptamer and QDNA, respectively. The ratio of C_{FDNA}:$C_{Aptamer}$:C_{QDNA} was 50 nM:100 nM:150 nM. Incubation at 4 °C for 60 min was applied.

2.5. Selectivity Test and Detection in Complex Sample Matrix

To assess the selectivity of the structure-switch aptasensor towards AFB1, we detected AFB1 (200 nM) along with a few non-target mycotoxins including OTA, OTB, FB1, FB2, and ZAE. As shown in Figure S6, non-target mycotoxins (1 μM) gave fluorescence intensity similar to the blank sample, while AFB1 showed higher fluorescence intensity. These results demonstrate the assay was selective, and co-existences of the tested non-target mycotoxins did not cause interference to the AFB1 detection.

To evaluate the performance of this structure-switch aptasensor in a complex sample matrix, we detected different concentrations of AFB1 spiked in 20-fold diluted beer and 20-fold diluted corn flour extraction. As shown in Figure S7, in both the diluted beer sample and diluted corn flour extraction sample, the fluorescence intensity increased with the increasing spiked amount of AFB1. Detection performances in 20-fold diluted beer and 20-fold diluted corn flour extraction were comparable to that in pure assay buffer. These results demonstrate the potential application of the developed structure-switch aptasensor in real samples analysis.

3. Materials and Methods

3.1. Chemicals and Reagents

All DNA strands listed in Table 1 were synthesized and purified by Sangon Biotech Inc. (Shanghai, China). Aflatoxin B1 (AFB1), ochratoxin A (OTA), ochratoxin B (OTB), fumonisin B1 (FB1), fumonisin B2 (FB2), and zearalenone (ZAE) were purchased from Pribolab (Qingdao, China). Beer and corn flour were bought from a local supermarket. Assay buffer was 10 mM Tris-HCl (pH 7.5) solution containing 50 mM NaCl, 50 mM $MgCl_2$, and 0.1% Tween20. Black 96-wells microplates were purchased from Thermo Fisher Scientific Inc. (Waltham, USA). All reagents used for experiments were analytical reagents. Solutions were prepared with ultrapure water (>18.2 MΩ·cm) from a Purelab Ultra Genetic Elga Labwater system (High Wycombe, UK).

Table 1. Sequences of different DNA strands.

Name	Function	Sequence (5′ to 3′)
FDNA	FDNA	FAM-TCACAGATGAGT
Af27	aptamer	ACTCATCTGTGATCACGTGTTGTCTCTCTGTGTCTCGTG
Af29	aptamer	ACTCATCTGTGATGCACGTGTTGTCTCTCTGTGTCTCGTGC
Af27-C12Q	QDNA	GACAACACGTGA-BHQ1
Af27-C13Q	QDNA	AGACAACACGTGA-BHQ1
Af27-C14Q	QDNA	GAGACAACACGTGA-BHQ1
Af29-C13Q	QDNA	GACAACACGTGCA-BHQ1
Af29-C14Q	QDNA	AGACAACACGTGCA-BHQ1
Af29-C15Q	QDNA	GAGACAACACGTGCA-BHQ1

The underlined section of aptamers (Af27 and Af29) are the sequence binding with AFB1.

3.2. Detection of AFB1

To detect AFB1, we mixed FDNA, QDNA, an aptamer, and AFB1 in the assay buffer. The final concentrations of FDNA, the aptamer, and QDNA were 50 nM, 100 nM, and 150 nM, respectively. After incubation for 60 min at 4 °C, 100 μL of the reaction mixture solution was transferred into a microplate well. Then, the fluorescence intensity was immediately measured by using a microplate reader (BioTek Synergy H1, Winooski, VT, USA), with an excitation wavelength of 485 nm and emission wavelength of 525 nm. Three repeated measurements for duplicate samples were carried out, and the average data were used.

3.3. Selectivity Test

To assess the selectivity of this structure-switch aptasensor towards AFB1, some non-target mycotoxins (OTA, OTB, FB1, FB2, and ZAE) were tested. The AFB1 concentration was 200 nM, and concentrations of the non-target mycotoxins were 1 μM. The assay procedures for non-target mycotoxins were the same as that for AFB1 detection.

3.4. Detection of AFB1 in Complex Sample Matrix

Beer was ultrasonicated to degas, and then filtered through a syringe filter (0.22 μm) before dilution with the assay buffer. Finally, different concentrations of AFB1 spiked in 20-fold diluted beer were detected using this structure-switch aptasensor following the same assay procedures. The corn flour extraction was prepared as follows: 3 mL of methanol/water (70:30, v/v) solution was added into corn flour (1 g), followed by a vortex for 20 min. After standing for 10 min, the supernatant was filtered through a syringe filter (0.22 μm) before dilution with the assay buffer. AFB1 concentrations spiked in 20-fold diluted extraction solution of corn samples were also detected by using the same assay procedures.

4. Conclusions

In summary, we reported a simple fluorescent switch aptasensor for AFB1 detection using FDNA with FAM label, QDNA with BHQ1 label, and an unlabeled aptamer. In the absence of AFB1, the three DNA sequences assembled to DNA duplex, showing low fluorescence due to fluorescence quenching. The addition of AFB1 caused dissociation of the DNA duplex consisting of FDNA, QDNA, and the unlabeled aptamer, resulting in a fluorescence increase. This aptasensor allowed the sensitive detection of AFB1, and a detection limit reached 61 pM. This aptasensor was selective and allowed to detect AFB1 in diluted beer and corn flour extraction. This method shows the advantages of a larger enhancement of fluorescence signal and use of an unlabeled aptamer. This sensor is promising in rapid, high-throughput, on-site, and cost-effective detection of AFB1.

Supplementary Materials: The following supporting information can be downloaded at: https://www.mdpi.com/article/10.3390/molecules27134257/s1, Table S1. Comparison of some aptamer-based fluorescence assays for AFB1; Figure S1. Effects of different FDNA concentrations on (A) fluorescence intensity of samples, (B) quenching efficiency, and (C) ratio of F_{AFB1}/F_{blank}; Figure S2. Effects of $MgCl_2$ concentrations in assay buffer on (A) fluorescence intensity of samples, (B) quenching efficiency, and (C) ratio of F_{AFB1}/F_{blank}; Figure S3. Effects of different NaCl concentrations in assay buffer on (A) fluorescence intensity of samples, (B) quenching efficiency, and (C) ratio of F_{AFB1}/F_{blank}; Figure S4. Effects of different incubation temperatures on (A) fluorescence intensity of samples, (B) quenching efficiency, and (C) ratio of F_{AFB1}/F_{blank}; Figure S5. Effects of incubation time on (A) fluorescence intensity of samples, (B) quenching efficiency, and (C) ratio of F_{AFB1}/F_{blank}; Figure S6. Selectivity of the fluorescent switch aptasensor towards AFB1; Figure S7. Detections of AFB1 spiked in 20-fold diluted beer and 20-fold diluted corn flour extraction.

Author Contributions: Writing—original draft preparation, C.W.; writing—review and editing, Q.Z., C.W. and H.Y.; investigation, C.W., H.Y.; conceptualization, resources, supervision, project administration, and funding acquisition, Q.Z. All authors have read and agreed to the published version of the manuscript.

Funding: This research was funded by National Natural Science Foundation of China (Grant No. 21874146, 22074156).

Conflicts of Interest: The authors declare no conflict of interest.

References

1. Nesbitt, B.F.; O'Kelly, J.; Sargeant, K.; Sheridan, A. Aspergillus flavus and turkey X disease: Toxic metabolites of aspergillus flavus. *Nature* **1962**, *195*, 1062–1063. [CrossRef]
2. Marroquin-Cardona, A.G.; Johnson, N.M.; Phillips, T.D.; Hayes, A.W. Mycotoxins in a changing global environment—A review. *Food Chem. Toxicol.* **2014**, *69*, 220–230. [CrossRef] [PubMed]
3. Da Rocha, M.E.B.; Freire, F.d.C.O.; Maia, F.E.F.; Guedes, M.I.F.; Rondina, D. Mycotoxins and their effects on human and animal health. *Food Control* **2014**, *36*, 159–165. [CrossRef]
4. Carnaghan, R.B.; Hartley, R.D.; O'Kelly, J. Toxicity and fluorescence properties of the aflatoxins. *Nature* **1963**, *200*, 1101. [CrossRef] [PubMed]
5. Llovet, J.M.; Zucman-Rossi, J.; Pikarsky, E.; Sangro, B.; Schwartz, M.; Sherman, M.; Gores, G. Hepatocellular carcinoma. *Nat. Rev. Dis. Primers* **2016**, *2*, 16018. [CrossRef]
6. Ostry, V.; Malir, F.; Toman, J.; Grosse, Y. Mycotoxins as human carcinogens-the IARC monographs classification. *Mycotoxin Res.* **2017**, *33*, 65–73. [CrossRef]
7. Koppen, R.; Koch, M.; Siegel, D.; Merkel, S.; Maul, R.; Nehls, I. Determination of mycotoxins in foods: Current state of analytical methods and limitations. *Appl. Microbiol. Biotechnol.* **2010**, *86*, 1595–1612.
8. Li, P.; Zhang, Q.; Zhang, W. Immunoassays for aflatoxins. *TrAC Trends Anal. Chem.* **2009**, *28*, 1115–1126. [CrossRef]
9. Chauhan, R.; Singh, J.; Sachdev, T.; Basu, T.; Malhotra, B.D. Recent advances in mycotoxins detection. *Biosens. Bioelectron.* **2016**, *81*, 532–545. [CrossRef]
10. Ellington, A.D.; Szostak, J.W. In vitro selection of RNA molecules that bind specific ligands. *Nature* **1990**, *346*, 818–822. [CrossRef]
11. Toh, S.Y.; Citartan, M.; Gopinath, S.C.; Tang, T.H. Aptamers as a replacement for antibodies in enzyme-linked immunosorbent assay. *Biosens. Bioelectron.* **2015**, *64*, 392–403. [CrossRef]
12. Citartan, M.; Gopinath, S.C.B.; Tominaga, J.; Tan, S.C.; Tang, T.H. Assays for aptamer-based platforms. *Biosens. Bioelectron.* **2012**, *34*, 1–11. [CrossRef]
13. Chen, L.; Wen, F.; Li, M.; Guo, X.; Li, S.; Zheng, N.; Wang, J. A simple aptamer-based fluorescent assay for the detection of aflatoxin B1 in infant rice cereal. *Food Chem.* **2017**, *215*, 377–382. [CrossRef]
14. Seok, Y.; Byun, J.Y.; Shim, W.B.; Kim, M.G. A structure-switchable aptasensor for aflatoxin B1 detection based on assembly of an aptamer/split DNAzyme. *Anal. Chim. Acta* **2015**, *886*, 182–187. [CrossRef]
15. Wang, C.; Liu, L.; Zhao, Q. Low temperature greatly enhancing responses of aptamer electrochemical sensor for aflatoxin B1 using aptamer with short stem. *ACS Sens.* **2020**, *5*, 3246–3253. [CrossRef]
16. Sun, L.; Wu, L.; Zhao, Q. Aptamer based surface plasmon resonance sensor for aflatoxin B1. *Microchim. Acta* **2017**, *184*, 2605–2610. [CrossRef]
17. Sun, L.; Zhao, Q. Direct fluorescence anisotropy approach for aflatoxin B1 detection and affinity binding study by using single tetramethylrhodamine labeled aptamer. *Talanta* **2018**, *189*, 442–450. [CrossRef]
18. Danesh, N.M.; Bostan, H.B.; Abnous, K.; Ramezani, M.; Youssefi, K.; Taghdisi, S.M.; Karimi, G. Ultrasensitive detection of aflatoxin B1 and its major metabolite aflatoxin M1 using aptasensors: A review. *TrAC Trends Anal. Chem.* **2018**, *99*, 117–128. [CrossRef]

19. Nutiu, R.; Li, Y. Structure-switching signaling aptamers: Transducing molecular recognition into fluorescence signaling. *Chem. A Eur. J.* **2004**, *10*, 1868–1876. [CrossRef]
20. Feng, C.; Dai, S.; Wang, L. Optical aptasensors for quantitative detection of small biomolecules: A review. *Biosens. Bioelectron.* **2014**, *59*, 64–74. [CrossRef]
21. Jia, Y.; Wu, F.; Liu, P.; Zhou, G.; Yu, B.; Lou, X.; Xia, F. A label-free fluorescent aptasensor for the detection of Aflatoxin B1 in food samples using AIEgens and graphene oxide. *Talanta* **2019**, *198*, 71–77. [CrossRef]
22. Nutiu, R.; Li, Y.F. Structure-switching signaling aptamers. *J. Am. Chem. Soc.* **2003**, *125*, 4771–4778. [CrossRef]
23. Ma, X.; Li, H.; Qiao, S.; Huang, C.; Liu, Q.; Shen, X.; Geng, Y.; Xu, W.; Sun, C. A simple and rapid sensing strategy based on structure-switching signaling aptamers for the sensitive detection of chloramphenicol. *Food Chem.* **2020**, *302*, 125359. [CrossRef]
24. Wang, C.; Sun, L.; Zhao, Q. A simple aptamer molecular beacon assay for rapid detection of aflatoxin B1. *Chin. Chem. Lett.* **2019**, *30*, 1017–1020. [CrossRef]
25. Sabet, F.S.; Hosseini, M.; Khabbaz, H.; Dadmehr, M.; Ganjali, M.R. FRET-based aptamer biosensor for selective and sensitive detection of aflatoxin B1 in peanut and rice. *Food Chem.* **2017**, *220*, 527–532. [CrossRef]
26. Xia, X.; Wang, Y.; Yang, H.; Dong, Y.; Zhang, K.; Lu, Y.; Deng, R.; He, Q. Enzyme-free amplified and ultrafast detection of aflatoxin B1 using dual-terminal proximity aptamer probes. *Food Chem.* **2019**, *283*, 32–38. [CrossRef]
27. Lu, Z.; Chen, X.; Wang, Y.; Zheng, X.; Li, C.M. Aptamer based fluorescence recovery assay for aflatoxin B1 using a quencher system composed of quantum dots and graphene oxide. *Microchim. Acta* **2014**, *182*, 571–578. [CrossRef]
28. Zhao, Z.; Yang, H.; Deng, S.; Dong, Y.; Yan, B.; Zhang, K.; Deng, R.; He, Q. Intrinsic conformation response-leveraged aptamer probe based on aggregation-induced emission dyes for aflatoxin B1 detection. *Dye. Pigment.* **2019**, *171*, 107767. [CrossRef]
29. Goud, K.Y.; Sharma, A.; Hayat, A.; Catanante, G.; Gobi, K.V.; Gurban, A.M.; Marty, J.L. Tetramethyl-6-carboxyrhodamine quenching-based aptasensing platform for aflatoxin B1: Analytical performance comparison of two aptamers. *Anal. Biochem.* **2016**, *508*, 19–24. [CrossRef]
30. Li, Y.; Wang, J.; Zhang, B.; He, Y.; Wang, J.; Wang, S. A rapid fluorometric method for determination of aflatoxin B1 in plant-derived food by using a thioflavin T-based aptasensor. *Microchim. Acta* **2019**, *186*, 214. [CrossRef]
31. Li, Y.P.; Sun, L.L.; Zhao, Q. Development of aptamer fluorescent switch assay for aflatoxin B1 by using fluorescein-labeled aptamer and black hole quencher 1-labeled complementary DNA. *Anal. Bioanal. Chem.* **2018**, *410*, 6269–6277. [CrossRef] [PubMed]
32. Wang, L.; Zhu, F.; Chen, M.; Zhu, Y.; Xiao, J.; Yang, H.; Chen, X. Rapid and visual detection of aflatoxin B1 in foodstuffs using aptamer/G-quadruplex DNAzyme probe with low background noise. *Food Chem.* **2019**, *271*, 581–587. [CrossRef] [PubMed]
33. Wang, B.; Zheng, J.; Ding, A.; Xu, L.; Chen, J.; Li, C.M. Highly sensitive aflatoxin B1 sensor based on DNA-guided assembly of fluorescent probe and TdT-assisted DNA polymerization. *Food Chem.* **2019**, *294*, 19–26. [CrossRef] [PubMed]

molecules

MDPI

Article

Evolution of Interferon-Gamma Aptamer with Good Affinity and Analytical Utility by a Rational In Silico Base Mutagenesis Post-SELEX Strategy

Lianhui Zhao [1], Qionglin Wang [2], Yingai Yin [1], Yan Yang [1], Huifang Cui [3] and Yiyang Dong [1,*]

[1] College of Life Science and Technology, Beijing University of Chemical Technology, Beijing 100029, China
[2] Henan Key Laboratory of Children's Genetics and Metabolic Diseases, Children's Hospital Affiliated to Zhengzhou University, Zhengzhou 450018, China
[3] College of Life Sciences, Zhengzhou University, Zhengzhou 450001, China
* Correspondence: yydong@mail.buct.edu.cn; Tel.: +86-010-64446260

Abstract: The Systematic Evolution of Ligands by EXponential enrichment (SELEX) is conventionally an effective method to identify aptamers, which are oligonucleotide sequences with desired properties to recognize targets specifically and sensitively. However, there are some inherent limitations, e.g., the loss of potential high-affinity sequences during biased iterative PCR enrichment processes and the limited structural diversity of the initial library, which seriously restrict their real-world applications. To overcome these limitations, the in silico base mutagenesis post-SELEX strategy based on the low Gibbs free energy (ΔG) and genetic algorithm was developed for the optimization of the interferon-gamma aptamer (B1-4). In the process of evolution, new sequences were created and the aptamer candidates with low ΔG values and advanced structures were produced. After five rounds of selection, systematic studies revealed that the affinity of the newly developed evolutionary aptamer (M5-5) was roughly 10-fold higher than that of the parent aptamer (B1-4), and an aptasensor detection system with a limit-of-detection (LOD) value of 3.17 nM was established based on the evolutionary aptamer. The proposed approach provided an efficient strategy to improve the aptamer with low energy and a high binding ability, and the good analytical utility thereof.

Keywords: aptamer; in silico; base mutagenesis; post-SELEX; interferon-gamma

check for updates

Citation: Zhao, L.; Wang, Q.; Yin, Y.; Yang, Y.; Cui, H.; Dong, Y. Evolution of Interferon-Gamma Aptamer with Good Affinity and Analytical Utility by a Rational In Silico Base Mutagenesis Post-SELEX Strategy. *Molecules* **2022**, *27*, 5725. https://doi.org/10.3390/molecules27175725

Academic Editors: Ramon Eritja, Sai Wang, Wei Wu and Long Wu

Received: 12 July 2022
Accepted: 1 September 2022
Published: 5 September 2022

1. Introduction

An aptamer, a kind of oligonucleotide sequence with a specific three-dimensional structure to selectively bind to a specific target, has many merits, such as a small size, customizable modification, long half-life, low cost, and low variability between batches. An aptamer is conventionally screened and isolated from the Systematic Evolution of Ligands by EXponential enrichment (SELEX) [1,2]. Although the SELEX is an effective screening method and many improvements have been made by researchers, it still has some inherent limitations. The structural diversity of oligonucleotide sequences in the initial library is practically limited. In addition, the biased iterative polymerase chain reaction (PCR) process in the SELEX makes it difficult to amplify sequences with advanced structures [3]. Therefore, the aptamers screened by traditional SELEX techniques are not always strong enough to detect targets.

Several post-SELEX strategies [4] have been proposed to make up for the shortcomings of the traditional SELEX, such as splitting or truncation, bivalent or multivalent construction, and chemical modification. Recently, the in silico post-SELEX strategy has emerged as a potential strategy to improve aptamers through an iterative process followed by in vitro experimental evaluation [5,6]. The computer-assisted method could efficiently optimize the aptamer through three main steps, including constructing the secondary and tertiary structures of the aptamer, predicting the structural conformation of the aptamer binding to the target, and optimizing the binding of the aptamer by mutation or modification. For instance, Song et al. engineered a high-affinity aptamer mutant that specifically binds to marine toxin gonyautoxin 1/4 [7]. Hsieh et al. successfully found one new sequence that could produce a relatively superior binding reaction with a prostate-specific antigen based on the known aptamers [8]. Wang et al. identified two aptamer mutants against the carcinoembryonic antigen exhibiting significantly high binding ability [9]. Thus far, the in silico post-SELEX strategy has been performed to engineer high-affinity aptamers for several targets. However, the thermodynamical stability of the aptamer has not been considered during the in silico post-SELEX program. Low ΔG was an important factor in the identification of an aptamer with high affinity [10,11]. Normally, the aptamer with the most thermodynamically stable structure has the lowest ΔG value [6,12,13]. In other words, as conformational flexibility is a key factor in limiting the affinity and specificity of aptamers, aptamers with a good binding ability mainly depend on stable conformation, especially in complex samples [14,15].

In the field of biomedical analysis, aptamers provide an alternative research platform for rapid detection and precise diagnosis [16]. Interferon-gamma (IFN-γ), an important cytokine, can act as an early indicator of infectious diseases, such as tuberculosis. Tuberculosis is one of the most catastrophic diseases and its transmission is very rapid [17,18]. Traditional antibody-based tuberculosis detection techniques commonly include the enzyme-linked immunosorbent assay (ELISA) and interferon-γ release assay (IGRAs) [19]. However, these methods have certain disadvantages, such as complicated operations and high costs. In contrast to antibodies, aptamers are still practically in a nascent stage without widespread use [20]. To further screen individuals efficiently for point-of-care testing, a model parent DNA aptamer (B1-4) against interferon-gamma was selected to optimize [21].

With the purpose of improving the performance of aptamers and promoting their application, as shown in Scheme 1, an innovative post-SELEX screening method was proposed. The scheme mainly involved the following steps: (1) the generation of aptamer mutants from B1-4 using a lab-written program; (2) the analysis of the minimum free energy of a secondary structure to select the aptamer candidates using the batch calculation function of Mfold; and (3) the model simulation and prediction of the binding ability of the aptamer candidates for IFN-γ using a molecular docking program. In this way, all sequences were screened and generated in silico without bias, and the scheme was advantageous for efficiently exploring the evolutionary aptamer with low ΔG and high affinity. Furthermore, the affinity values of the aptamer mutant with the highest docking score and B1-4 were tested by biolayer interferometry (BLI), which is an optical technique to measure the surface–biomolecule interactions by analyzing the interference patterns [22]. Furthermore, a fluorescent assay based on the evolutionary aptamer for IFN-γ detection with high sensitivity and selectivity was developed.

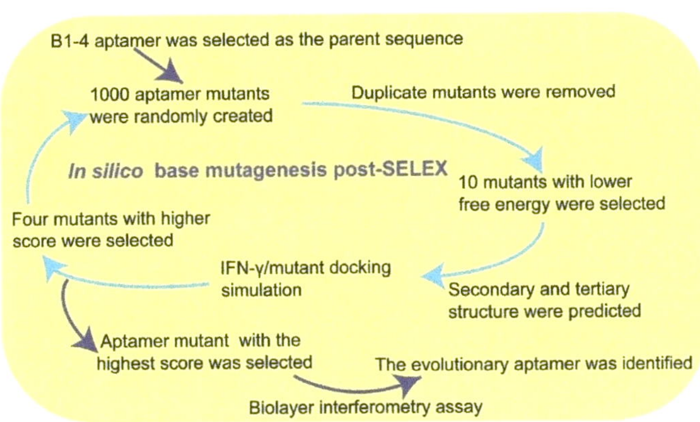

Scheme 1. Production of evolutionary aptamer against IFN-γ by in silico base mutagenesis post-SELEX.

2. Results and Discussion

2.1. Production of Evolutionary Aptamer against IFN-γ by In Silico Base Mutagenesis Post-SELEX

The in silico base mutagenesis post-SELEX process is an efficient method for the selection of an evolutionary aptamer against IFN-γ. Using an in silico base mutagenesis program and Mfold batch process, 10 mutants with low ΔG values were produced in each round, as shown in Tables S2–S7. The secondary structure of B1-4 was predicted using the Mfold and its ΔG was −1.03 kcal/mol (Figure 1a). The docking score of the B1-4/IFN-γ complex was 1246.537 (Figure 1b). Furthermore, the key binding sites of B1-4 with IFN-γ were analyzed by a PLIP, and the results showed that the bases of G31, A32, C33, A34, T35, and C54 bind to the IFN-γ by hydrogen bonding (Table S8). In the first round, the top 10 mutants (ΔG ranging from −3.18 to −4.76 kcal/mol) had lower ΔG than B1-4 and they were docked with IFN-γ, in which 4 mutants (M1-2, M1-5, M1-8, M1-9) with high ZDOCK scores indicated good docking results (Table S2). For the four sequences, each was mutated to 250 sequences, which were used as the library for the next round of screening. In the second round, the ΔG values of the top 10 mutants ranged from −5.92 to −7.11 kcal/mol, in which four sequences (M2-6, M2-8, M2-9, and M2-10) with high ZDOCK scores were used for the third round of screening (Table S3). After the programmed process, the ΔG values of the top 10 mutants were in the range of −8.44 to −9.44 kcal/mol in the third generation, −10.76 to −11.42 kcal/mol in the fourth generation, −12.46 to −14.13 kcal/mol in the fifth generation, and −14.57 to −16.29 kcal/mol in the sixth generation, indicating that each round of base mutagenesis could evolve sequences with lower ΔG than the previous round (Tables S4–S7). In the fifth generation, the ΔG of the mutant M5-5 was −12.98 kcal/mol (Figure 1c), much lower than that of B1-4, and the ZDOCK score of 1543.334 was the highest score among all the mutants. The bases involved in the interactions of the M5-5/IFN-γ are mainly distributed in C2, G3, A9, T10, C11, C12, G14, and A20 (Table S8). In the sixth generation, the ZDOCK score of the top mutant was 1406.824, much lower than that of M5-5, indicating that further mutation probably caused a worsening in the docking results. In addition, one aptamer mutant with positive ΔG (CCGCCCAAATCCCTAAGAGAA-GACTATAATGACATCAAACCAGACACACTACACACGCA, 0.04 kcal/mol) was used as a negative control to be docked to IFN-γ. The ZDOCK score was 1124.301, much lower than that of B1-4. The result further indicated that low ΔG was an important factor for the aptamer binding IFN-γ. Thus, the mutant M5-5 with the highest ZDOCK score was determined to be the best aptamer candidate, binding the IFN-γ in all the mutants.

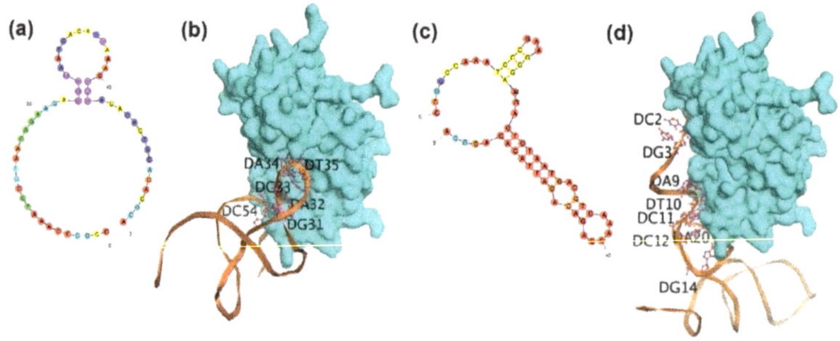

Figure 1. Predicted secondary structures of (**a**) B1-4 and (**c**) M5-5, using the Mfold. Predicted docking results of (**b**) B1-4/IFN-γ complex and (**d**) M5-5/IFN-γ complex, using ZDOCK online software. Three-dimensional images have been visualized by MOE.

Furthermore, although there exist many bioinformatics tools for predicting secondary structure of aptamer sequences, further improvement in accuracy is still awaited. To demonstrate the change in the aptamer conformation in practice, the conformations of B1-4 and M5-5 before and after binding with IFN-γ in the PBS buffer were investigated by circular dichroism spectroscopy. As shown in Figure S2, the CD spectrum of B1-4 and M5-5 both have a negative peak around 245 nm and a positive peak around 275 nm, suggesting that they both have B-type DNA conformation. Obviously, the positive peak of M5-5 is higher than that of B1-4, and the negative peak is lower than that of B1-4, indicating that the evolutionary aptamer has more complementary base pairings and a folded structure, which is consistent with the predicted results. In addition, the peak intensity of the B1-4 circular dichroism suggested that its secondary structure also tends to form a folded structure, indicating that the aptamer with poor thermodynamic stability may not exist in the predicted form in practice, further indicating the necessity of optimization. Upon IFN-γ binding, the absorption intensity of the positive and the negative peaks has no significant change for B1-4 and M5-5. The results reveal that IFN-γ cannot induce the changes in the aptamers' B-type conformation. Therefore, the structure of the evolutionary aptamer evolved in a way that was more favorable for binding to the target.

2.2. Generation of Evolutionary Aptamer with High Affinity

In order to verify the validity of the in silico screening result, the affinities of the evolutionary aptamer M5-5 and the parent aptamer B1-4 were characterized by BLI. As shown in Figure 2a,b, the BLI signal responses increased as the IFN-γ concentration increased from 125 to 2000 nM. The K_d values of M5-5 and B1-4 were 105.6 and 913.9 nM, respectively, which revealed that the affinity of M5-5 was roughly 10-fold higher than that of B1-4. In addition, as shown in Figure S3, the K_d values of the top mutants for each round of selection (M1-8, M2-6, M3-6, and M4-10) were determined to be 439.5, 481.6, 267.1, and 276.6 nM, respectively. The results showed that M5-5 has the best affinity for IFN-γ, which was consistent with the simulation results and indicated that the in silico base mutagenesis post-SELEX we proposed could be used to remarkably improve the aptamer binding ability.

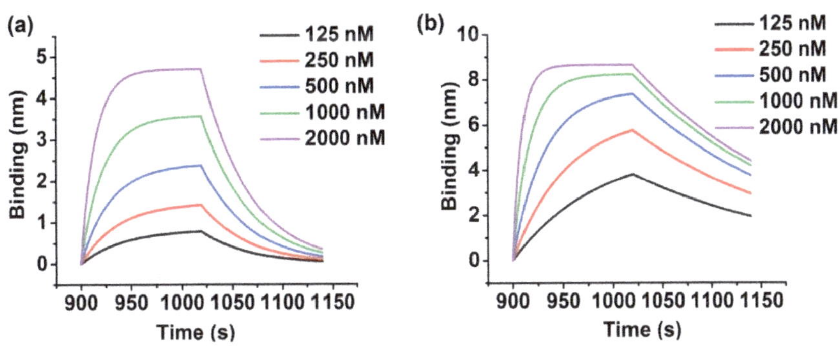

Figure 2. BLI characterization of the affinity of (**a**) B1-4 and (**b**) M5-5.

2.3. Detection of IFN-γ Using Evolutionary Aptamer

In order to investigate the applicability of the evolutionary aptamer M5-5, a fluorescent assay for the detection of IFN-γ was constructed. As shown in Figure 3a, the molecular beacon (MB) was first designed to contain three parts, including a loop that hybridizes with the aptamer, a stem that contains five base pairs of a complementary sequence, and a fluorophore–quencher pair. Carboxyfluorescein (FAM) as a fluorophore was attached to one terminus of MB, and Black Hole Quencher 1 (BHQ1) was located at the other terminus. As shown in Figure 3b, in the absence of IFN-γ, the MB hybridized with the M5-5, resulting in the separation of the FAM and BHQ1 pairs and enhanced fluorescence. After introducing IFN-γ, due to the strong affinity of the aptamer with IFN-γ, the MB was released from the duplex of the M5-5/MB to form the terminal hybridization structure (i.e., the fluorophore and quencher were in close proximity). Thus, the fluorescence was efficiently quenched. To verify the successful hybridization of the M5-5/MB, as shown in Figure S4, the fluorescence emission spectra of the M5-5/MB duplex and MB were tested, respectively. The results showed that the fluorescence value of the M5-5/MB duplex recovered significantly.

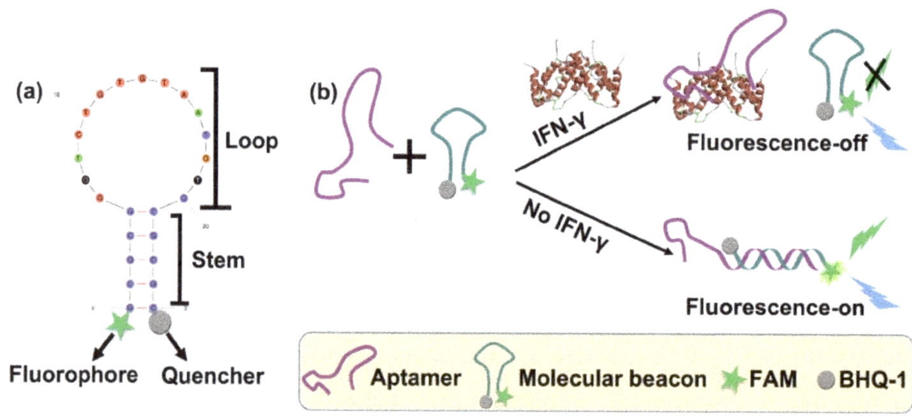

Figure 3. (**a**) Molecular beacon (MB) designed to contain three parts. (**b**) Schematic illustration of the aptamer-based fluorescent assay for detection of IFN-γ.

To optimize the length of the MB, 250 nM M5-5 was mixed with MBs of different lengths (MB-10, MB-15, MB-20, and MB-25) at a molar ratio of 1:3. By measuring the fluorescence intensity of the M5-5/MB duplex before and after adding 250 nM IFN-γ, the fluorescence reduction rate was obtained. As shown in Figure 4a, the highest rate was obtained at MB-20. The fluorescence intensity of the M5-5/MB-10 duplex was weak owing to its short length. A longer loop in the MB sequence (MB-25) resulted in a lower fluorescence reduction rate. MB-25 had more nucleotides in the loop to promote the formation of the duplex with M5-5, thus limiting the M5-5 structure switching after the addition of the IFN-γ. Thus, the MB-20 was chosen for the IFN-γ analysis. Then, in order to ensure M5-5 fully hybridized with the MB, the ratio of M5-5 to MB was also investigated. In this experiment, the M5-5 at a final concentration of 250 nM was mixed with MB-20 at 1:1, 1:2, 1:3, and 1:4 molar ratios. As shown in Figure 4b, the fluorescence intensity increased with the increase in the molecular beacon concentration, indicating that more aptamers were bound with MBs, and few unbound aptamers were in the solution. When the concentration ratio was 1:3, the fluorescence intensity almost reached its maximum.

Under the optimal conditions, the responses of the proposed method based on the evolutionary aptamer M5-5 were investigated by varying the IFN-γ concentrations. As shown in Figure 4c,d, the value of $(F_0 - F)/F_0$ increased with an increasing concentration of IFN-γ and the fluorescence reduction rate performed a good linear correlation with the concentration of IFN-γ within a linear range of 1−250 nM. The calibration curve of $(F_0 - F)/F_0$ against the IFN-γ concentration was $y = 0.106\ C + 5.758$ ($R^2 = 0.998$), in which y was $(F_0 - F)/F_0$ and C was the concentration of IFN-γ. The limit of detection (LOD) of the aptasensor was defined as the concentration that corresponds to a three-fold standard deviation of the blank value of $(F_0 - F)/F_0$, calculated as 3.17 nM. This LOD is comparable to the previous method for IFN-γ detection based on fluorescence resonance energy transfer [23].

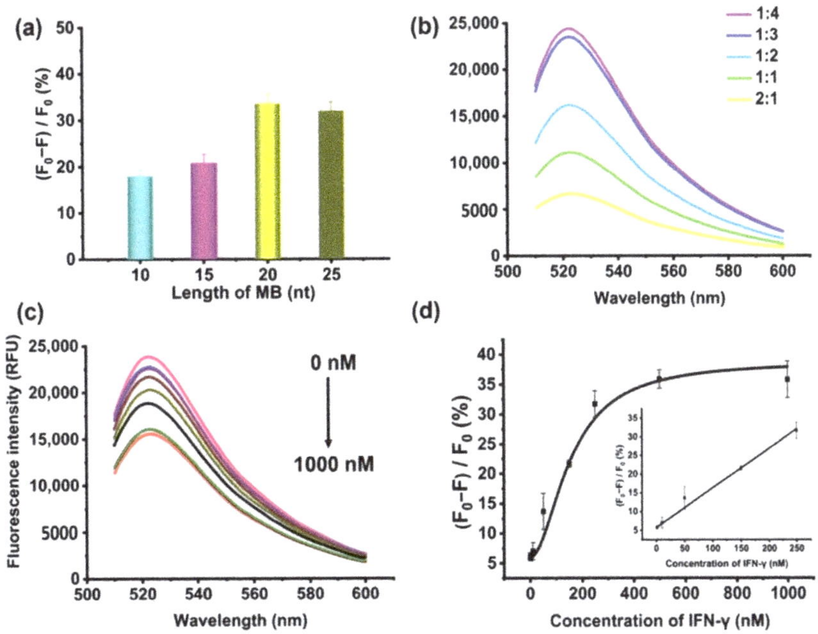

Figure 4. Optimization of experimental parameters, including (**a**) the length of MB and (**b**) the molar ratio between M5-5 and MB. (**c**) Fluorescence emission spectra of the detection solutions in the addition of IFN-γ at various concentrations (from top to bottom): 0, 0.1, 1, 10, 50, 150, 250, 500, 1000 nM. (**d**) Calibration curve of the $(F_0-F)/F_0$ and the concentration of IFN-γ. Error bars represent standard deviations of the three experiments.

2.4. High Selectivity of the Assay Based on Evolutionary Aptamer

In order to determine the selectivity of this assay based on the evolutionary aptamer M5-5, the fluorescence intensity was measured in the presence of different interfering targets (Myoglobin (Mb), human serum albumin (HSA), and immunoglobulin G (IgG)) with a concentration of 250 nM under the same experiment condition as IFN-γ. As shown in Figure 5, only IFN-γ induced a significant fluorescence reduction, whereas less value was observed with the addition of other interferences, indicating that the assay based on the evolutionary aptamer exhibited high selectivity in the detection of IFN-γ. The three interfering proteins have also been used as interfering substances in the previous methods for the detection of IFN-γ and the results also confirmed that these methods had good selectivity [24,25].

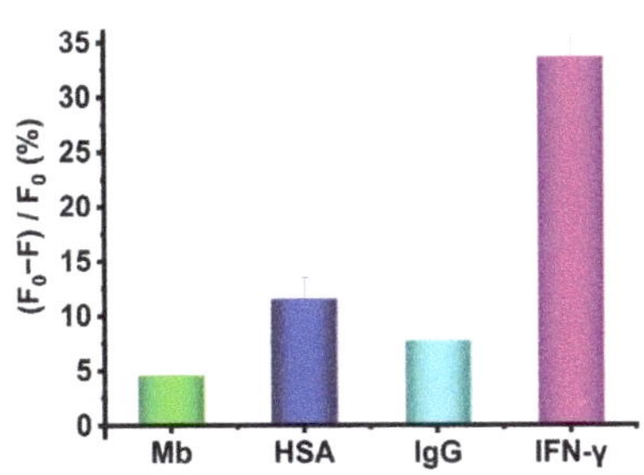

Figure 5. The selectivity of the assay based on evolutionary aptamer in the presence of various interference.

3. Materials and Methods

3.1. Materials and Instruments

Human interferon-gamma (IFN-γ, greater than 95% purity) was purchased from Sino-Bio Biotech Ltd. (Shanghai, China). Human serum albumin (HSA), immunoglobulin G (IgG), and myoglobin (Mb) were purchased from Beijing Solarbio Science & Technology Co., Ltd. (Beijing, China). Phosphate-buffered saline (PBS, pH 7.4, 10×) was purchased from Beijing Kehua Jingwei scientific Co., Ltd. (Beijing, China). The synthesis and HPLC (high-performance liquid chromatography) purification of all oligonucleotides in this study were performed by Sangon Biotechnology Co., Ltd. (Shanghai, China). B1-4 (CCGCCCAAATCC-CTAAGAGAAGACTGTAATGACATCAAACCAGACACACTACACACGCA) was screened by Cao et al. [21] and other sequences were listed in Table S1.

The affinity of the aptamer was characterized by Octet RED96e System (Fortebio, CA, USA) (accessed on 10 July 2022). The fluorescent signal was measured by Synergy H1 Microplate Reader (Biotek, Winooski, VT, USA). The structure of aptamer was analyzed by J-815 spectropolarimeter (Jasco, TKY, Japan).

3.2. Production of Aptamer Mutants by In Silico Mutagenesis

The B1-4, as a parent aptamer, was randomly mutated through an in silico base mutagenesis program designed by our team. The number of mutants in this study was about 1000 in each round, much higher than before [9]. The higher number of mutants means more possibilities of obtaining the sequence with better performance. Duplicate mutants were removed. Then, the secondary structures of aptamer mutants were predicted by the Mfold web server (http://www.unafold.org/mfold/applications/dna-folding-form.php) (accessed on 10 July 2022), based on a free-energy minimization algorithm and thermodynamic-based approach [26,27]. Using the shell script provided by Mfold, a batch process was written to fold secondary structures and arrange the sequences in ascending order of ΔG, generating a text file containing the sequences and secondary structures of dot–bracket information. The matched brackets represent base pairs, and dots represent unpaired bases. The top ten mutants were selected, and their three-dimensional (3D) structures were automatically modeled by RNA Composer (http://rnacomposer.ibch.poznan.pl/) (accessed on 10 July 2022) in batch mode [28,29]. Because the outcome was 3D RNA form, the modification was needed to ensure the structure of mutants in DNA coordinate. Molecular Operating Environment (MOE) DNA/RNA builder was used to convert the RNA to DNA, and energy minimization was used for determining low energy conformation. Finally, the prepared 3D DNA structures were docked with prepared IFN-γ (PDB: 1FG9) through the ZDOCK program (http://zdock.umassmed.edu) (accessed on 10 July 2022) [30,31]. The top four mutants with high docking scores were selected for the second cycle of base mutagenesis. The same steps of random mutagenesis and molecular docking were repeated 6 cycles. Finally, the aptamer mutant with the highest score was produced.

3.3. Model Simulation of Aptamer/IFN-γ

The prepared aptamer mutants and IFN-γ were docked using ZDOCK online server. The ZDOCK docking program, based on the Fast Fourier Transform, was an automated tool to search all possible binding modes between the aptamer mutants and IFN-γ, producing top 10 predictions with a high score. The best docking result of each mutant was selected to be compared with others. The interaction sites and interaction types of aptamer and IFN-γ were analyzed by automated protein–ligand interaction profiler server (PLIP, https://plip-tool.biotec.tu-dresden.de/plip-web/plip/index) (accessed on 10 July 2022) [32].

3.4. Biolayer Interferometry Assay

The equilibrium binding affinities (K_d values) of B1-4 and evolutionary aptamer were characterized by biolayer interferometry (BLI) through the following steps. First, the BLI chips coated with streptavidin (SA) were pretreated in baseline solution (Phosphate Buffered Saline, PBS, pH 7.4) for equilibration. Then, the BLI chips were immersed in solution containing biotin-modified aptamers with concentration of 500 nM to immobilize aptamers. Then, the chips functionalized with aptamer were immersed in PBS buffer for washing. Moreover, the IFN-γ in sample solution was captured by aptamers and the thickness of the biolayer increased. Finally, the chips were immersed in PBS solution for dissociation, and the thickness of the biolayer decreased. The reaction time of the above five steps was 120, 600, 180, 120, and 120 s, respectively. All steps were performed at 25 °C in a 96-well plate containing 200 μL solution in each well. The dissociation constant (K_d) could be obtained by monitoring the biolayer thickness in real time.

3.5. Circular Dichroism Measurement

CD measurements were performed to investigate the conformations of B1-4 and M5-5 before and after binding with IFN-γ in the PBS buffer. The CD spectrum of aptamer (5 μM) or with IFN-γ (5 μM) was measured three times at a scanning rate of 200 nm/min in the range of 220–320 nm. The background signals of the PBS buffer were also measured and subtracted from the CD spectrum.

3.6. Fluorescence Detection of IFN-γ Using Evolutionary Aptamer

The complementary sequence as molecular beacon (MB) was designed to partially hybridize with evolutionary aptamer. Firstly, aptamer and MB were dissolved in PBS solution, respectively. Then, to allow the hybridization, the mixture of aptamer and MB was incubated at 95 °C for 5 min and cooled at 25 °C for 30 min. Then, different concentrations of IFN-γ were added to the mixture, thoroughly mixed, and then the obtained solutions were incubated at 25 °C for 30 min. Finally, 50 μL of the obtained solution was analyzed by fluorescence, with excitation wavelength at 480 nm and emission wavelength in the range of 510–600 nm. The fluorescence reduction rate was defined as $y = (F_0 - F)/F_0 \times 100\%$ (F_0 and F are fluorescence intensity in the absence and presence of IFN-γ).

4. Conclusions

In this study, a novel in silico base mutagenesis post-SELEX strategy, based on minimum free energy and a genetic algorithm, was proposed to make up for the shortcomings of the traditional SELEX technology. The obtained evolutionary aptamer has a low ΔG, −12.98 kcal/mol, and a high ZDOCK score, 1543.334, indicating that it was conducive to binding with the IFN-γ. Moreover, it could also prove that the evolutionary aptamer had a good binding ability characterized by BLI. Then, a fluorescent assay was developed to detect IFN-γ by using an M5-5/MB duplex with an LOD as low as 3.17 nM. Our rational design of the in silico base mutagenesis post-SELEX strategy showed that the evolutionary approach was very promising for improving an aptamer, and the evolutionary aptamer combined with different technology could be further used to promote the application of early disease diagnosis and the prevention of disease transmission.

Supplementary Materials: The following supporting information can be downloaded at: https://www.mdpi.com/article/10.3390/molecules27175725/s1, Table S1: List of oligonucleotides; Table S2: B1-4 and mutants with low free energy in the first round library; Table S3: Mutants with low free energy in the second round library; Table S4: Mutants with low free energy in the third round library; Table S5: Mutants with low free energy in the fourth round library; Table S6: Mutants with low free energy in the fifth round library; Table S7: Mutants with low free energy in the sixth round library; Table S8: The binding residues of B1-4 and M5-5 based on the Mfold prediction interacted with IFN-γ through hydrogen bonding; Figure S1: The phylogeny of the M5-5 aptamer predicted by the Mfold. The mutation sites are shaded in red; Figure S2: Circular dichroism (CD) analysis of B1-4 and M5-5 with or without IFN-γ; Figure S3: BLI characterization of the affinity of (a) M1-8, (b) M2-6, (c) M3-6, and (d) M4-10; Figure S4: Fluorescence emission spectra of the M5-5/MB duplex and MB.

Author Contributions: Conceptualization, Y.D.; methodology, L.Z.; software, Q.W. and H.C.; validation, Y.Y. (Yingai Yin); formal analysis, L.Z.; investigation, L.Z.; resources, Y.D.; data curation, L.Z.; writing—original draft preparation, L.Z.; writing—review and editing, Y.D. and Y.Y. (Yan Yang);

visualization, L.Z.; supervision, Y.D.; project administration, Y.D.; funding acquisition, Y.D. All authors have read and agreed to the published version of the manuscript.

Funding: This research was supported by the Beijing University of Chemical Technology–China-Japan Friendship Hospital Biomedical Transformation Engineering Research Center Joint Project (Grant No. RZ2020-02) and the National Key Research and Development Program of China (Grant No. 2016YFF0203703).

Institutional Review Board Statement: Not applicable.

Informed Consent Statement: Not applicable.

Data Availability Statement: Not applicable.

Acknowledgments: The authors gratefully thank Professor Jingyang Gao, Department of Information Science and Technology, Beijing University of Chemical Technology, for her help in this work.

Conflicts of Interest: The authors declare no conflict of interest.

References

1. Tuerk, C.; Gold, L. Systematic evolution of ligands by exponential enrichment: RNA ligands to bacteriophage T4 DNA polymerase. *Science* **1990**, *249*, 505–510. [CrossRef] [PubMed]
2. Ellington, A.D.; Szostak, J.W. In vitro selection of RNA molecules that bind specific ligands. *Nature* **1990**, *346*, 818–822. [CrossRef] [PubMed]
3. Klug, S.J.; Famulok, M. All you wanted to know about SELEX. *Mol. Biol. Rep.* **1994**, *20*, 97–107. [CrossRef] [PubMed]
4. Gao, S.X.; Zheng, X.; Jiao, B.H.; Wang, L.H. Post-SELEX optimization of aptamers. *Anal. Bioanal. Chem.* **2016**, *408*, 4567–4573. [CrossRef] [PubMed]
5. Ahmad, N.A.; Zulkifli, R.M.; Hussin, H.; Nadri, M.H. In silico approach for Post-SELEX DNA aptamers: A mini-review. *J. Mol. Graph. Model.* **2021**, *105*, 107872. [CrossRef] [PubMed]
6. Navien, T.N.; Thevendran, R.; Hamdani, H.Y.; Tang, T.H.; Citartan, M. In silico molecular docking in DNA aptamer development. *Biochimie* **2021**, *180*, 54–67. [CrossRef]
7. Song, M.H.; Li, G.; Zhang, Q.; Liu, J.P.; Huang, Q. De novo post-SELEX optimization of a G-quadruplex DNA aptamer binding to marine toxin gonyautoxin 1/4. *Comput. Struct. Biotechnol. J.* **2020**, *10*, 3425–3433. [CrossRef]
8. Hsieh, P.C.; Lin, H.T.; Chen, W.Y.; Tsai, J.J.P.; Hu, W.P. The combination of computational and biosensing technologies for selecting aptamer against prostate specific antigen. *Biomed. Res. Int.* **2017**, *2017*, 5041683. [CrossRef]
9. Wang, Q.L.; Cui, H.F.; Du, J.F.; Lv, Q.Y.; Song, X.J. In silico post-SELEX screening and experimental characterizations for acquisition of high affinity DNA aptamers against carcinoembryonic antigen. *RSC. Adv.* **2019**, *9*, 6328–6334. [CrossRef]
10. Mousivand, M.; Anfossi, L.; Bagherzadeh, K.; Barbero, N.; Mirzadi-Gohari, A.; Javan-Nikkhah, M. In silico maturation of affinity and selectivity of DNA aptamers against aflatoxin B1 for biosensor development. *Anal. Chim. Acta* **2020**, *1105*, 178–186. [CrossRef]
11. Setlem, K.; Mondal, B.; Ramlal, S.; Kingston, J. Immuno affinity SELEX for simple, rapid, and cost-effective aptamer enrichment and identification against aflatoxin B1. *Front. Microbiol.* **2016**, *1*, 1909. [CrossRef] [PubMed]
12. Thiel, W.H.; Bair, T.; Wyatt Thiel, K.; Dassie, J.P.; Rockey, W.M.; Howell, C.A.; Liu, X.Y.; Dupuy, A.J.; Huang, L.; Owczarzy, R.; et al. Nucleotide Cbserved with a short SELEX RNA aptamer library. *Nucleic Acid Ther.* **2011**, *21*, 253–263. [CrossRef] [PubMed]
13. Yue, H.; Chen, J.; Chen, X.; Wang, X.; Zhou, N. Systematic screening and optimization of single-stranded DNA aptamer specific for N-acetylneuraminic acid: A comparative study. *Sens. Actuators B—Chem.* **2021**, *344*, 130270. [CrossRef]
14. Kuai, H.L.; Zhao, Z.L.; Mo, L.T.; Liu, H.; Hu, X.X.; Fu, T.; Zhang, X.B.; Tan, W.H. Circular bivalent aptamers enable in vivo stability and recognition. *J. Am. Chem. Soc.* **2017**, *139*, 9128–9131. [CrossRef]
15. Zhao, L.H.; Qi, X.Y.; Yan, X.C.; Huang, Y.F.; Liang, X.G.; Zhang, L.Q.; Wang, S.; Tan, W.H. Engineering Aptamer with enhanced affinity by triple helix-based terminal fixation. *J. Am. Chem. Soc.* **2019**, *141*, 17493–17497. [CrossRef] [PubMed]
16. Dong, Y.Y.; Wang, S.; Zhao, L.H.; Yang, J.B.; Xu, Y. Some frontier technologies for aptamers in medical applications. In *Aptamers for Medical Applications: From Diagnosis to Therapeutics*; Dong, Y., Ed.; Springer: Singapore, 2021; pp. 375–403.
17. Boehm, U.; Klamp, T.; Groot, A.M.; Howard, J.C. Cellular responses to interferon-gamma. *Annu. Rev. Immunol.* **1997**, *15*, 749–795. [CrossRef]
18. Dhenadhayalan, N.; Sriram, M.I.; Lin, K.C. Aptamer-based fluorogenic sensing of interferon-gamma probed with ReS2 and TiS2 nanosheets. *Sens. Actuators B—Chem.* **2017**, *258*, 929–936. [CrossRef]
19. Januarie, K.C.; Uhuo, O.V.; Iwuoha, E.; Feleni, U. Recent advances in the detection of interferon-gamma as a TB biomarker. *Anal. Bioanal. Chem.* **2022**, *414*, 907–921. [CrossRef]
20. Kulabhusan, P.K.; Hussain, B.; Yüce, M. Current perspectives on aptamers as diagnostic tools and therapeutic agents. *Pharmaceutics* **2020**, *12*, 646. [CrossRef]
21. Cao, B.B.; Hu, Y.; Duan, J.H.; Ma, J.; Xu, D.K.; Yang, X.D. Selection of a novel DNA aptamer for assay of intracellular interferon-gamma. *PLoS ONE* **2014**, *9*, e98214. [CrossRef]

22. Li, Y.Y.; Wen, H.; Shi, X.B.; Li, H.T. Determination of protein-DNA (ZMYND11-DNA) interaction by a label-free biolayer interferometry assay. *Bio-Protocol* **2015**, *5*, e1402. [CrossRef]
23. Tuleuova, N.; Jones, C.N.; Yan, J.; Ramanculov, E.; Yokobayashi, Y.; Revzin, A. Development of an aptamer beacon for detection of interferon-gamma. *Anal. Chem.* **2010**, *82*, 1851–1857. [CrossRef]
24. Abnous, K.; Danesh, N.M.; Ramezani, M.; Alibolandi, M.; Hassanabad, K.Y.; Emrani, A.S.; Bahreyni, A.; Taghdisi, S.M. A triple-helix molecular switch-based electrochemical aptasensor for interferon-gamma using a gold electrode and Methylene Blue as a redox probe. *Microchim. Acta* **2017**, *184*, 4151–4157. [CrossRef]
25. Jin, H.; Gui, R.J.; Gao, X.H.; Sun, Y.J. An amplified label-free electrochemical aptasensor of γ-interferon based on target-induced DNA strand transform of hairpin-to-linear conformation enabling simultaneous capture of redox probe and target. *Biosens. Bioelectron.* **2019**, *145*, 111732. [CrossRef]
26. Zuker, M. Mfold web server for nucleic acid folding and hybridization prediction. *Nucleic Acids Res.* **2003**, *31*, 3406–3415. [CrossRef]
27. SantaLucia, J., Jr. A unified view of polymer, dumbbell, and oligonucleotide DNA nearest-neighbor thermodynamics. *Proc. Natl. Acad. Sci. USA* **1998**, *95*, 1460–1465. [CrossRef]
28. Maciej, A.; Mariusz, P.; Tomasz, Z.; Joanna, S.; Tomasz, R.; Katarzyna, T.; Adamiak, R.W.; Marta, S. New functionality of RNAComposer: An application to shape the axis of MiR160 precursor structure. *Acta Biochim. Pol.* **2016**, *63*, 737–744.
29. Mariusz, P.; Marta, S.; Maciej, A.; Purzycka, K.J.; Piotr, L.; Natalia, B.; Jacek, B.; Adamiak, R.W. Automated 3D structure composition for large RNAs. *Nucleic Acids Res.* **2012**, *40*, e112.
30. Thiel, D.J.; Du, M.; Walter, R.L.; Arcy, A.D.; Chène, C.; Fountoulakis, M.; Garotta, G.; Winkler, F.K.; Ealick, S.E. Observation of an unexpected third receptor molecule in the crystal structure of human interferon-γ receptor complex. *Structure* **2000**, *8*, 927–936. [CrossRef]
31. Pierce, B.G.; Kevin, W.; Howook, H.; Bong-Hyun, K.; Thom, V.; Weng, Z. ZDOCK server: Interactive docking prediction of protein–protein complexes and symmetric multimers. *Bioinformatics* **2014**, *30*, 1771–1773. [CrossRef]
32. Salentin, S.; Schreiber, S.; Haupt, V.J.; Adasme, M.F.; Schroeder, M. PLIP: Fully automated protein–ligand interaction profiler. *Nucleic Acids Res.* **2015**, *43*, W443–W447. [CrossRef] [PubMed]

molecules

MDPI

Article

Aptamer Selection Based on Microscale Electrophoretic Filtration Using a Hydrogel-Plugged Capillary Device

Junku Takao [1], Reina Nagai [1], Tatsuro Endo [1], Hideaki Hisamoto [1] and Kenji Sueyoshi [1,2,*]

[1] Department of Applied Chemistry, Graduate School of Engineering, Osaka Metropolitan University, Osaka 545-0051, Japan

[2] Precursory Research for Embryonic Science and Technology (PRESTO), Japan Science and Technology Agency (JST), Saitama 332-0012, Japan

* Correspondence: sueyoshi@omu.ac.jp

Abstract: This study reports a novel aptamer selection method based on microscale electrophoretic filtration. Aptamers are versatile materials that recognize specific targets and are attractive for their applications in biosensors, diagnosis, and therapy. However, their practical applications remain scarce due to issues with conventional selection methods, such as complicated operations, low-efficiency separation, and expensive apparatus. To overcome these drawbacks, a selection method based on microscale electrophoretic filtration using a capillary partially filled with hydrogel was developed. The electrophoretic filtration of model target proteins (immunoglobulin E (IgE)) using hydrogel, the electrokinetic injection of DNAs to interact with the trapped proteins, the elimination of DNAs with weak interactions, and the selective acquisition of aptamer candidates with strong interactions were successfully demonstrated, revealing the validity of the proposed concept. Two aptamer candidates for IgE were obtained after three selection cycles, and their affinity for the target was confirmed to be less than 1 nM based on their dissociation constant (K_D) values. Therefore, the proposed method allows for the selection of aptamers with simple operations, highly effective separation based on electrophoresis and filtration, and a relatively cheap apparatus with disposable devices.

Keywords: aptamer selection; microscale electrophoretic filtration; systematic evolution of ligands by exponential enrichment (SELEX)

Citation: Takao, J.; Nagai, R.; Endo, T.; Hisamoto, H.; Sueyoshi, K. Aptamer Selection Based on Microscale Electrophoretic Filtration Using a Hydrogel-Plugged Capillary Device. *Molecules* **2022**, *27*, 5818. https://doi.org/10.3390/molecules27185818

Academic Editor: Roberto Mandrioli

Received: 20 June 2022
Accepted: 2 September 2022
Published: 8 September 2022

1. Introduction

Aptamers are oligonucleotides that bind strongly to specific target molecules owing to their three-dimensional conformations [1–3]. Compared with an antibody that must be biosynthesized, the aptamer possesses many advantages, such as a low cost through artificial synthesis, a low cytotoxicity, and a high applicability to various target molecules, such as large targets (protein [4], virus [5–7], and cell [8]), as well as small targets (peptides [9], amino acids [10], and inorganic ions [11]). Thus, aptamers have attracted attention for applications in biosensors [12–14], diagnoses [15–17], and therapies [18–20]. However, for these applications, target-specific aptamers must be experimentally isolated from a large library of oligonucleotides with random sequences.

Aptamer selection from a random library was generally employed in four experimental steps: (i) mixing a target with a library, (ii) separation of the target-oligonucleotide complex from the free/unbound ones, (iii) dissociation of the isolated complexes, and (iv) amplification of the obtained oligonucleotides as aptamer candidates. During the mixing step, the target molecules were added to an oligonucleotide library solution. Oligonucleotides with different sequences showed various affinities for the target molecules under the experimental conditions. Thus, oligonucleotides can form complexes according to the association/dissociation equilibrium. In the separation step, the formed complexes were isolated from the unbound oligonucleotides. In the dissociation step, the addition of denaturants, such as acids, high-concentration salts, surfactants, and organic solvents, weakened the interaction between the target and oligonucleotides, resulting in the dissociation of bound oligonucleotides. After collecting the dissociated oligonucleotides, they were amplified by the polymerase chain reaction (PCR) in the amplification step. The amplified oligonucleotides were used as next-generation libraries containing aptamer candidates. By repeating 10–15 rounds of the selection cycle comprising the above four steps, target-specific aptamers were enriched. This selection strategy, called systematic evolution of ligands by exponential enrichment (SELEX), has become one of the golden standards for aptamer selection [21,22]. The dissociation constants (K_D) of aptamers obtained by SELEX reach the micromolar to picomolar scale, comparable to that of antibodies [23].

In SELEX, a better separation process can reduce the number of SELEX rounds by improving the quality of the amplified libraries, resulting not only in a decrease in the consumption of reagents and time but also in the experimental errors caused by PCR bias, non-specific adsorption, and manual handling [2]. However, insufficient separation leads to contamination by unbound oligonucleotides and greatly degrades the quality of the amplified libraries.

In conventional SELEX, target-immobilized beads are often used for the separation [1–3,24]. The target molecule must be immobilized on the bead surface before the beads could be dispersed in the solution containing the oligonucleotide library. After mixing, the oligonucleotides were bound to the immobilized target molecules, forming a complex based on each association/dissociation equilibrium. The complexes on the beads were isolated from the library solution using centrifugal or magnetic forces. Finally, the aptamer candidates were detached from the beads using eluent. Although SELEX techniques using beads have the advantages of simplifying the experimental procedures required to select aptamer candidates, they also require many selection cycles because of the effects of low separation efficiency and non-specific adsorption to the beads. To overcome these limitations, Duan et al. developed a SELEX method using magnetic beads coupled with a DNA library instead of target molecules to suppress the non-specific adsorption of unbound DNAs [25]. The aptamers for histamine and tryptamine were successfully obtained using the proposed method, which requires 17 rounds due to the limited diversity of the library. Wang et al. reported an aptamer-screening device that integrated positive and negative selection units [26]. The developed microfluidic device has two integrated units for packing beads: the upstream or downstream unit is packed with microbeads immobilizing the negative (non-target) proteins or the target protein, respectively. During the introduction of a library solution into the device, only DNAs without binding to non-target proteins at the upstream unit reach the positive-selection beads immobilizing target molecules, which allows successive negative and positive selections in one device. The target-specific aptamer candidates with strong binding could be concentrated at the positive selection unit by introducing only the library solution into the device, and the myoglobin-specific aptamer could be selected only for the seven-round selection.

As described above, SELEX using beads can be achieved through simple procedures, which are still evolving; however, there are numerous limitations. The first is that the binding reaction mainly depends on their diffusion, resulting in a method that requires a long time for binding. Another issue is that they require the immobilization of the target molecules on the bead surface, which leads to structural changes, affecting the binding affinity. Furthermore, non-specific adsorption of oligonucleotides to the beads degrades the quality of the amplified library, which can result in a simple but laborious and costly procedure that consumes large quantities of valuable biological samples.

To overcome these limitations, capillary electrophoresis (CE) has recently been utilized for separation in aptamer selection due its great potential to separate various molecules in a liquid phase with a minimal consumption of reagents [2,27]. Mendonsa et al. first applied CE to the separation step in the SELEX cycles [28,29]. The target molecules were mixed with the DNA library solution and incubated to facilitate binding. Then, the mixture was introduced into the capillary as a short plug, and the complexes were rapidly separated from the free DNAs by electrophoresis. Based on the high separation efficiency in CE, the number of SELEX cycles could be drastically reduced from to 10–15 to 2–4 rounds. Wakui et al. explored a high-affinity and specific aptamer using a separation method combining CE with target-immobilized magnetic beads [30]. The proposed method has the advantages of high recovery using magnetic beads and highly effective separation based on electrophoresis, which demonstrate the selection of aptamers with a high affinity within a few cycles. Saito et al. combined the CE separation of complexes with online sample preconcentration based on transient isotachophoresis [31]. In this method, a large-volume mixture solution containing a DNA library and target microbes are concentrated in the capillary as a narrow zone based on isotachophoresis. Then, the complexes of DNA and microbes are completely resolved from the free DNAs by CE. During electrophoretic migration, the DNAs weakly bound to the target microbes were dissociated, whereas those that were strongly bound remained as complexes. Thus, high-affinity aptamers could be successfully isolated using the proposed method with only one cycle of selection.

This advance in CE-SELEX has markedly reduced the number of rounds of SELEX cycles, although premixing of the target and DNA library in each cycle is required for the equilibrium of association/dissociation. Premixing requires not only a long time owing to the slow molecular diffusion, but also a large-volume consumption of both targets and the DNA library because the optimization of the binding conditions requires many mixtures with different solutions. Zhu et al. studied aptamer selection in the SELEX process without pre-mixing [32]. In the proposed method, plugs of the library and target solutions were individually introduced into the capillary. When a high separation voltage was applied to the capillary, the injected target molecules and DNAs were mixed owing to differences in their electrophoretic mobility. After on-column mixing, the complex and unbound DNAs were rapidly separated by CE. The proposed on-column mixing can easily change the various experimental conditions, which allows the determination of the effects of the aptamer selection environment, such as pH, buffer components, and temperature. In contrast, the number of aptamer candidates obtained decreased with a decrease in the diversity of DNAs that attracted the target molecules in the capillary. Overall, although CE-SELEX is an ideal tool to obtain the desired aptamers, serious problems that require expensive CE equipment and highly skilled technicians remain. Unfortunately, despite their great utility, the number of researchers and technicians able to utilize CE-SELEX is limited, hindering the application of aptamers. Thus, to facilitate aptamer usage as a molecular recognition tool worldwide, a methodology for easier, shorter, more efficient, and less expensive selection is urgently required.

To address this, we propose a novel SELEX method based on microscale electrophoretic filtration (MEF) using a capillary partially plugged with a hydrogel. Our group previously reported on the filtration of enzymes to analyze their kinetics [33]. As previously reported, enzymes incorporated in a hydrogel maintain their enzymatic activity [34,35]; thus, we proposed that the filtrated enzymes also maintain their activity even when they are located inside the hydrogel. The developed technology successfully trapped the target enzymes, alkaline phosphatase (ALP, M_W = ~140 kDa, pI = 4.4~5.8), through the sieving effect of the plugged hydrogel. The activity of the trapped ALP at the interface of the hydrogel could be measured by introducing the substrate molecules by the electrophoretic migration. This revealed that the activity of the trapped ALP did not change significantly compared to that of the ALP in the solution. These results indicated that the conformation of the trapped enzyme was maintained in the proposed selection method, which is based on electrophoretic filtration. Therefore, we applied this technology for aptamer selection (Figure 1). The target molecule was electrokinetically introduced into the capillary partially filled with hydrogel as a trapping step. Based on the molecular sieving effect, target molecules were trapped near the upstream interface of the hydrogel (Figure 1a). After trapping the target molecules, the DNA library was introduced via electrophoresis (the binding step). The DNAs migrated toward the anode and then interacted with the trapped target molecules. According to the association/dissociation equilibrium, DNAs with strong interactions form stable complexes, whereas those with weak or no interactions cannot. Thus, DNAs without strong binding migrated through the hydrogel and were rapidly removed from the capillary (Figure 1b). Only DNAs with binding affinities higher than the threshold were enriched near the interface during the binding step. To obtain surviving DNAs with moderate interactions, the trapped complexes were treated with a washing solution (the washing step). The electrostatic and hydrophobic interactions between the DNAs and target molecules were moderately suppressed by the addition of a salt and surfactant to the washing solution introduced into the capillary. Consequently, the moderately bound DNAs were also dissociated from the complex during the washing step and removed from the boundary (Figure 1c). In the elution step, the eluent containing a higher concentration of salt and surfactant was introduced into the capillary after the washing step. The strongly bound DNAs were dissociated by the eluent before moving toward the downstream reservoir.

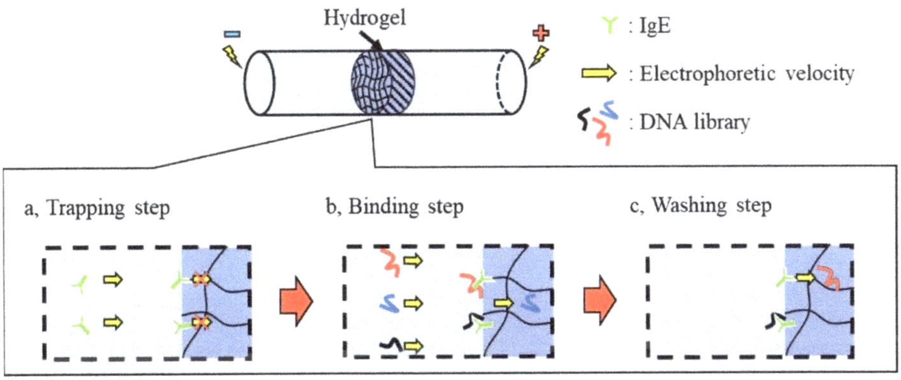

Figure 1. Schematic illustration of aptamer selection in the proposed method. (**a**) Trapping target molecules by the molecular sieving effect of the plugged hydrogel. (**b**) Binding DNAs electrokinetically introduced into the capillary after the trapping targets. The unbound DNA is eliminated to the anodic reservoir solution. (**c**) Washing the complexes using a high-ionic strength buffer to dissociate the weakly/moderately bound DNAs. Finally, the remaining DNAs forming stable complexes is eluted by introducing the eluent containing higher concentrations of additives.

In the proposed method, the unbound, weakly interacting, and moderately bound DNAs can be easily eliminated to the anodic reservoir solution during the binding and washing steps, while the strongly bound DNAs remain as a complex near the upstream interface of the hydrogel. In addition, the random DNA library was continuously supplied during the binding step, enriching the amount of bound DNAs to the trapped targets. Moreover, the performance of the obtained aptamers can be easily modulated by adjusting the additive/composition of the washing/eluting solutions. This method will enable aptamer selection with simple experimental procedures and inexpensive systems compared to conventional methods. In this study, a combination of microscale electrophoretic filtration and SELEX (MEF-SELEX) was demonstrated to confirm the proposed methodology.

2. Results

2.1. *Confirmation of Electrophoretic Filtration of Target Molecules*

In a previous study, the filtration of the target protein (alkaline phosphatase, ALP; M_W = approximately 140 kDa) was confirmed using a capillary partially plugged with a hydrogel (15 wt% acrylamide (AAm)/bis-acrylamide solution (bis-AAm) [33]. The molecular weight of the model target immunoglobulin G (IgG; M_W = approximately 150 kDa), was larger than that of ALP. Thus, the trapping of IgG using the capillary was tested under weakly basic conditions for the electrophoretic migration of IgG. After the fluorescently labeled IgG (F-IgG) was introduced into the hydrogel-plugged capillary by applying the voltage, the fluorescence intensity near the upstream interface of the hydrogel was significantly increased with increasing time for the electrokinetic injection (Figure S1). This indicates that IgG was successfully filtered and trapped at the cathodic interface of the hydrogel in the capillary using the proposed method. Because the molecular weight of immunoglobulin E (IgE; M_W = approximately 190 kDa) is larger than that of IgG, it is assumed that this method also allows for the trapping of IgE, and an experiment related to aptamer selection was conducted.

2.2. Confirmation of the Interaction of DNAs with IgG

After the plugged hydrogel had trapped the IgG, 5′-fluorescein isothiocyanate-labeled DNAs (FITC-DNA) were electrokinetically introduced into the capillary by applying a voltage during the binding step (Figure 2a). Consequently, the fluorescence of FITC-DNAs was observed near the upstream interface of the hydrogel immediately after applying the voltage (Figure 2b,c). The fluorescence intensity profile along the red horizontal dotted line centered in the capillary was estimated using ImageJ software ver. 1.53r (Figure 2d). The fluorescence intensity near the interface increased upon increasing the injection time of FITC-DNA, whereas weak fluorescence in the hydrogel remained during the binding step. This indicated that negatively charged FITC-DNAs migrated toward the anodic side of the capillary, and FITC-DNAs bound to the trapped IgG remained near the upstream interface of the hydrogel, whereas FITC-DNAs without binding passed through the interface. The fluorescence intensity reached a plateau after 10 min of injection with FITC-DNAs (Figure 2d), suggesting that almost the entirety of the IgG could bind with FITC-DNAs. Thus, it was confirmed that DNAs bound to the trapped target were separable from the unbound one using the proposed method.

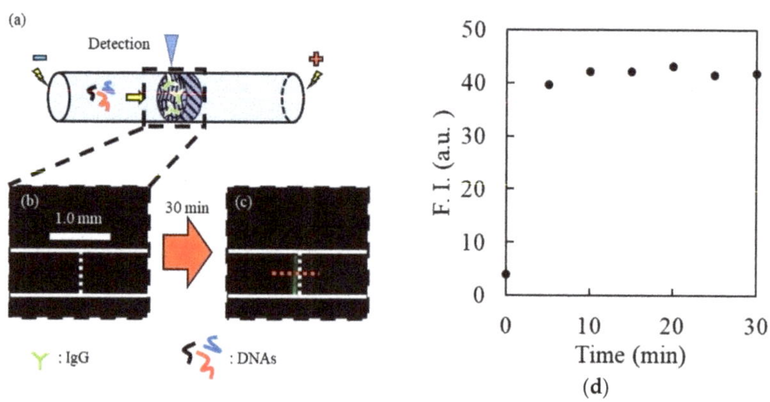

Figure 2. Confirmation of the binding DNAs with IgG. Schematic illustration (**a**) and fluorescence images obtained near the upstream interface of the gel before (**b**) and after (**c**) the introduction of FITC-DNAs (10 nM). The white vertical broken lines represent the cathodic interface of the plugged hydrogel. (**d**) Relationship between time for binding step and fluorescence intensity near the interface of the hydrogel.

2.3. Evaluation of Effects of Additives on DNA Binding

After the binding step, the anodic reservoir solution was exchanged with one containing the background solution, and a washing voltage was applied for 30 min. Consequently, the fluorescence intensity near the upstream interface of the hydrogel continued to decrease to a steady value (black circle in Figure 3a), indicating the dissociation of some FITC-DNAs and the migration of the released FITC-DNAs toward the anodic side of the capillary. In contrast, many FITC-DNAs still remained near the interface as complexes.

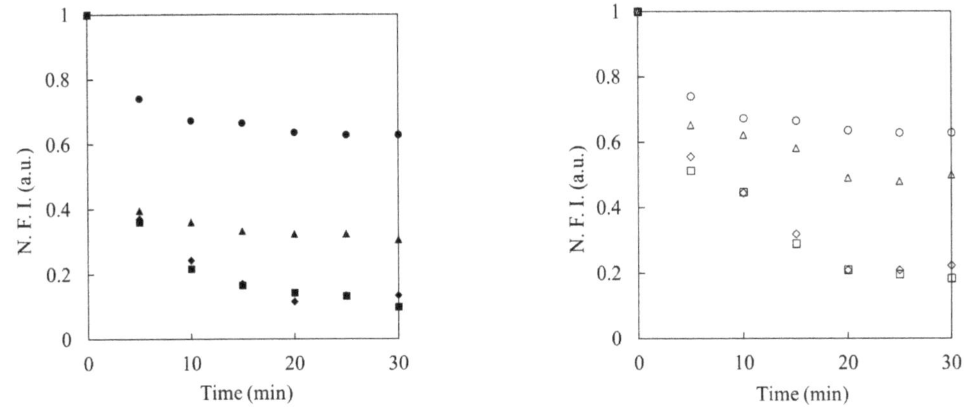

(**a**) (**b**)

Figure 3. Effects of the additives on the interaction between DNAs and trapped IgE during the washing step. (**a**) The concentrations of added NaCl were 0 mM (black circle), 5 mM (black triangle), 25 mM (black rhombus), and 100 mM (black square). (**b**) The concentration of added SDS were 0 mM (white circle), 0.5 mM (white triangle), 0.8 mM (white rhombus), and 1.0 mM (white square). The value of normalized fluorescence intensity (N. F. I.) was defined as 1 when the fluorescence intensity of FITC-DNAs located near the upstream interface of the hydrogel immediately after introducing 10 nM (**a**) or 20 nM DNAs (**b**) for 30 min, respectively.

To study the contribution of the additives to electrostatic and hydrophobic interactions during complexation, a background solution containing various concentrations of sodium chloride (NaCl) or sodium dodecyl sulfate (SDS) was used as the washing solution. When NaCl or SDS was added to the washing solution, the decrease in the fluorescence intensity accelerated as the concentration of the additives increased, compared to that without additives (Figure 3a,b). These results indicate that the addition of salt or surfactant suppressed the electrostatic or hydrophobic interactions between the DNAs and target molecules, respectively, under the experimental conditions. Thus, it was clarified that the strongly bound DNAs could survive by adding an appropriate amount of salt and surfactant to the washing solution.

2.4. Confirmation of Binding DNAs with IgE

To confirm the effect of the trapped amount of target molecules on DNA binding, IgE and FITC-labeled IgE aptamer [29] (FITC-IgE-Apt) were employed as model samples in the proposed method. After trapping 0.0, 1.0, 1.5, and 2.0 μg/mL IgE for 30 min, the cathodic side of the reservoir solution was exchanged with that containing the background solution, and then 100 V was applied for 5 min to prevent IgE from reaching the hydrogel. FITC-IgE-Apt was then introduced into the IgE-trapped capillaries for 30 min. Thus, a strong fluorescence was also observed near the cathodic-side interface of the hydrogel plug (Figure 4), as described above. After the binding step, the values of fluorescence intensity increased as higher concentrations of IgE were introduced, as shown by the binding step in Figure 4. During the washing step, weakly or moderately adsorbed FITC-IgE-Apt was eliminated by additives. The higher concentration of IgE introduced into the capillary also caused stronger fluorescence to remain after the washing step. This indicates that the higher the introduced concentration of IgE, the greater the amount of trapped IgE and, accordingly, the greater the number of binding sites. When FITC-IgE-Apt was introduced without trapped IgE, on the other hand, the fluorescence increased slightly during the binding step, which indicated that the weak stacking effect was due to the molecular sieving effect of the hydrogel, which decelerated the FITC-IgE-Apt. The observed fluorescence decreased until it was nearly the same as the background fluorescence intensity during the washing step because the FITC-IgE-Apt located near the interface migrated toward the anodic side of the capillary without binding/adsorption. Therefore, it was concluded that the non-specific adsorption of the DNAs to the hydrogel was negligible under the experimental conditions.

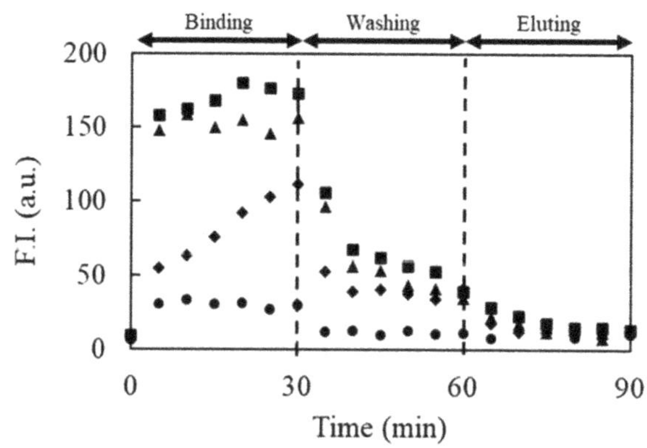

Figure 4. Effect of the concentration of introduced IgE on the fluorescence intensity of bound F-IgE-Apt. Concentrations of the trapped IgE were 0 µg/mL (black circle), 1.0 µg/mL (black rhombus), 1.5 µg/mL (black triangle), and 2.0 µg/mL (black square).

When the eluent containing a higher concentration of NaCl than that in the washing solution was introduced into the capillary (elution step in Figure 4), the fluorescence intensities decreased near the background. The disappearance of fluorescence indicated that the strongly bound FITC-IgE-Apt could be eluted from the trapped interface owing to further suppression of the electrostatic interactions. These results confirm that the proposed method allows aptamer candidates to be obtained only by simple electrophoresis, by tuning the concentrations of the additives in the washing solution and eluent.

2.5. Selection of IgE Aptamer Candidates by the Proposed Method

Based on the above results and discussion, a 2.0 µg/mL IgE solution prepared with the background solution was employed to increase the amount of bound DNAs. A library solution containing 10 nM random DNAs and 50 mM NaCl was also prepared with the background solution to suppress the binding of DNAs with weak interactions with IgE during the binding step. A background solution containing 100 mM NaCl and 1 mM SDS was prepared as the washing solution to eliminate the DNAs moderately bound to IgE during the washing step. To elute the aptamer candidates effectively, a higher concentration of NaCl (300 mM) was added to the background solution containing 1 mM SDS to suppress both electrostatic and hydrophobic interactions.

To enrich DNAs with strong binding to IgE, the proposed selection process was repeated for three cycles, as shown in Scheme 1. After each elution step, PCR was performed to confirm the collection of eluted DNAs. Consequently, the band of the PCR product was observed between the 60 and 80 bp ladders, indicating the successful amplification of the aptamer candidates with a number of their sequence being 78 bp in each PCR product (Figure 5a). This confirmed that the introduction of the eluent allowed for the collection of DNAs that survived the washing step. In each cycle of the proposed method, next-generation sequencing (NGS) using the MiSeq system was carried out after labeling each product (1st, 2nd, and 3rd products in Scheme 1) by PCR for sequencing. The obtained sequence data were analyzed using AptamCORE software to explore the aptamer candidates against IgE. As a result, more than two thousand sequences were output by NGS, and the number of major sequences was listed using AptamCORE. In this experiment, the top two sequences, S1 and S2 (Table 1), were specifically enriched during the three-cycle selection, and the ratio of the number of S1 and S2 sequences to that of the total sequence reached over 20% and 10%, respectively (Figure 5b). Repeating the selection also increased the level of S1 and S2, as shown in Figure 5c. Unfortunately, the sequence S2 was not found in the NGS data in the first round because only 5 μL of the eluent (total volume, 500 μL) was used for the NGS analysis in the first round of the experimental procedure. The enrichment factor in each round is defined as follows:

(enrichment factor) = (number S1 or S2 in round n)/(number of S1 or S2 in round $(n-1)$).

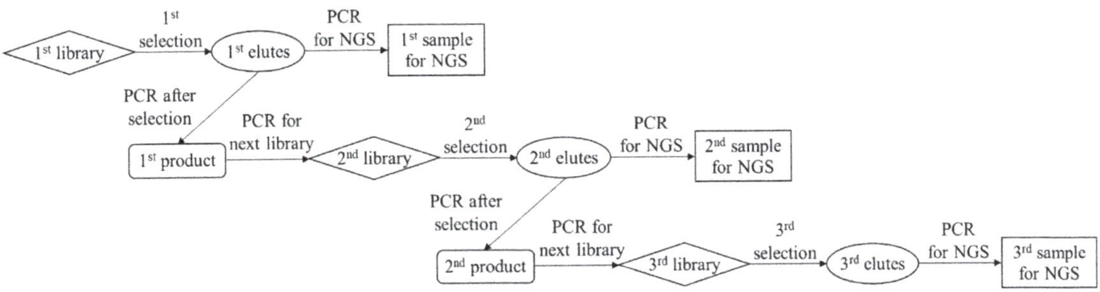

Scheme 1. Schematic illustration of the amplification of the obtained DNAs by the proposed method for the aptamer selection and NGS.

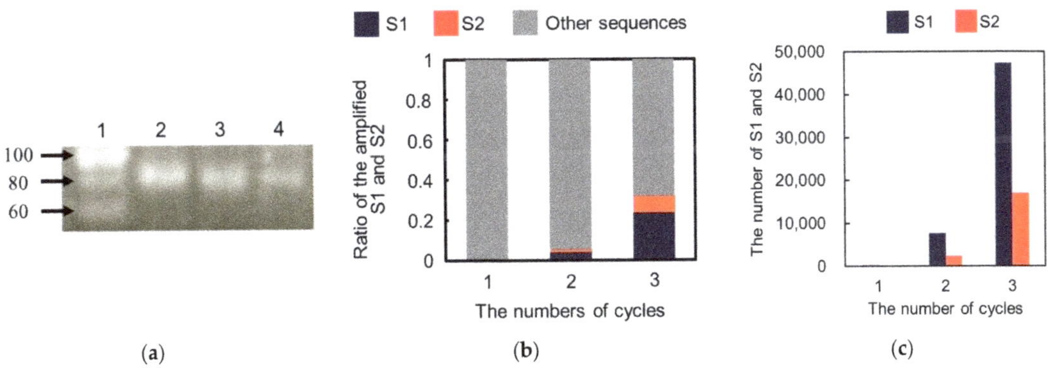

Figure 5. (**a**) Confirmation of the PCR amplification by agarose gel electrophoresis. (Lane 1:20 bp ladder, lanes 2–4: 1st, 2nd, and 3rd products in Scheme 1) (**b**) Ratio of specifically enriched S1 and S2 to the number of total sequences in each cycle. (**c**) The number of enriched S1 and S2 sequences in each cycle.

Table 1. The most amplified sequences after three cycles.

	Sequences (5′–3′)
S1	**GACTACCCGGGTATCTAATCC**GCGTATGGTCATGGATG-TGCATACACCCCGCCTATCAGAACTG**CTGCCGCCCGTAGG**
S2	**GACTACCCGGGTATCTAATCC**GACCATTTTTCGTCT-CCTTTGTACGAGCAGTGTGCTCGAC**CTGCCGCCCGTAGG**

Bold text indicates the primer sequence.

The enrichment factor values for the S1 sequence were calculated to be 26- and 6.2-fold in the 2nd and 3rd rounds, respectively. The enrichment value for S2 was calculated to be 7.2-fold in the 3rd round (Figure 5c). As described above, only strongly bound DNAs, owing to their sequence, are able to survive the proposed selection scheme. Thus, DNAs with different sequences were assumed aptamer candidates and their affinity for IgE was verified.

2.6. Evaluation of the Sequences and Affinity of the Obtained Aptamer Candidates

The binding affinities of the two aptamer candidates, S1 and S2, were evaluated using CE-LIF to verify the concept of aptamer selection based on MEF. As shown in Figure 6, 5′-FITC-labeled S1 or S2 (FITC-S1, FITC-S2) was detected as a single peak with strong intensity at 100 s. When FITC-S1 or FITC-S2 was mixed with IgE, the signal of each aptamer candidate disappeared, and new peaks were observed at a detection time of 570 s. These results indicate that almost all free FITC-S1 and FITC-S2 were bound to IgE, resulting in the variation of electrophoretic mobility by complex formation. Focusing on the shapes of the newly appeared peaks of the complexes in Figure 6a,b, two clearly separated small signals were observed in the cases of the mixture of FITC-S1 or FITC-S2 and IgE. These results suggest that one or two molecules of S1 or S2 can bind to similar structures in IgE, such as Fab fragments. In addition, the peak areas of these signals were smaller than those of the free FITC-S1 or FITC-S2. The decrease in the total peak area may be due to the interaction of labeled FITC with aromatic amino acids. However, the fundamental reasons for the variations in both the peak shapes and areas were not elucidated in this study, which made it difficult to evaluate the dissociation constants of S1 and S2 under experimental conditions. Experiments using non-equilibrium capillary electrophoresis of equilibrium mixtures (NECEEM) experiments were conducted to evaluate the binding ability of S1 or S2 to IgE. Based on the results of these experiments, the K_D values of S1 and S2 were estimated to be 0.83 and 1.55 nM, respectively (see Supporting Information). This indicated that the selected aptamer candidates, S1 and S2, exhibited strong binding abilities to IgE. Mixtures of FITC-S1 or FITC-S2 and IgG or BSA were also analyzed by CE-LIF to evaluate the specificity of S1 and S2. No significant changes were observed in the electropherograms compared to those obtained by injecting FITC-S1 or FITC-S2 alone (Figure 6). These results indicate that S1 and S2 exhibited specificity to IgE; at least there were no noticeable interactions with IgG or BSA. Consequently, the aptamer candidates selected by the proposed method exhibited both a strong binding ability and specificity to IgE.

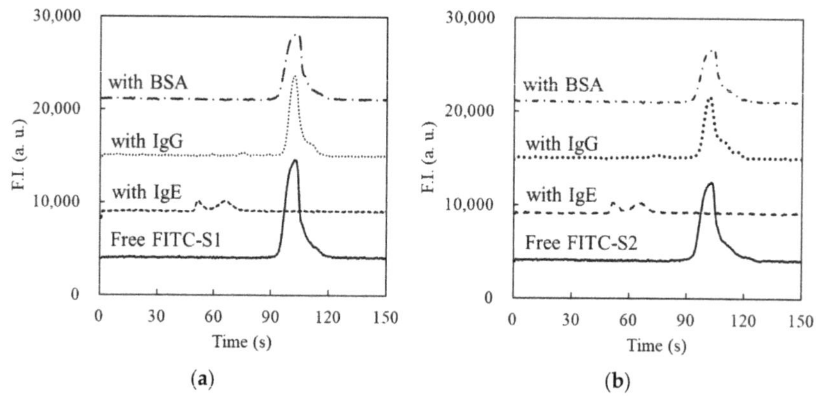

Figure 6. Confirmation of the affinity of S1 (**a**) or S2 (**b**) by CE-LIF. The separation voltage of 15 kV was applied to the fused silica capillary (total length: 30.2 cm, effective length: 20.0 cm). The lines represent the results in the CE analyses of 10 nM FITC-S1 or FITC-S2 (solid line), mixture of 10 nM FITC-S1 or FITC-S2 with 100 nM IgE (dashed line), mixture of 10 nM FITC-S1 or FITC-S2 with 100 nM IgG (dotted line), and mixture of 10 nM FITC-S1 or FITC-S2 with 100 nM BSA (dashed–dotted line), respectively.

3. Materials and Methods

3.1. Chemical Reagents

IgE purified from human myeloma was purchased from Athens Research & Technology, Inc. (Athens, GA, USA). NaOH was obtained from Kanto Chemical Co., Inc. (Tokyo, Japan). A prepolymer solution containing 30% (*w*/*w*) acrylamide (AAm)/bis-acrylamide solution (bis-AAm) (AAm:bis-AAm = 37.5:1) and IgG from human serum was purchased from Sigma-Aldrich (St. Louis, MO, USA). The reagents included in the polymerase and dNTPs (KAPA HiFi HS ReadyMix) were purchased from Nippon Genetics Co., Ltd. (Tokyo, Japan). Hydrochloric acid, SDS and 2-amino-2-hydroxymethyl-1,3-propanediol were purchased from FUJIFILM Wako Pure Chemical Corporation (Osaka, Japan). The reagents, 2-hydroxy-2-methylpropiophenone (HOMPP) and 3-(trimethoxysilyl)propyl methacrylate, were purchased from Tokyo Chemical Industry (Tokyo, Japan). Xpress Micro Dialyzer MD300 was purchased from Scienove GmbH (Jena, Germany). VB-G25Slide-A-Lyzer dialysis cassettes (2 K MWCO, 0.5 mL) were obtained from Thermo Fisher Scientific Inc. (Waltham, MA, USA). Sodium chloride (NaCl) was obtained from Nacalai Tesque, Inc. (Kyoto, Japan). The HiLyte Fluor^{TM} 555 Labeling Kit-NH_2 was purchased from Cosmo Bio Co., Ltd. (Tokyo, Japan). The PCR clean-up kit was purchased from Takara Bio, Inc. (Shiga, Japan). A DNA library of random sequences and fluorescently labeled IgE aptamers reported by Shuan et al. (5′-FITC-GACTACCCGGGTATCTAATCCTGAAACATAGCATATTTACTTATGTCGCCTTGCCGGT-TCTGCTGCCGCCCGTAGG-3′, K_D = 23 ± 12 nM) was synthesized by Tsukuba Oligo Service Co., Inc. (Ibaraki, Japan). Forward and reverse primers for sequencing and forward and reverse primers for selection were synthesized by BEX Co., Ltd. (Aichi, Japan) and are listed in Table S1. Deionized water (18×10^7 S/cm) was prepared using a Milli-Q system (direct-Q UV 3; Merck, KGaA, Darmstadt, Germany).

3.2. Apparatus

A fluorescence microscope (VB-S20; Keyence Corp., Osaka, Japan) equipped with a light setup (120 W) (VB-L10; Keyence Corp., Osaka, Japan) and a battery (PA250-0.25B; TEXIO Corp., Kanagawa, Japan) was used to observe the fluorescence of the labeled IgE and DNAs in the capillary. A mask aligner (250 W) (UV-CL251S; San-Ei Electric, Osaka, Japan) was used for photopolymerization of the hydrogel. The thermal cycler (TurboCycler) was purchased from Blue-Ray Biotech Corp. (Xindian, Taiwan). A capillary electrophoresis apparatus equipped with a laser-induced fluorescence detector (CE-LIF) (PA800 plus; AB SCIEX LLC, Framingham, MA, USA) was used to evaluate the affinities of the aptamer candidates.

3.3. Fabrication of Electrophoresis Device for Aptamer Selection

The preparation of the hydrogel-plugged capillary is described in a previous study [33]. Briefly, a glass capillary (Drummond Scientific Company, Broomall, PA, USA) was cut into 2 cm pieces, and their inner surfaces were activated by sinking them in 0.1 M NaOH solution for 30 min. After washing the capillaries with deionized water and drying them, the inner surface of the capillary was modified by sinking in a solution mixed with 3-(triethoxysilyl)propyl methacrylate and 100 mM HCl (4:1, *v*/*v*) for 3 h. After the modification and washing of the remaining solutions, the modified capillary was dried using a vacuum pump.

To prepare a capillary partially filled with a hydrogel, a 0.2 μL prepolymer solution (50 mM Tris-HCl buffer (pH 9.0) containing 15 wt% AAm/bis-AAm and 1 vol% HOMPP) was introduced into the modified capillary, and the solution was moved to the central part of the capillary by pneumatic pressure using a micropipette. After forming the hydrogel (length: ~2 mm) by UV light irradiation for 2 min, the hollow parts of the capillary were filled with 10 mM Tris-HCl buffer (pH 9.0) as a background solution. In the experiments, a capillary with an inner diameter of 500 or 350 μm was used for the measurements. A thicker capillary was used to observe electrophoretic migration and the filtration of fluorescently labeled proteins and DNA. A narrower one was used for the selection of aptamers to suppress Joule heating during the experiments. As disposable reservoirs, contact holes were punched into the flat top and curved side surfaces of the sampling tubes (650 μL). The fluidic adapter was made of a silicon elastomer (Fukoku Bussan Co., Ltd. Tokyo, Japan) and was attached to the flat top of the disposable reservoirs to suppress solution leakage. The capillary was connected to the reservoirs via connectors, and then each platinum electrode was inserted into the connecting hole located on the curved side surface of the reservoirs. To prevent misalignment during the experiments, the assembled devices were placed on a jig fabricated using a 3D printer (Figure S2).

3.4. *Confirmation of Electrophoretic Filtration of Target Molecules*

To confirm the trapping of target molecules by microscale electrophoretic filtration, IgG (M_W, approximately 150 kDa) was used as a model protein instead of the rare and expensive IgE (M_W, approximately 190 kDa). IgG was fluorescently labeled using a HiLyte Fluor^{TM} 555 labeling kit according to a previous work [33].

After labeling, a fluorescently labeled IgG (F-IgG) solution was prepared with a 10 mM background solution (pH 9.0). The cathodic/anodic reservoirs were filled with a buffer solution with or without F-IgG. F-IgG was then electrokinetically introduced into the hydrogel-plugged capillary by applying a voltage of 100 V for 30 min. The fluorescence near the upstream interface of the hydrogel plug was observed using a fluorescence microscope equipped with excitation/emission filters (RFP for excitation, 540 ± 25 nm; RFP for emission, >572 nm). The fluorescence intensity was calculated from the obtained fluorescence images using the ImageJ software (NIH, Bethesda, MD, USA) (Figure S1a).

3.5. *Confirming the Interaction of DNAs with IgG*

To confirm the interaction of DNAs with the target molecules trapped near the hydrogel, IgG and 5′-fluorescein isothiocyanate-labeled random DNAs (FITC-DNAs) were employed as model samples. After the electrokinetic injection of 100 nM IgG at a voltage of 100 V for 30 min, the cathodic reservoir was replaced with that filled with background solution, and a voltage of 100 V was added for 5 min to focus the IgG remaining in the IgG-trapped capillary. During the electrokinetic injection of 10 nM FITC-DNA by applying 100 V for 30 min (binding step), fluorescence near the hydrogel interface was monitored using a fluorescence microscope. After exchanging the anodic and cathodic reservoirs containing the background solution, a washing voltage of 100 V was applied for 30 min (the washing step). During the washing step, fluorescence from F-DNA was observed at the upstream interface of the capillary. Fluorescence imaging during the binding and washing steps was performed using a microscope equipped with filters (GFP-B: 470 ± 40 nm; GFP-B: >535 ± 50 nm). The fluorescence intensities were also evaluated using ImageJ software.

3.6. *Evaluation of Effects of Additives on DNA Binding*

To assess the effects of the additives in the washing solutions on the interactions between DNAs and IgG, washing solutions containing various concentrations of NaCl (0, 5, 25, 50, and 100 mM) or SDS (0, 0.5, 0.8, and 1 mM) were prepared to control the electrostatic and hydrophobic interactions on complexation, respectively. After the electrokinetic injection of 10 or 20 nM FITC-DNAs into the IgG-trapped capillary, as described above, both reservoirs were exchanged with those filled with the washing solution containing the additives. The fluorescence imaging and estimation of the fluorescence intensity were the same as described in the previous section.

3.7. *Confirmation of Binding DNA with IgE*

The target protein solutions containing IgE were prepared by diluting commercially available IgE solution (1550 μg/mL) with 10 mM Tris-HCl buffer solution (pH 9.0). FITC-labeled IgE aptamer solution (FITC-IgE-Apt, 10 nM) was prepared with 10 mM Tris-HCl buffer solution (pH 9.0). Under the experimental conditions, it was confirmed that the IgE aptamer could specifically bind to IgE by CE-LIF (Figure S3). Each solution with a different concentration of IgE (0.0, 1.0, 1.5, and 2.0 μg/mL) was electrokinetically introduced into each capillary by applying voltages of 100 V for 30 min, resulting in the focusing of IgE onto the upstream interface of each hydrogel. After washing with the background buffer solution for 5 min, the cathodic reservoir was exchanged with that containing FITC-IgE-Apt solution. FITC-IgE-Apt was then introduced into the capillary by applying a voltage of 100 V for 30 min. A washing solution (10 mM Tris HCl (pH 9.0) containing 20 mM NaCl) was employed in the washing step. After the washing step, FITC-IgE-Apt was eluted using an eluent (10 mM Tris HCl (pH 9.0) containing 50 mM NaCl). Among the above processes, fluorescence by FITC-IgE-Apt was observed and evaluated, as described in the experiment using F-DNAs.

3.8. Selection of IgE Aptamer Candidates by the Proposed Method

For the selection of aptamer candidates against IgE, 2.0 μg/mL IgE was prepared with the background solution to increase the amount of bound DNAs. A solution containing DNAs with successive 40 random sequences between the forward and reverse primer regions was selected as the library solution. The reagent (10 nM) was prepared with a 10 mM background solution containing 50 mM NaCl without any fluorescent labeling. In the washing step, anodic and cathodic reservoirs filled with 10 mM Tris-HCl buffer solution (pH 9.0) containing 100 mM NaCl and 1 mM SDS were employed to eliminate unbound or weakly bound DNAs by reducing both electrostatic and hydrophobic interactions, respectively. In the elution step, the anodic and cathodic reservoirs filled with 10 mM Tris-HCl buffer solution (pH 9.0) containing 300 mM NaCl and 1 mM SDS were employed for the elution of the strongly bound DNAs to collect the potential aptamer candidates. The other experimental conditions were identical to those described above.

3.9. Amplification of DNAs for IgE Aptamer Candidates and Sequence Analysis

As shown in Scheme 1, PCR amplification of the eluted DNAs and purification of the amplified products were performed to enrich the aptamer candidates with strong binding. Briefly, the PCR solution was prepared by mixing the eluent (2.5 μL), 2 × KAPA HiFi HS ReadyMix (12.5 μL), 1 μM solution of reverse primer for selection (5 μL), and 1 μM solution of forward primer for selection (5 μL). PCR tubes containing the mixed solutions were placed on a thermal cycler. After the first heating step at 95 °C for 3 min, PCR amplification was conducted by repeating the following steps: denaturation at 95 °C for 30 s, annealing at 60 °C for 30 s, and extension at 72 °C for 10 s. These steps were repeated for 25 cycles, and unreacted sequences were extended completely at 72 °C for 5 min. After PCR amplification, the solution containing the PCR products was purified using a PCR clean-up kit, resulting in the first generation of purified PCR products (1st product). Then, the 1st products solutions were amplified and purified again to increase the amount of the next-generation library required for the proposed selection protocol, providing a second generation of the library (2nd library) solution. As shown in Scheme 1, the 2nd products, 3rd libraries, and 3rd products were also obtained by repeating the described cycle, including selection, amplification, and purification. In each cycle, gel electrophoresis of the PCR products was performed using planar agarose gels (3% (w/v)) to confirm the appropriate amplification. Additionally, each generation of the products was amplified as per the described PCR protocol by using the forward and reverse primers for sequencing instead of those for the selection, resulting in the 1st, 2nd, and 3rd samples for NGS.

3.10. Evaluation of the Sequences and Affinity of the Obtained Aptamer Candidates

After preparation of the 1st, 2nd, and 3rd samples, NGS was performed using the Miseq system (Illumina, San Diego, CA, USA) for common use within the Osaka Metropolitan University. The NGS data were analyzed using AptamCORE software to list the aptamer candidates.

To evaluate the binding affinity of the aptamer candidates targeting IgE, listed in Table 1 (S1 and S2), CE-LIF analyses of the fluorescently labeled aptamer candidates were performed. A fused silica capillary (total length: 30.2 cm, effective separation length: 20.0 cm) was used for the analyses. The solutions of the FITC-labeled IgE aptamer candidates, FITC-S1 and FITC-S2, were prepared and mixed with solutions of IgE, IgG, or BSA in 10 mM Tris-HCl buffer solution (pH 9.0). The final concentration of FITC-S1 and FITC-S2 was 10 nM, whereas those of the proteins were 0 (not added) or 100 nM. Before the analyses, 1 M NaOH solution was introduced at a pressure of 20 psi for 3 min to refresh the inner surface of the capillary, and deionized water was then introduced to remove the NaOH solution at a pressure of 20 psi for 3 min. The capillary was then filled with 10 mM Tris-HCl buffer solution (pH 9.0) by applying a pressure of 20 psi for 1 min to precondition the capillary. After preconditioning, each sample solution was injected into the capillary at a pressure of 0.5 psi for 8 s. Both ends of the capillary were inserted into the anodic and cathodic reservoirs containing the background solution, and a separation voltage of 15 kV was applied to the capillary via the reservoir solutions. The fluorescence of the free candidates or complexes was observed at the detection point, revealing their affinity for IgE.

4. Discussion

In the proposed method, only three cycles were required for selection, which was less than that obtained by the conventional method using IgE-immobilized beads [36]. Although the number of cycles was comparable to that using capillary electrophoresis [29], CE-based selection requires a much more expensive apparatus and sophisticated knowledge and expertise, whereas the proposed method can be conducted using inexpensive apparatus and disposable devices with simple experimental procedures. Thus, the proposed method is expected to become an ideal tool for the establishment of aptamers that target the desired molecules for capture using a hydrogel.

5. Conclusions

In this study, a novel method for aptamer selection based on microscale electrophoretic filtration was developed. The basic concept of the proposed method was confirmed by using IgG and IgE as model targets. The target molecules were successfully trapped by the sieving effect of the plugged hydrogel, which overcame the issues in SELEX using target-immobilized beads. It was also confirmed that the interaction between DNAs and the trapped targets could be controlled by the electrokinetic injection of the DNA library. Online mixing due to fast electrophoresis allowed tuning of the experimental conditions without labor-intensive premixing, which is generally required for conventional SELEX methods owing to slow molecular diffusion. Additionally, the efficiency of the binding was expected to improve because the non-binding DNAs were immediately removed from the interface, trapping the target molecules by electrophoresis. Furthermore, the amount of DNAs that interact with target molecules could be easily increased by extending the injection time.

In the washing step, weakly and moderately bound DNAs could be easily eliminated by tuning the washing solution containing additives. Thus, only the DNAs that can survive during the washing step can be obtained in the elution step, which enhances the quality of the aptamer selection based on the proposed method in each cycle for selection. After three cycles for selecting the IgE aptamer, two types of DNAs with different sequences, S1 and S2, were selectively enriched as aptamer candidates. Both aptamers were found to be specific for IgE and to strongly interact with it, with K_D values around 1 nM. These findings establish the validity of the proposed method. Compared to other reports, the proposed method provides not only an effective selection of aptamers but also simple experimental procedures using disposable and cheap devices. However, the inner diameter of the capillary (350 μm) was larger than that of the conventional CE-SELEX (50 μm). Thus, the electrophoretic migration of the molecules in the developed device was slower owing to the lower applicable voltage, requiring a short time for binding, washing, and elution for 30 min. In addition, the higher electric current caused Joule heating and electrolysis, such that large-volume reservoirs (500 μL) had to be used to reduce their effects. Although issues remain and the proposed method is not yet ideal for aptamer selection, these can be overcome easily by integrating the proposed method on a microfluidic chip. Aptamer selection using a microfluidic device integrating MEF will simplify the experimental procedure, reduce the consumption of reagents, and shorten the total time, which will also facilitate the improvement of the specificity of the selected aptamer by further integration to connect the negative selection device, as previously reported.

Supplementary Materials: The following supporting information can be downloaded at: https://www.mdpi.com/article/10.3390/molecules27185818/s1, Table S1: The sequences of the primers and the library which used with measurements., Figure S1: Confirmation of the trapping IgG., Figure S2: Photographs of the experimental setup., Figure S3: Analysis of the affinity of the IgE aptamer for IgE in 10 mM Tris-HCl buffer solution (pH 9.0) using CE-LIF. Section SI1: Calculating LOD for S1 on electropherogram results. Section SI2: Evaluation of the binding ability of the aptamer candidates using nonequilibrium capillary electrophoresis of equilibrium mixtures (NECEEM) experiments [37].

Author Contributions: Conceptualization and methodology, J.T., T.E., H.H., and K.S.; validation, J.T. and R.N.; formal analysis and investigation, resources, and data curation, J.T. and K.S.; writing-original draft preparation, J.T.; writing-review and editing, visualization, and supervision, J.T., T.E., H.H. and K.S.; project administration and funding acquisition, K.S. All authors have read and agreed to the published version of the manuscript.

Funding: This study was partially supported by JST, PRESTO, Japan (Grant No. JPMJPR19H7).

Institutional Review Board Statement: Not applicable.

Informed Consent Statement: Not applicable.

Data Availability Statement: Not applicable.

Acknowledgments: The authors greatly appreciate Junko Muramatsu (Institute for Advanced Genomics, Osaka Metropolitan University, Osaka, Japan) for performing next-generation sequencing using MiSeq. The authors also thank Eri Matsui (KONICA MINOLTA Inc., Tokyo, Japan) for evaluating the affinity of the aptamer candidates obtained using CE-LIF.

Conflicts of Interest: The authors declare no conflict of interest.

Sample Availability: Samples of the compounds are available from the companies listed.

References

1. Darmostuk, M.; Rimpelova, S.; Gbelcova, H.; Ruml, T. Current approaches in SELEX: An update to aptamer selection technology. *Biotechnol. Adv.* **2015**, *33*, 1141–1161. [CrossRef] [PubMed]
2. Saito, S. SELEX-based DNA Aptamer Selection: A Perspective from the Advancement of Separation Techniques. *Anal. Sci.* **2021**, *37*, 17–26. [CrossRef] [PubMed]
3. Wang, T.; Chen, C.; Larcher, L.M.; Barrero, R.A.; Veedu, R.N. Three decades of nucleic acid aptamer technologies: Lessons learned, progress and opportunities on aptamer development. *Biotechnol. Adv.* **2019**, *37*, 28–50. [CrossRef]
4. Bogná, Z.; Gyurcsányi, R.E. Aptamers against Immunoglobulins: Design, Selection and Bioanalytical Applications. *Int. J. Mol. Sci.* **2020**, *21*, 5748. [CrossRef]
5. Hong, S.-L.; Xiang, M.-Q.; Tang, M.; Pang, D.-W.; Zhang, Z.-L. Ebola Virus Aptamers: From Highly Efficient Selection to Application on Magnetism-Controlled Chips. *Anal. Chem.* **2019**, *91*, 3367–3373. [CrossRef]
6. Sun, M.; Liu, S.; Wei, X.; Wan, S.; Huang, M.; Song, T.; Lu, Y.; Weng, X.; Lin, Z.; Chen, H.; et al. Aptamer Blocking Strategy Inhibits SARS-CoV-2 Virus Infection. *Angew. Chem. Int. Ed.* **2021**, *60*, 10266–10272. [CrossRef]
7. Zhang, Y.; Yu, Z.; Jiang, F.; Fu, P.; Shen, J.; Wu, W.; Li, J. Two DNA Aptamers against Avian Influenza H9N2 Virus Prevent Viral Infection in Cells. *PLoS ONE* **2015**, *10*, e0123060–e0123072. [CrossRef]
8. Kordasht, H.K.; Hasanzadeh, M. Aptamer based recognition of cancer cells: Recent progress and challenges in bioanalysis. *Tlanta* **2020**, *220*, 121436–121452. [CrossRef]
9. Zamanian, J.; Khoshbin, Z.; Abnous, K.; Taghdisi, S.M.; Hosseinzadeh, H.; Danesh, N.M. Current progress in aptamer-based sensing tools for ultra-low level monitoring of Alzheimer's disease biomarkers. *Biosens. Bioelectron.* **2022**, *197*, 113789–113804. [CrossRef]
10. McKeague, M.; DeRosa, M.C. Challenges and Opportunities for Small Molecule Aptamer Development. *J. Nucleic Acids.* **2012**, *2012*, 748913. [CrossRef]

11. Guo, W.; Zhang, C.; Ma, T.; Liu, X.; Chen, Z.; Li, S.; Deng, Y. Advances in aptamer screening and aptasensors' detection of heavy metal ions. *J. Nanobiotechnol.* **2021**, *19*, 166–184. [CrossRef] [PubMed]
12. Singhal, C.; Bruno, J.G.; Kaushal, A.; Sharma, T.K. Recent Advances and a Roadmap to Aptamer-Based Sensors for Bloodstream Infections. *ACS Appl. Bio Mater.* **2021**, *4*, 3962–3984. [CrossRef] [PubMed]
13. Sharma, T.K.; Bruno, J.G.; Dhiman, A. ABCs of DNA aptamer and related assay development. *Biotechnol. Adv.* **2017**, *35*, 275–301. [CrossRef]
14. Yu, H.; Alkhamis, O.; Canoura, J.; Liu, Y.; Xiao, Y. Advances and Challenges in Small-Molecule DNA Aptamer Isolation, Characterization, and Sensor Development. *Angew. Chem. Int. Ed.* **2021**, *60*, 16800–16823. [CrossRef] [PubMed]
15. Radom, F.; Jurek, P.M.; Mazurek, M.P.; Otlewski, J.; Jeleń, F. Aptamers: Molecules of great potential. *Biotechnol. Adv.* **2013**, *31*, 1260–1274. [CrossRef]
16. Dzuvor, C.K.O.; Tettey, E.L.; Danquah, M.K. Aptamers as promising nanotheranostic tools in the COVID-19 pandemic era. *Wiley Interdiscip. Rev. Nanomed. Nanobiotechnol.* **2022**, *14*, e1785–e1813. [CrossRef]
17. Sánchez-Báscones, E.; Parra, F.; Lobo-Castañón, M.J. Aptamers against viruses: Selection strategies and bioanalytical applications. *Trends Anal. Chem.* **2021**, *143*, 116349–116370. [CrossRef]
18. Banerjee, J.; Nilsen-Hamilton, M. Aptamers: Multifunctional molecules for biomedical research. *J. Mol. Med.* **2013**, *91*, 1333–1342. [CrossRef]
19. Zhuo, Z.; Yu, Y.; Wang, M.; Li, J.; Zhang, Z.; Liu, J.; Wu, X.; Lu, A.; Zhang, G.; Zhang, B. Recent Advances in SELEX Technology and Aptamer Applications in Biomedicine. *Int. J. Mol. Sci.* **2017**, *18*, 2142. [CrossRef]
20. Vandghanooni, S.; Barar, J.; Eskandani, M.; Omidi, Y. Aptamer-conjugated mesoporous silica nanoparticles for simultaneous imaging and therapy of cancer. *Trends Anal. Chem.* **2020**, *123*, 115759–115772. [CrossRef]
21. Tuerk, C.; Gold, L. Systematic Evolution of Ligands by Exponential Enrichment: RNA Ligands to Bacteriophage T4 DNA Polymerase. *Science* **1990**, *249*, 505–510. [CrossRef] [PubMed]
22. Ellington, A.D.; Szostak, J.W. In vitro selection of RNA molecules that bind specific ligands. *Nature* **1990**, *346*, 818–822. [CrossRef]
23. Kimoto, M.; Yamashige, R.; Matsunaga, K.; Yokoyama, S.; Hirao, I. Generation of high-affinity DNA aptamers using an expanded genetic alphabet. *Nat. Biotechnol.* **2013**, *31*, 453–457. [CrossRef] [PubMed]
24. Bruno, J.G. In Vitro Selection of DNA to Chloroaromatics Using Magnetic Microbead-Based Affinity Separation and Fluorescence Detection. *Biochem. Biophys. Res. Commun.* **1997**, *234*, 117–120. [CrossRef]
25. Duan, A.N.; Li, C.; Zhu, X.; Qi, S.; Wang, Z.; Wu, S. Simultaneous coupled with Separate SELEX for heterocyclic biogenic amine-specific aptamers screening and their application in establishment of an effective aptasensor. *Sens. Actuators B Chem.* **2022**, *352*, 130985–130993. [CrossRef]
26. Wang, Q.; Liu, W.; Xing, Y.; Yang, X.; Wang, K.; Jiang, R.; Wang, P.; Zhao, Q. Screening of DNA Aptamers against Myoglobin Using a Positive and Negative Selection Units Integrated Microfluidic Chip and Its Biosensing Application. *Anal. Chem.* **2014**, *86*, 6572–6579. [CrossRef]
27. Zhu, C.; Yang, G.; Ghulam, M.; Li, L.; Qu, F. Evolution of multi-functional capillary electrophoresis for high-efficiency selection of aptamers. *Biotechnol. Adv.* **2019**, *37*, 107432–107447. [CrossRef]
28. Mendonsa, S.D.; Bowser, M.T. In Vitro Evolution of Functional DNA Using Capillary Electrophoresis. *J. Am. Chem. Soc.* **2004**, *126*, 20–21. [CrossRef]
29. Mendonsa, S.D.; Bowser, M.T. In Vitro Selection of High-Affinity DNA Ligands for Human IgE Using Capillary Electrophoresis. *Anal. Chem.* **2004**, *76*, 5387–5392. [CrossRef]
30. Wakui, K.; Yoshitomi, T.; Yamaguchi, A.; Tsuchida, M.; Saito, S.; Shibukawa, M.; Furusho, H.; Yoshimoto, K. Rapidly Neutralizable and Highly Anticoagulant Thrombin-Binding DNA Aptamer Discovered by MACE SELEX. *Mol. Ther. Nucleic. Acids.* **2019**, *16*, 348–359. [CrossRef]
31. Saito, S.; Hirose, K.; Tsuchida, M.; Wakui, K.; Yoshimoto, K.; Nishiyama, Y.; Shibukawa, M. Rapid acquisition of high-affinity DNA aptamer motifs recognizing microbial cell surfaces using polymer-enhanced capillary transient isotachophoresis. *Chem. Commun.* **2016**, *52*, 461–464. [CrossRef] [PubMed]
32. Zhu, C.; Wang, X.; Li, L.; Hao, C.; Hu, Y.; Rizvi, A.S.; Qu, F. Online reaction based single-step CE for Protein-ssDNA complex obtainment to assist aptamer selection. *Biochem. Biophys. Res. Commun.* **2018**, *506*, 169–175. [CrossRef] [PubMed]
33. Takao, J.; Endo, T.; Hisamoto, H.; Sueyoshi, K. Direct Measurement of Initial Rate of Enzyme Reaction by Electrokinetic Filtration Using a Hydrogel-plugged Capillary Device. *Anal. Sci.* **2021**, *37*, 1439–1446. [CrossRef] [PubMed]
34. Hughes, A.J.; Herr, A.E. Quantitative Enzyme Activity Determination with Zeptomole Sensitivity by Microfluidic Gradient-Gel Zymography. *Anal. Chem.* **2010**, *82*, 3803–3811. [CrossRef]
35. Sakai-Kato, K.; Kato, M.; Toyo'oka, T. On-line drug–metabolism system using microsomes encapsulated in a capillary by the sol–gel method and integrated into capillary electrophoresis. *Anal. Biochem.* **2002**, *308*, 278–284. [CrossRef]
36. Wiegand, T.W.; Williams, P.B.; Dreskin, S.C.; Jouvin, M.-H.; Kinet, J.-P.; Tasset, D. High-affinity oligonucleotide ligands to human IgE inhibit binding to Fc epsilon receptor I. *J. Immunol.* **1996**, *157*, 221–230.
37. Victor, O.; Svetlana, M.K.; Sergey, N.K. Nonequilibrium Capillary Electrophoresis of Equilibrium Mixtures, Mathematical Model. *Anal. Chem.* **2004**, *76*, 1507–1512. [CrossRef]

molecules

Article

A Dual-Mode Method Based on Aptamer Recognition and Time-Resolved Fluorescence Resonance Energy Transfer for Histamine Detection in Fish

Xin Wang [1,†], Fu Yang [1,†], Chengfang Deng [1], Yujie Zhang [1], Xiao Yang [1,2], Xianggui Chen [1,2], Yukun Huang [1,2,*], Hua Ye [3,*], Jianjun Zhong [3] and Zhouping Wang [4]

1 School of Food and Biological Engineering, Xihua University, Chengdu 610039, China
2 Chongqing Key Laboratory of Speciality Food Co-Built by Sichuan and Chongqing, Chengdu 610039, China
3 School of Grain Science and Technology, Jiangsu University of Science and Technology, Zhenjiang 212100, China
4 School of Food Science and Technology, Jiangnan University, Wuxi 214122, China
* Correspondence: hyk_dianal@163.com (Y.H.); huaye@just.edu.cn (H.Y.)
† These authors contributed equally to this work.

Abstract: Histamine produced via the secretion of histidine decarboxylase by the bacteria in fish muscles is a toxic biogenic amine and of significant concern in food hygiene, since a high intake can cause poisoning in humans. This study proposed a fluorometric and colorimetric dual-mode specific method for the detection of histamine in fish, based on the fluorescence labeling of a histamine specific aptamer via the quenching and optical properties of gold nanoparticles (AuNPs). Due to the fluorescence resonance energy transfer phenomenon caused by the proximity of AuNPs and NaYF4:Ce/Tb, resulting in the quenching of the fluorescence signal in the detection system, the presence of histamine will compete with AuNPs to capture the aptamer and release it from the AuNP surface, inducing fluorescence recovery. Meanwhile, the combined detection of the two modes showed good linearity with histamine concentration, the linear detection range of the dual-mode synthesis was 0.2–1.0 μmol/L, with a detection limit of 4.57 nmol/L. Thus, this method has good selectivity and was successfully applied to the detection of histamine in fish foodstuffs with the recoveries of 83.39~102.027% and 82.19~105.94% for *Trichiurus haumela* and *Thamnaconus septentrionalis*, respectively. In addition, this method was shown to be simple, rapid, and easy to conduct. Through the mutual verification and combined use of the two modes, a highly sensitive, rapid, and accurate dual-mode detection method for the analysis of histamine content in food was established, thereby providing a reference for the monitoring of food freshness.

Keywords: histamine; time-resolved fluorescence; aptamer; colormetric; dual-mode detection

Citation: Wang, X.; Yang, F.; Deng, C.; Zhang, Y.; Yang, X.; Chen, X.; Huang, Y.; Ye, H.; Zhong, J.; Wang, Z. A Dual-Mode Method Based on Aptamer Recognition and Time-Resolved Fluorescence Resonance Energy Transfer for Histamine Detection in Fish. *Molecules* **2022**, *27*, 8711. https://doi.org/10.3390/molecules27248711

Academic Editors: Sai Wang, Wei Wu and Long Wu

Received: 9 November 2022
Accepted: 6 December 2022
Published: 9 December 2022

1. Introduction

As a heterocyclic biogenic amine produced by histidine under the action of decarboxylase, histamine is commonly found in dark meat and fish products containing Morganella morganii and Photobacterium phosphoreum [1–3]. Since it is highly toxic, histamine content is often used as an important index to evaluate the degree of food spoilage. Studies have proven that histamine is a potential precursor of carcinogenic N-nitroso compounds, and the excessive intake of food containing histamine can cause symptoms of poisoning such as conjunctivitis, dizziness, nausea, chest tightness, and in cases of severe reaction, even death [1,4]. Consequently, numerous countries have prescribed limits to the permissible content of histamine, depending on the specific characteristics of the food type. The United States Food and Drug Administration (FDA), for example, stipulates that the histamine content in fish should not exceed 50 mg/kg, while the European Food Safety Authority (EFSA) prescribes a histamine content limit of 100–200 mg/kg in fresh fish products.

China's food safety supervision and management department has set the limit of histamine including high histamine fish, which is not permitted to exceed 400 mg/kg, while in other sea fish, it may not exceed 200 mg/kg [5]. With the development of the foreign trade of aquatic products in China, along with the focus on improving the quality of life, food safety has gained increased attention including an awareness that the development of a sensitive and rapid technology to detect food histamine is crucial to ensure food safety.

Current histamine detection methods traditionally involve high performance liquid chromatography [6,7], capillary electrophoresis technology [8,9], enzyme-linked immunosorbent assay technology [1], and electrochemical sensing technology [10,11], among others. Although these methods do have the advantages of high sensitivity and accuracy, they also have some disadvantages such as expensive chromatographic equipment, long detection cycles, and relatively complex operation, which are not suitable for rapid detection in the field; electrophoretic techniques and enzyme-linked immunosorbent assays (ELISA) have limitations in detection volume and operational techniques and time. The portability of rapidly evolving electrochemical instruments has been applied to rapid detection in the field, but their limited temperature range and lifetime limit the popularity of electrochemical sensing techniques. The more recently developed aptamer, a type of biological molecular probe produced through a process of artificial ligand screening called the systematic evolution of ligands by exponential enrichment (SELEX) from a single-stranded oligonucleotide with a high affinity and specificity to target molecules, has been shown to have excellent potential as an alternative to antibodies [12,13]. Aptamers are currently widely used in the field of food safety detection. Sharma et al. [14] developed an aptamer assay for aflatoxin B1 (AFB1) and ochratoxin A (OTA). The developed assay platform showed high specificity and was practically applied in the detection of milk samples. Huang et al. [15] selected and characterized an aptamer against Streptococcus pyogenes and developed an aptamer-based fluorescent detection method for Streptococcus pyogenes in cooked chickens. AuNPs have attracted extensive attention in the field of chemical and biosensing because of their surface plasmon resonance, which can quickly respond to and mediate colorimetric signals [5]. Lerga et al. [16,17] used SELEX (Systematic Evolution of Ligands by Exponential Enrichment) technology to screen and obtain the histamine aptamer H_2 and constructed a colorimetric detection method based on AuNPs in order to realize the detection of histamine in saliva, urine, blood, and food extracts.

Time-resolved fluorescence is a common detection method that uses lanthanide labeling to identify target molecules, monitors the deactivation process of excited molecules in the form of light radiation, and uses time-resolved technology to resolve signals, thereby realizing the microscopic dynamic analysis of molecules [18]. This technology avoids the background interference caused by sample autofluorescence, has the advantages of high sensitivity and easy operation, and plays an important role in food processing, food detection, and food composition analysis, among other fields [19–21]. In recent years, the visual and rapid detection by AuNPs based on their aggregation effect has been increasingly reported, and has been applied in fields ranging from food safety and life science analysis to cancer diagnosis and treatment. In our previous work, the time-resolved fluorescent nanomaterial $NaYF_4$:Ce/Tb was synthesized using the one-step solvothermal method, followed by the successful construction of a time-resolved fluorescence-sensitive nanoprobe via the addition of aptamers to its surface [22]. However, fluorescence/colorimetric dual-mode detection technology based on AuNPs has thus far not been reported and, hence, this study aimed to establish a time-resolved fluorescence/colorimetric dual-mode detection method for histamine based on the multiple optical effects of AuNPs. This approach not only utilizes the high sensitivity of fluorescence analysis, but also the colorimetric detection advantages of simple operation and rapid reaction. Moreover, compared with the single detection mode, dual-mode detection has been shown to have higher sensitivity, a wider range of detection, and a faster detection speed, thereby providing a new mode of technical support for the rapid and accurate detection of histamine content in aquatic products.

2. Results

2.1. Feasibility of Fluorometric and Colorimetric Dual-Mode Detection of Histamine

The fluorescence results and UV–Vis absorption spectra of the histamine were used to verify the feasibility of the novel sensing strategy [23,24]. As shown in Figure 1a, the Aptamer@$NaYF_4$:Ce/Tb exhibited a high initial fluorescence value. In the blank sample without histamine, the probe was adsorbed on the surface of the AuNPs. Due to the fluorescence resonance energy transfer, the fluorescence of the Aptamer@$NaYF_4$:Ce/Tb was quenched, thereby greatly reducing the fluorescence signal of the reaction system. However, in the presence of histamine, the Aptamer@$NaYF_4$:Ce/Tb bonded to the histamine, undergoing conformational change, and then fell away from the surface of the AuNPs, whereafter the fluorescence intensity was significantly increased.

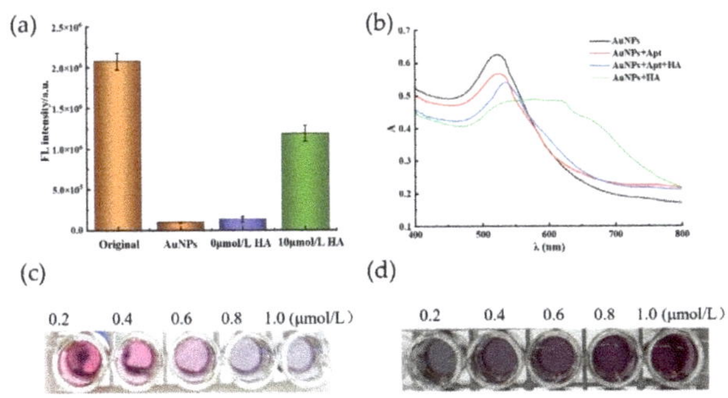

Figure 1. (**a**) Fluorescence signal of the histamine aptamer in different systems. (**b**) The UV–Visible absorption spectrogram of detection principle feasibility (histamine concentration of 10 μmol/L). (**c**) Colorimetric results of different histamine concentrations in the same volume of AuNP solution at the same aptamer concentration (from left to right histamine concentrations in the order of 0.2–1.0 μmol/L). (**d**) Colorimetric results of histamine at 2.0 μmol/L concentration in solutions with different aptamer concentrations and AuNPs (from left to right aptamer concentration of 20–100 nmol/L).

In Figure 1b, the characteristic surface plasmon resonance (SPR) peak band of the pure AuNP solution could be observed at approximately 540 nm. Compared with the pure AuNP solution, the AuNP solution adsorbed with the aptamer showed basically no change in the position of the peak band, absorbance and color (which remained red), as shown in Figure 1c. However, after the addition of histamine, the color of the AuNP solution changed from red to blue-gray, and a strong absorption peak appeared at 640 nm, indicating the serious aggregation of the AuNPs (Figure 1b). Furthermore, the surface of the AuNPs adsorbed with the aptamer exhibited a red shift at 540 nm, and the absorption peak at 640 nm was slightly increased, implying that the aptamer could protect the AuNPs from aggregation while, at the same time, the histamine could replace citrate ions on the surface of the AuNPs through the imidazole ring, leading to partial aggregation [25,26]. In conclusion, these two effects are clear indications that the dual-mode detection method based on the AuNP aptamer is feasible.

2.2. Optimization of Dual-Mode Sensing System

2.2.1. Optimization of Aptamer Concentration

In this experiment, different concentrations of inducer were added to 180 μL of AuNPs and incubated for 30 min at 25 °C, and then 2.0 μmol/L of histamine was added to observe the change in colorimetric signal. The results are shown in Figure 1d, and it can be seen that the absorption value of the AuNP solution and the corresponding color change of the solution were not obvious with the increase in the inducer concentration, indicating that the aggregation effect of histamine on the AuNPs was relatively weak. When the concentration of the inducer was greater than 40 nmol/L, histamine had no aggregation effect on AuNPs and the color was no longer gray, confirming the protective effect of the inducer on AuNPs, and partial histamine was needed to induce the aggregation of AuNPs for colorimetric detection. Meanwhile, the effect of different concentrations of Aptamer@NaYF4:Ce/Tb on the fluorescence recovery of AuNPs was measured using a microplate reader. As can be seen in Figure 2a, the fluorescence recovery value (ΔF) gradually decreased as the concentration of Ap-tamer@NaYF4:Ce/Tb increased with the same amount of AuNPs and histamine involved in changing the fluorescence intensity of the substance. When the inducer concentration was 40 nmol/L, ΔF reached a maximum value, and then decreased and stabilized. Therefore, based on the combined results of these two methods, the inducer concentration of 40 nmol/L was selected as the optimal dose concentration.

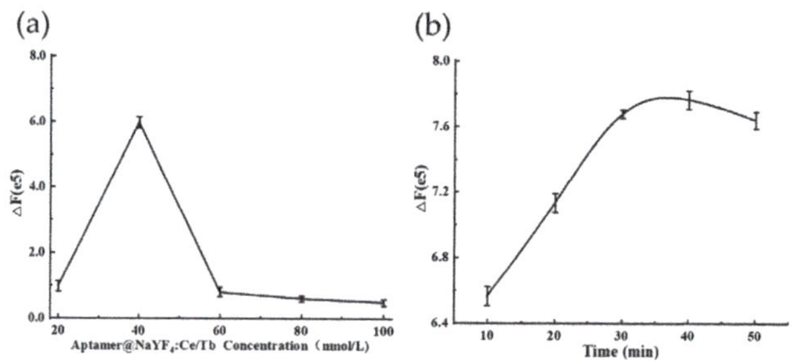

Figure 2. (**a**) Fluorescence recovery values of histamine at different Aptamer@NaYF$_4$:Ce/Tb concentrations. (**b**) Fluorescence recovery value of Aptamer@NaYF$_4$:Ce/Tb after different periods of reaction with histamine.

2.2.2. Optimization of Reaction Time

The effects of different incubation times on the fluorescence recovery of the detection system after the addition of histamine were investigated. The results are presented in Figure 2b, in which it can be seen that with the addition of histamine, the fluorescence recovery value gradually increased as the reaction time extended. When the period of reaction reached 40 min, the fluorescence recovery value reached its maximum and subsequently stabilized, indicating that a stable fluorescence signal could be obtained after 40 min incubation with histamine.

2.3. *Establishment of Fluorometric/Colorimetric Dual-Mode Method for Detection of Histamine*

Based on the preliminary optimization experiments, two detection methods, the fluorescence method and colorimetric method, were used to detect histamine, and the relationship between fluorescence recovery and absorbance ratio and histamine concentration were obtained, respectively. The detection results were synthesized according to Equation (1) [27], as shown in Figure 3, with a linear range of 0.2–1.0 μmol/L, a correlation coefficient of R = 0.9996, a linear regression equation of Y = 841376X−1592, and a detection limit of 4.57 nmol/L.

$$K = \frac{\Delta F}{A_{640}/A_{540}} \quad (1)$$

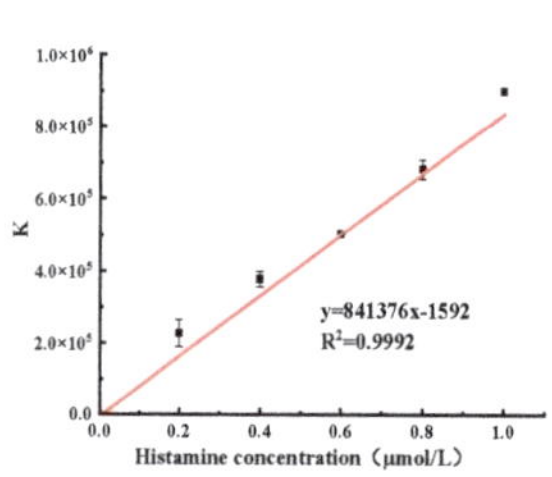

Figure 3. Fluorescence colorimetric dual-mode integrated value (*K*) and histamine concentration.

2.4. Precision Analysis of the Measurement Results

Precision is a general term indicating the precision of the measurement methods and results. Herein, the precision of the measurement method is denoted by the relative standard deviation of the sample (the ratio of sample standard deviation S to the sample mean $\overline{x}$) with the symbol *RSD* (relative standard deviation of the sample) = $S/\overline{x}$. According to the experimental method described in Section 3.4.5, the results are shown in Tables 1 and 2.

Table 1. Precision analysis of the fluorometric (ΔF) determination results.

Histamine Concentration (μmol)	Measurement Results (n = 3) (10^5)			$\overline{x}$ (10^5)	S (10^5)	RSD/%
0	0	0	0	0	0	0
0.2	2.0	2.1	2.2	2.2	7.5	3.4
0.4	3.3	3.4	3.6	3.5	9.7	2.7
0.6	5.1	5.0	5.3	5.2	2.0	3.8
0.8	6.5	6.7	6.6	6.7	9.1	1.4
1.0	8.3	8.1	8.4	8.2	2.1	2.6

Table 2. Precision analysis of the absorbance value (A) determination results.

Histamine Concentration (μmol)	Measurement Results (n = 3)			$\overline{x}$	S (10^{-3})	RSD/%
0	0.50	0.49	0.50	0.49	9.1	1.8
0.2	0.58	0.56	0.59	0.58	16.4	2.8
0.4	0.64	0.65	0.66	0.65	0.4	0.1
0.6	0.73	0.73	0.75	0.74	16.8	2.2
0.8	0.81	0.85	0.87	0.86	14.1	1.6
1.0	0.87	0.87	0.89	0.88	7.8	0.8

The results showed that the precision range of the fluorometric recovery value was 1.4–3.8%, while the precision of the UV absorbance value was 0.1–2.8%, indicating that the precision, stability, and repeatability of the method were good.

2.5. Analysis of Specificity

Muscle tissue usually contains a large amount of free histidine, which, under certain conditions, is converted into histamine under the action of decarboxylase. In order to further verify the specificity and selectivity of this method, dimethylamine, trimethylamine, β-phenylethylamine, ethylenediamine, spermidine, tryptamine, and other histamine analogs (all with concentrations of 10 μmol/L) were selected for the specificity test. As can be seen from the experimental results presented in Figure 4, dimethylamine, trimethylamine, β-phenylethylamine, ethylenediamine, spermidine, and tryptamine produced no notable signal value, while the relative fluorescence intensities of histamine and histidine, dead amine and spermine were 1.75%, 7.31%, and 56.22% of the histamine, respectively. These results not only indicate that the method has good selectivity, but also that the presence and interference of spermine in the detection system should be avoided.

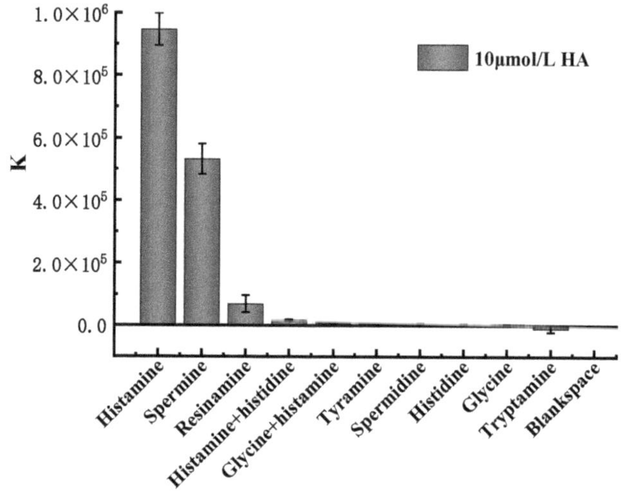

Figure 4. Specificity evaluation of the detection system based on the K signal value.

2.6. Comparison with Other Methods

Table 3 lists the detection effects of different histamine detection methods, wherein it can be seen that the detection limit of the dual-mode method constructed by this method was 4.57 nmol/L, with a linear range of 0.2–1.0 μmol/L. Furthermore, compared with both the single fluorescence method and the colorimetric method, it showed a higher degree of sensitivity. Thus, in comparison with other methods, the proposed method was found to have a wider linear range and higher detection sensitivity.

Table 3. Comparison of different histamine detection methods.

Methods	Linear Range	Detection Limit	Reference
Indirect competitive ELISA method	20~1350 μmol/L	4500 nmol/L	[28]
Non-enzymatic unmodified graphene electrode based method	40~900 μmol/L	5580 nmol/L	[29]
Molecularly imprinted polymer-based fluorescence assay	1.8~44.98 μmol/L	1800 nmol/L	[30]
Plasmonic nanoparticle-based surface plasmon resonance electrochemiluminescence of polyacid-bipyridyl ruthenium	1~1000 μmol/L	100 nmol/L	[31]
Colorimetric	0.2~1.0 μmol/L	69.37 nmol/L	This work
Fluorescence	0.2~1.0 μmol/L	9.21 nmol/L	This work
Dual mode (K-value)	0.2~1.0 μmol/L	4.57 nmol/L	This work

2.7. Sample Detection Analysis and Recovery Experiment

According to the national standard method (high performance liquid chromatography, HPLC) and the histamine detection experiment described by Huang et al. [32], the standard curve of correlation between the peak area and histamine concentration was Y = 30.09031X + 20.09641, based on a series of standard samples (histamine standard with gradient concentrations of 1–50 mg/L), while the detection limit of the national standard method was 0.0370 mg/kg. Based on the same series of samples, the concentrations of histamine in *Trichiurus haumela* and *Thamnaconus septentrionalis* determined by the national standard method were 0.0696 mg/kg and 0.0466 mg/kg, respectively, while the concentrations of histamine in the *Trichiurus haumela* and *Thamnaconus septentrionalis* determined by the repeatability condition were 0.07395 mg/kg and 0.0485 mg/kg, respectively. The absolute difference between the two independent measurements was not more than 10% of the arithmetic mean value, indicating that the national standard method had good precision. The concentrations of histamine detected by aptamer identification time-resolved fluorescence resonance energy transfer were 0.0720 mg/kg and 0.0508 mg/kg, respectively. For the same sample, the absolute difference between the two independent determination results under the condition of repeatability did not exceed 10% of the arithmetic mean value, and the two methods had a good correlation.

In order to evaluate the practical application of this method, real samples of *Trichiurus haumela* and *Thamnaconus septentrionalis* were collected for the experiment. The detection results are shown in Table 4. At the concentration of 0–10 mol/L, the recovery rates of the *Trichiurus haumela* and *Thamnaconus septentrionalis* samples were 83.41–101.94% and 82.24–105.92%, respectively. These results indicate that the dual-mode detection system developed in this study has a good recovery rate and can be used to detect histamine in *Trichiurus haumela* and *Thamnaconus septentrionalis*.

Table 4. Recovery rate of the fish samples.

Samples	Amount Added (μmol)	Detection Amount	Recovery (%)
Trichiurus haumela	0	-	0
	0.2	0.166 ± 0.003	83.41 ± 0.016
	0.4	0.371 ± 0.021	92.87 ± 0.051
	0.6	0.564 ± 0.044	94.02 ± 0.072
	0.8	0.816 ± 0.070	101.94 ± 0.087
	1.0	0.949 ± 0.069	94.98 ± 0.068
Thamnaconus septentrionalis	0	0.075 ± 0.0138	-
	0.2	0.164 ± 0.010	82.24 ± 0.047
	0.4	0.389 ± 0.005	97.41 ± 0.012
	0.6	0.636 ± 0.010	105.92 ± 0.017
	0.8	0.759 ± 0.30	94.97 ± 0.036
	1.0	0.919 ± 0.04	91.88 ± 0.040

Although high performance liquid chromatography has high reproducibility, good sensitivity, and column efficiency, it requires pretreatment such as the derivatization of histamine, which is not conducive to rapid detection in situ. Based on aptamer identification time-resolved fluorescence resonance energy transfer, this experiment determined histamine simply and quickly. Moreover, this experiment verified that the performance indices met the requirements, the comparison results with the reference method GB 5009.208–2016 were satisfactory, and the established method was feasible and can be used for the determination of the histamine residue in the actual samples.

3. Materials and Methods

3.1. Materials and Reagents

Avidin, ampicillin sodium salt, $Y(NO_3)_3 \bullet 6H_2O$, $Ce(NO_3)_3 \bullet 6H_2O$, $Tb(NO_3)_3 \bullet 5H_2O$ were purchased from Sigma Aldrich (Shanghai, China) Trading Co., Ltd.; O-phosphorylethanolamine (AEP) was purchased from Tishi (Shanghai, China) Chemical Industrial Development Co., Ltd.; $HAuCl_4 \bullet 4H_2O$ (purity $\geq$ 99.9%) was purchased from Shanghai Aladdin Biochemical Technology Co., Ltd.; ethylene glycol solution containing appropriate amounts of NH_4F, ethylene glycol, pentanediol (25% in v/v), HCl, NaCl, trisodium citrate ($Na_3C_6H_5O_7 \bullet 2H_2O$), Sinopharm Chemical Reagent Co. histamine (HTA), dimethylamine (DMA), trimethylamine (TMA), β-phenylethylamine (β-PA), ethylenediamine (ELA), spermidine (SMD), and tryptamine (TPA) were purchased from Shanghai Maclean Biochemical Technology Co., Ltd.; ultrapure water prepared by a Milli Direct-Q water purification system (Millipore Corp., Bedford, MA, USA) was used in all experiments (18.2 MΩ•cm); *Trichiurus haumela* and *Thamnaconus septentrionalis* (local supermarket).

The biotin-modified histamine aptamer sequence (5′-AGCTCCAGAAGATAAATTACAG GGAACGTGTTGGTGCGGTTCTTCCGATCTGCTGTTCTCTCTATCTGTGCCAT GCAAC TAGGATACTATGACCCCGG-3′) was synthesized by Bioengineering Ltd. (Shanghai, China) and purified by high performance liquid chromatography.

3.2. Instruments and Equipment

The following instruments and equipment were used in the analysis: A laser particle size meter (ZSN3600), Malvern Instruments Ltd., Malvern, UK; fluorescence spectrometer (FLUOROMAX-4CP), HORIBA Corporation, Irvine, CA, USA; double beam ultraviolet spectrophotometer (SDPTOPUV2800PC), Shanghai Sunyu Hengping Scientific Instruments Co., Ltd. (Shanghai, China); small high-speed centrifuge (5424), Eppendorf AG, Hamburg, Germany; vacuum drying oven (BZF-50), Shanghai Boxun Industrial Co., Ltd., Shanghai, China; pure/ultrapure water all-in-one machine (Direct-Q), Millipore Corporation, Burlington, MA, USA; and JEM-2100 high-resolution transmission electron microscope (Nippon Electron Co., Ltd. Tokyo, Japan).

3.3. Detection Principle

The principle of histamine detection applied in this study is shown in Figure 5. First, the histamine aptamer was immobilized on $NaYF_4$:Ce/Tb nanofluorescent material through biotin–avidin interaction. The subsequent addition of unmodified AuNPs to the system triggered the fluorescence resonance energy transfer phenomenon, resulting in the quenching of the fluorescence signal. When the target histamine was added, it competed with the AuNPs to bind to the aptamer because of the high affinity of its aptamer to the target histamine [33]. The AuNPs were, thus, separated from the aptamer and released into the solution, enabling the recovery of the fluorescence signal. Finally, the dual-mode detection of histamine was realized via the measurement of the degree of time-resolved fluorescence recovery and linear analysis of the histamine concentration. The colorimetric method replaces the citrate ion on the surface of synthetic AuNPs by a special imidazole ring on the surface of the histamine. The combination of positively charged histamine and negatively charged citrate ions reduces the net surface charge of AuNPs, which destabilizes AuNPs and triggers aggregation, and the color of the solution changes from red to blue-gray, so the presence of histamine can also be detected by visual observation. Moreover, AuNPs have a characteristic absorption peak at 540 nm, and when aggregation occurs, the intensity of the absorption peak at 540 nm decreases and a new absorption peak appears at 640 nm, which is due to the red-shift of the surface plasmon resonance peak of the aggregates. Therefore, the concentration of the target histamine was analyzed by measuring the absorbance at 540 and 640 nm and determining the degree of aggregation of AuNPs by its ratio (A640/A540). Accurate quantification and simple, rapid determination of the histamine content was conducted by dual-mode fluorescence and colorimetric methods.

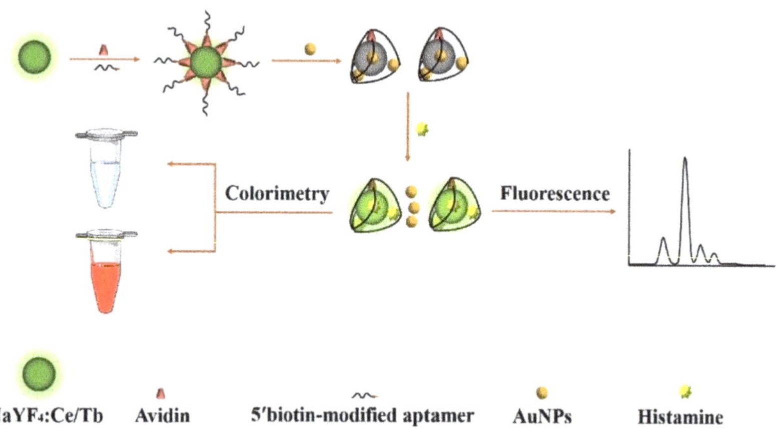

Figure 5. Schematic diagram of histamine detection based on the aptamer/AuNPs fluorometric and colorimetric dual-mode detection.

3.4. Experimental Methods

3.4.1. Preparation of AuNPs

Add 1 mL $HAuCl_4 \bullet 4H_2O$ solution (1 wt%) and 99 mL ultrapure water to a round bottom flask, heat to boiling under magnetic stirring, then add 2 mL 1% aqueous sodium citrate solution (weigh 0.5 g sodium citrate solid into 100 mL beaker and add 50 mL deionized water, stir to dissolve); the reaction is kept boiling with continuous stirring for 10 min, after which it is turned off. The heating power was turned off to stop the heating, and then the stirring was continued for 15 min, and the solution was cooled to room temperature to finally obtain a transparent wine red solution, which was placed in a brown reagent bottle and stored at 4 °C for use (the prepared AuNPs were burgundy in color and characterized with transmission electron microscopy) (Supplementary Material).

3.4.2. Preparation of NaYF4:Ce/Tb Nanoparticles

The nanomaterials of $NaYF_4$:Ce/Tb were synthesized by referring to the method of Wang et al. [34]. In a round bottom flask, 1.0 mmol AEP and a certain stoichiometric ratio of NaCl, $Y(NO_3)_3 \bullet 6H_2O$, $Ce(NO_3)_3 \bullet 6H_2O$, and $Tb(NO_3)_3 \bullet 5H_2O$ were added and dissolved with ethylene glycol solution containing an appropriate amount of NH_4F and aged for 30 min, transferred to an autoclave at 180 °C for 4 h, and removed and cooled naturally to room temperature. After washing and drying in a vacuum drying oven at 60 °C for 10 h, the solid particles were stored at room temperature and protected from light (scanning time-resolved fluorescence of the synthesized NaYF4:Ce/Tb nanofluorescent materials and characterization by transmission electron microscopy).

3.4.3. Preparation of NaYF4:Ce/Tb Nanoparticles Modified with Avidin

The classical glutaraldehyde method was used to couple the avidin with the amino groups on the surface of $NaYF_4$:Ce/Tb. A total of 5 mg of $NaYF_4$:Ce/Tb nanoparticles were weighed and dispersed in 1 mL of phosphate buffered saline (PBS), sonicated for 10 min, the solution was reacted with 5 mL of glutaraldehyde aqueous solution, and shaken slowly at room temperature for 2 h. The remaining glutaraldehyde reaction solution was discarded and washed three times with phosphate buffer, and a certain amount of avidin was added overnight at 37 °C. The reaction was followed by centrifugation and washing with PBS three times. Finally, the samples were resuspended in 5 mL PBS and stored at 4 °C for later use.

3.4.4. Assembly of Aptamer@NaYF4:Ce/Tb Nanoparticles

The quantified aptamers were added into the avidin-based nanoparticle suspension and incubated at 37 °C for 3 h with slow shaking. After the reaction, it was centrifuged and washed three times with PBS, resuspended in PBS, and stored at 4 °C.

3.4.5. Construction of a Fluorescent Colorimetric Dual-Mode Based Method for the Detection of Histamine

In the detection system, different concentrations of histamine solutions were detected with a certain concentration of fluorescent probe. The target histamine was incubated with the fluorescent detection probe at 25 °C with shaking for 0.5 h. The fluorescence intensity of the solution was measured after the reaction. The measurement conditions were set as follows: excitation wavelength 273 nm, delay time 0.1 ms, detection time 1 ms.

The synthesized AuNPs (180 μL) were mixed with the histamine aptamer (30 μL, 40 nmol/L) so that the aptamer was uniformly adsorbed on the surface of the AuNPs. Different concentrations of histamine solutions were added, mixed well, and incubated at room temperature (25 °C) for 30 min before UV–Vis spectroscopy scanning to evaluate the degree of AuNP aggregation by the ratio of the absorbance value at 640 nm to that at 540 nm. Meanwhile, the fluorescence intensity of the supernatant was measured using an equal volume of PBS buffer instead of histamine as a blank control with an excitation wavelength of 273 nm and emission wavelength of 545 nm, and the fluorescence difference (ΔF) between the experimental and blank groups was calculated. Finally, the measurement values of the two methods were combined to produce the final data K, which is the ratio of the fluorescence recovery value to the signal change in absorbance at wavelengths of 640 nm and 540 nm [27].

3.4.6. Spiked Recoveries of Actual Samples with Detection Applications

Referring to the method of Lerga et al. [16], the flesh of *Trichiurus haumela* and *Thamnaconus septentrionalis* was picked, and the fish was minced for subsequent experiments. A total of 0.5 g of fish mince was weighed, 25 mL ultrapure water was added, and the supernatant was collected after 1~2 min of vortex oscillation, and then centrifuged at 10,000 r/min for 5 min. Histamine standard solutions of different concentrations were added to the fish extract supernatant, and three standard addition and recovery experiments were performed for each solution according to the method in Section 3.4.4, and the recoveries were calculated.

4. Conclusions

A simple and rapid dual-mode method for the detection of histamine was proposed by using a histamine aptamer modified with fluorophore (aptamer@NaYF4:Ce/Tb), which was conducted in fluorometric and colorimetric dual-mode method through taking the AuNPs as carriers and the time-resolved fluorescence nanoparticle modified aptamer as the recognition element. The results showed that the changes in fluorescence intensity and absorbance ratio were linearly related to the histamine concentration, showed good selectivity for histamine, and was successfully applied to histamine in fish meat. In addition, the method does not require the action of enzymes, the instrumentation and operating procedures used are relatively simple and largely independent of the environment, and the unique biocompatibility and optical properties of AuNPs can be utilized for their powerful signal generation and transduction capabilities, which contribute to the rapid and efficient reactions. By mutual validation comparison, this dual-mode achieved a satisfactory analytical performance for complex food matrix samples, avoiding the single detection. The dual mode has great potential for development and application in the field of food decontamination by avoiding possible interference in a single detection mode. The implementation of multiple detection methods with joint application and mutual validation of the diversified detection mode is important to achieve real-time, simultaneous monitoring of hazardous substances in complex samples in the field. Currently, this dual-mode detection method can only detect a single histamine, and has limitations in detecting multiple biogenic amines with the widely varying types and levels in food.

Supplementary Materials: The following supporting information can be downloaded at: https://www.mdpi.com/article/10.3390/molecules27248711/s1. Figure S1. The prepared AuNPs object; Figure S2. (a) Ultraviolet peak spectrum of AuNPs; (b) Electron microscopic picture of AuNPs; Figure S3. (a) Time-resolved fluorescence spectra; (b) Time-resolved fluorescence nanoparticle electron microscope; Figure S4. (a) Relationship between fluorescence recovery value (ΔF) and histamine concentration; (b) Absorbance ratio (A_{640}/A_{540}) and histamine concentration; Supplementary Material for HPLC method operation for Sample Analysis and Detection.

Author Contributions: Conceptualization, Y.H. and H.Y.; Methodology, F.Y. and C.D.; Software, J.Z.; Validation, X.W., Y.Z. and X.Y.; Formal analysis, H.Y.; Investigation, Y.H.; Resources, X.C.; Data curation, Z.W.; Writing—original draft preparation, X.W. and F.Y.; Writing—review and editing, X.W., Y.H. and H.Y.; Visualization, X.W.; Supervision, Z.W.; Project administration, Y.H.; Funding acquisition, X.C. All authors have read and agreed to the published version of the manuscript.

Funding: This research was funded by the National Natural Science Foundation of China (31801647), the Sichuan Science and Technology Program (2020YFN0153, 2020YFN0151, 2022JDTD0028), the Innovation and Entrepreneurship Training Program of Sichuan Province (S202110650035), Sichuan Institute of International Science and Technology Cooperation (Australia and New Zealand) (AXYJ2022-006) and the Natural Science Foundation of Jiangsu Province (BK20210882).

Institutional Review Board Statement: Not applicable.

Informed Consent Statement: Not applicable.

Data Availability Statement: Not applicable.

Conflicts of Interest: The authors declare no conflict of interest.

Sample Availability: Samples of the compounds Aptamer@NaYF4:Ce/Tb Nanoparticles are available from the authors.

References

1. Karovicova, J.; Kohajdova, Z. Biogenic amines in food. *Chem. Pap.* **2005**, *59*, 70–79. [CrossRef]
2. Emborg, J.; Dalgaard, P. Formation of histamine and biogenic amines in cold-smoked tuna: An investigation of psychrotolerant bacteria from samples implicated in cases of histamine fish poisoning. *J. Food Prot.* **2006**, *69*, 897–906. [CrossRef]
3. Kanki, M.; Yoda, T.; Ishibashi, M.; Tsukamoto, T. Photobacterium phosphoreum caused a histamine fish poisoning incident. *Int. J. Food Microbiol.* **2004**, *92*, 79–87. [CrossRef]
4. Papageorgiou, M.; Lambropoulou, D.; Morrison, C.; Modzinska, E.; Namiesnik, J.; Plotka-Wasylka, J. Literature update of analytical methods for biogenic amines determination in food and beverages. *TrAC-Trends Anal. Chem.* **2018**, *98*, 128–142. [CrossRef]
5. Lerga, T.M.; Skouridou, V.; Bermudo, M.C.; Bashammakh, A.S.; El-Shahawi, M.S.; Alyoubi, A.O.; O'Sullivan, C.K. Gold nanoparticle aptamer assay for the determination of histamine in foodstuffs. *Microchim. Acta* **2020**, *187*, 452. [CrossRef] [PubMed]
6. Hwang, B.S.; Wang, J.T.; Choong, Y.M. A rapid gas chromatographic method for the determination of histamine in fish and fish products. *Food Chem.* **2003**, *82*, 329–334. [CrossRef]
7. Tao, Z.; Sato, M.; Han, Y.; Tan, Z.; Yamaguchi, T.; Nakano, T. A simple and rapid method for histamine analysis in fish and fishery products by TLC determination. *Food Control* **2011**, *22*, 1154–1157. [CrossRef]
8. Nishiwaki, F.; Kuroda, K.; Inoue, Y.; Endo, G. Determination of histamine, 1-methylhistamine and N-methylhistamine by capillary electrophoresis with micelles. *Biomed. Chromatogr. BMC* **2000**, *14*, 184–187. [CrossRef]
9. Spichiger, U.E.; Fakler, A. Potentiometric microelectrodes as sensor and detectors. Magnesium-selective electrodes as sensors, and hofmeister electrodes as detectors for histamine in capillary electrophoresis. *Electrochim. Acta* **1997**, *42*, 3137–3145. [CrossRef]
10. Veseli, A.; Vasjari, M.; Arbneshi, T.; Hajrizi, A.; Svorc, L.; Samphao, A.; Kalcher, K. Electrochemical determination of histamine in fish sauce using heterogeneous carbon electrodes modified with rhenium(IV) oxide. *Sens. Actuators B-Chem.* **2016**, *228*, 774–781. [CrossRef]
11. Chen, f.; Huang, y.; Yuan, y. Non-destructive identification of adulterated honey based on electronic nose. *J. Xihua Univ.* **2018**, *37*, 56–60.
12. Wu, L.; Zhou, M.; Wang, Y.; Liu, J. Nanozyme and aptamer- based immunosorbent assay for aflatoxin B1. *J. Hazard. Mater.* **2020**, *399*, 123154. [CrossRef] [PubMed]
13. Ye, H.; Yang, Z.; Khan, I.M.; Niazi, S.; Guo, Y.; Wang, Z.; Yang, H. Split aptamer acquisition mechanisms and current application in antibiotics detection: A short review. *Crit. Rev. Food Sci. Nutr.* **2022**, 1–12. [CrossRef]
14. Sharma, A.; Catanante, G.; Hayat, A.; Istamboulie, G.; Ben Rejeb, I.; Bhand, S.; Marty, J.L. Development of structure switching aptamer assay for detection of aflatoxin M1 in milk sample. *Talanta* **2016**, *158*, 35–41. [CrossRef] [PubMed]
15. Huang, Y.; Wang, X.; Duan, N.; Xia, Y.; Wang, Z.; Che, Z.; Wang, L.; Yang, X.; Chen, X. Selection and characterization, application of a DNA aptamer targeted to Streptococcus pyogenes in cooked chicken. *Anal. Biochem.* **2018**, *551*, 37–42. [CrossRef]
16. Mairal Lerga, T.; Jauset-Rubio, M.; Skouridou, V.; Bashammakh, A.S.; El-Shahawi, M.S.; Alyoubi, A.O.; O'Sullivan, C.K. High Affinity Aptamer for the Detection of the Biogenic Amine Histamine. *Anal. Chem.* **2019**, *91*, 7104–7111. [CrossRef] [PubMed]
17. Shoji, S.; Sun, H.B.; Kawata, S. Photofabrication of wood-pile three-dimensional photonic crystals using four-beam laser interference. *Appl. Phys. Lett.* **2003**, *83*, 608–610. [CrossRef]
18. Huang, Y. *Study of a Nucleic Acid Aptamer Recognition-Based Time-Resolved Fluorescent Nanoprobe for Biotoxin Detection*; Jiangnan University: Wuxi, China, 2015.
19. Brandao, M.P.; dos Anjos, V.d.C.; Bell, M.J.V. Time resolved fluorescence of cow and goat milk powder. *Spectrochim. Acta Part A-Mol. Biomol. Spectrosc.* **2017**, *171*, 193–199. [CrossRef]
20. Kulpakko, J.; Kopra, K.; Hanninen, P. Time-resolved fluorescence-based assay for rapid detection of Escherichia coli. *Anal. Biochem.* **2015**, *470*, 1–6. [CrossRef]
21. Mu, T.; Chen, S.; Zhang, Y.; Chen, H.; Guo, P. Characterization of edible oils using time-resolved fluorescence. *Anal. Methods* **2014**, *6*, 940–943. [CrossRef]
22. Huang, Y.; Wang, C.; Huo, Q.; Song, Y.; Du, G.; Wang, L.; Yang, X.; Chen, X. A time-resolved luminescence aptasensor of ofloxacin based on rolling circle amplification and magnetic separation. *Anal. Bioanal. Chem.* **2020**, *412*, 4555–4563. [CrossRef] [PubMed]
23. Wang, Y.; Li, H.; Zhou, J.; Qi, Q.; Fu, L. A colorimetric and fluorescent gold nanoparticle-based dual-mode aptasensor for parvalbumin detection. *Microchem. J.* **2020**, *159*, 105413. [CrossRef]
24. Hu, X.; Shi, J.; Shi, Y.; Zou, X.; Tahir, H.E.; Holmes, M.; Zhang, W.; Huang, X.; Li, Z.; Xu, Y. A dual-mode sensor for colorimetric and fluorescent detection of nitrite in hams based on carbon dots-neutral red system. *Meat Sci.* **2019**, *147*, 127–134. [CrossRef] [PubMed]

25. Ling, B.; Ma, Y.; Chen, H.; Wang, L. A SPR aptamer sensor for mercury based on AuNPs@NaYF4: Yb, Tm, Gd upconversion luminescent nanoparticles. *Anal. Methods* **2017**, *9*, 6032–6037. [CrossRef]
26. Liu, Q.; Tang, P.; Xing, X.; Cheng, W.; Liu, S.; Lu, X.; Zhong, L. Colorimetry/SERS dual-sensor of H_2O_2 constructed via TMB-Fe_3O_4@ AuNPs. *Talanta* **2022**, *240*, 123118. [CrossRef]
27. Yuan, M.; Wang, M.; Zheng, Y.; Cao, H.; Xu, F.; Ye, T.; Yu, J. Dual-mode fluorescent colorimetric detection of As(III) based on aptamer and gold nanoparticles. *Anal. Chem.* **2021**, *49*, 76–84. [CrossRef]
28. Li, Y.-F.; Lin, Z.-Z.; Hong, C.-Y.; Huang, Z.-Y. Histamine detection in fish samples based on indirect competitive ELISA method using iron-cobalt co-doped carbon dots labeled histamine antibody. *Food Chem.* **2021**, *345*, 128812. [CrossRef]
29. Nakthong, P.; Kondo, T.; Chailapakul, O.; Siangproh, W. Development of an unmodified screen-printed graphene electrode for nonenzymatic histamine detection. *Anal. Methods* **2020**, *12*, 5407–5414. [CrossRef]
30. Feng, X.; Ashley, J.; Zhou, T.; Halder, A.; Sun, Y. A facile molecularly imprinted polymer-based fluorometric assay for detection of histamine. *RSC Adv.* **2018**, *8*, 2365–2372. [CrossRef]
31. Wang, H.; Xu, G.; Dong, S. Electrochemiluminescence of tris(2,2′-bipyridine)ruthenium(II) immobilized in poly(p-styrenesulfonate)-silica-Triton X-100 composite thin-films. *Analyst* **2001**, *126*, 1095–1099. [CrossRef]
32. Huang, Y.; Song, Y.; Chen, F. Analysis of 8 biogenic amines in traditional Pixian douban based on response surface methodology optimized by precolumn derivatization-high performance liquid chromatography. *J. Xihua Univ.* **2020**, *39*, 61–71.
33. Song, Y.; Hu, C. Detection of pefloxacin veterinary drug residues in milk based on nucleic acid aptamer recognition with time-resolved fluorescence resonance energy transfer. *Food Ferment. Ind.* **2021**, *47*, 244–250. [CrossRef]
34. Wang, X.; Huang, Y.; Wu, S.; Duan, N.; Xu, B.; Wang, Z. Simultaneous detection of Staphylococcus aureus and Salmonella typhimurium using multicolor time-resolved fluorescence nanoparticles as labels. *Int. J. Food Microbiol.* **2016**, *237*, 172–179. [CrossRef] [PubMed]

molecules

Article

The SEB1741 Aptamer Is an Efficient Tool for Blocking CD4+ T Cell Activation Induced by Staphylococcal Enterotoxin B

Leslie Chavez-Galan [1,†], Andy Ruiz [1,†], Lucero A. Ramón-Luing [1], Alejandro Escamilla-Gutiérrez [2,3], Anahí Sánchez-Monciváis [4], Brenda Tecuatzi-Cadena [4], Karen Medina-Quero [4,*] and María Guadalupe Córdova-Espinoza [2,4,*]

1 Laboratory of Integrative Immunology, Instituto Nacional de Enfermedades Respiratorias Ismael Cosío Villegas, Mexico City 14080, Mexico
2 Laboratory of Medical Bacteriology, Department of Microbiology, Escuela Nacional de Ciencias Biológicas, Instituto Politécnico Nacional, Mexico City 11350, Mexico
3 Hospital General "Dr. Gaudencio González Garza", Centro Médico Nacional La Raza, Instituto Mexicano del Seguro Social IMSS, Mexico City 02990, Mexico
4 Laboratory of Immunology, Escuela Militar de Graduados de Sanidad, SEDENA, Mexico City 11200, Mexico
* Correspondence: kmq.kmq5@gmail.com (K.M.-Q.); mixtlipp@yahoo.com.mx (M.G.C.-E.)
† These authors contributed equally to this work.

Abstract: Staphylococcal enterotoxin B (SEB) is a protein produced by *Staphylococcus aureus*, which is toxic to humans. It is well known for its ability to stimulate the exacerbated activation of proinflammatory CD4+ T cells (Th1 profile), and in vitro studies have been conducted to understand its mechanism of action and its potential use as an immune therapy. However, the efficiency of the SEB1741 aptamer in blocking SEB has not been experimentally demonstrated. Methods: Enrichment CD4+ T cells were stimulated with SEB, and as a blocker, we used the SEB1741 aptamer, which was previously synthesised by an "in silico" analysis, showing high affinity and specificity to SEB. The efficiency of the SEB1741 aptamer in blocking CD4+ T cell activation was compared with that of an anti-SEB monoclonal antibody. Flow cytometry and Bio-Plex were used to evaluate the T-cell function. Results: In vitro, SEB induced the activation of CD4+ T cells and favoured a Th1 profile; however, the SEB1741 aptamer was highly efficient in decreasing the frequency of CD4+ T cells positive to ki-67 and CD69 cells, this means that proliferation and activation of CD4+ T cells was decreased. Moreover, the production of interleukin 2 (IL-2) and interferon-gamma (IFN-γ) was affected, suggesting that the Th1 profile is not present when the SEB1441 aptamer is used. Thus, the SEB1741 function was similar to that of anti-SEB. Conclusions: The SEB1741 aptamer is a valuable tool for blocking CD4+ T cell activation and the subsequent release of proinflammatory cytokines by SEB stimulation.

Keywords: SEB; anti-SEB; SEB1741 aptamer; CD4+ T cells; inflammation

check for updates

Citation: Chavez-Galan, L.; Ruiz, A.; Ramón-Luing, L.A.; Escamilla-Gutiérrez, A.; Sánchez-Monciváis, A.; Tecuatzi-Cadena, B.; Medina-Quero, K.; Córdova-Espinoza, M.G. The SEB1741 Aptamer Is an Efficient Tool for Blocking CD4+ T Cell Activation Induced by Staphylococcal Enterotoxin B. *Molecules* **2023**, *28*, 3480. https://doi.org/10.3390/molecules28083480

Academic Editors: Long Wu, Wei Wu and Sai Wang

Received: 7 March 2023
Revised: 1 April 2023
Accepted: 11 April 2023
Published: 14 April 2023

1. Introduction

Staphylococcus aureus (*S. aureus*) is a Gram-positive bacterium, and approximately 30% of the adult population is colonised by this bacterium. However, it plays a larger role than just being a commensal organism. *S. aureus* induces various human infections, including skin infections, upper respiratory tract infections, foodborne illnesses, and more severe illnesses, such as toxic shock syndrome [1–3]. Approximately 80% of *S. aureus* strains clinically isolated are enterotoxin producers [4], and staphylococcal enterotoxin B (SEB) is the main toxin produced by this bacterium. SEB is a superantigen (SAg) that activates T-cells in a non-specific manner, causing an exaggerated and potentially harmful immune response [3–7].

SAgs bind to the major histocompatibility complex class II (MHC-II) to stimulate the T-cell receptor (TCR) in many T cells, resulting in toxic shock syndrome due to the exacerbated release of proinflammatory cytokines [7–10]. In addition, a low concentration of SAgs in the bloodstream is enough to elevate toxic levels of proinflammatory cytokines, such as interleukin-2 (IL-2), interferon-gamma (IFN-γ), and tumour necrosis factor (TNF) [8,10–12].

The rapid onset of fever, organ failure, and, in some cases, death is a characteristic of toxic shock syndrome. Patient survival depends on the level of preexisting neutralising antibodies that prevent the binding of SAg/MHC-II, preventing the release of proinflammatory cytokines [10,13].

Treatment of toxic shock syndrome consists of antibiotics such as penicillin, cephalosporin, vancomycin, or methicillin, intravenous fluids, and immunoglobulin to neutralise the toxin; however, some of these drugs increase blood pressure, and cleaning and removal of infected tissue and antibiotic resistance may occur [14]. Thus, it is necessary to develop new treatments that focus on blocking T cell activation to avoid hyperinflammation in these patients.

The development of aptamers represents a therapeutic alternative with significant advantages; they are small RNA or DNA sequences, which by forming three-dimensional structures, can act as antibodies capable of binding to a given antigen with high recognition and affinity [15,16]. Thus, the design of a functional aptamer to inhibit the superantigenic activity of SEB could be an optimal alternative treatment. One study showed that in vitro, this aptamer significantly inhibited the production of proinflammatory cytokines, and in vivo, the survival rate was increased [15].

Recently, we reported the synthesis of an aptamer against SEB (hereafter called SEB1741), which by an "in silico" analysis, showed a prediction of high affinity and specificity in the binding site of SEB. The interactions are as follows: Cytosine 25-Lysine 57, Guanine 14-Glutamine 210, Adenine 30-Lysine 109 and 111, and Cytosine 23-Aspartic acid 108 [7]. The aptamer obtention was based on a prediction of binding affinity using three research algorithms (rDock, patchDock, and HDOCK). The binding affinity value between SEB and SEB1741 is only at a theoretical level, and algorithms provided the values "in silico." Moreover, when the conventional AUTODOCK program calculates the dissociation constant (Kd) directly using small molecules with rotatable bonds, both molecules (SEB and SEB1741) have a considerable size, making them unpredictable to obtain the Kd.

Thus, this study aimed to evaluate the efficiency of the SEB1741 aptamer in blocking CD4+ T cell activation and the consequent proinflammatory cytokines released by SEB stimulation. Our data confirm experimentally that SEB1741 is efficient in blocking the interaction of SEB/CD4+ T cells, which helps regulate the production of IFN-γ but not that of TNF.

2. Results

2.1. The Aptamer SEB1741 Interacts with the Active Site of SEB

Previously, to select the best aptamer that could block the binding site of MHC-II molecules, we selected and characterised some aptamers "in silico" and identified SEB1741 as a potential blocking aptamer [7]. Figure 1a shows how SEB1741 interacts with the active site of SEB. Figure 1b shows the electrostatic surface of the SEB. According to the colour scale, red represents the hydrophobic character, and blue represents the hydrophilic character. The beta-alpha helix of SEB interacts by hydrogen bonds of SEB1741 with Cytosine 25-Lysine 57, Guanine 14-Glutamine 210, Adenine 30-Lysine 109 and 111, and Cytosine 23-Aspartic acid 108. In general, most strategies for blocking the immune response due to SEB SAg have focused on blocking the binding site that partially occupies the MHC-II binding site; however, in this case, the effort was to directly block the SEB binding site, suggesting that the SEB1741 aptamer could be an optimal option to avoid SEB/cell interactions.

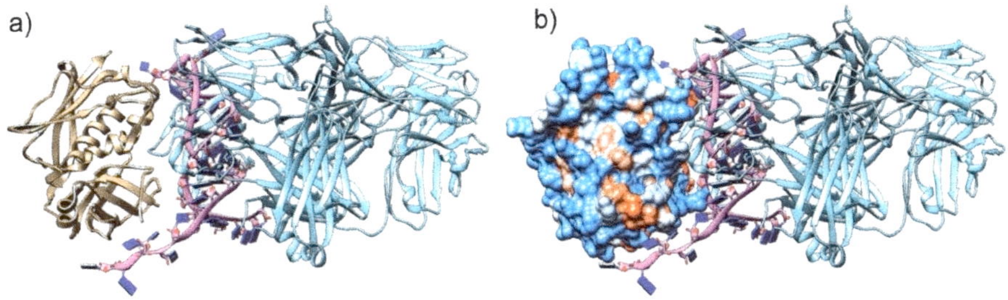

Figure 1. Molecular interactions between the tertiary structure of staphylococcal enterotoxin B (SEB), SEB1741, and monoclonal antibody (mAb) anti-SEB: (**a**) The beta-alpha helix SEB (brown) interacting with the hydrogen bonds of SEB1741 (purple) with Cytosine 25-Lysine 57, Guanine 14-Glutamine 210, Adenine 30-Lysine 109 and 111, and Cytosine 23-Aspartic acid 108; (**b**) SEB1741 (purple) and the mAb anti-SEB (light blue) interacting in the active site of the SEB hydrophobic region.

2.2. The SEB1741 Aptamer Inhibits the CD4+ T-Cell Proliferation Induced by SEB

An increase in the rate of CD4+ cell proliferation may indicate an active immune response against diverse infections. Ki-67 is routinely evaluated as a proliferation marker [17,18]. Figure 2 shows that SEB induced a similar frequency of CD4+Ki-67+ T cells as the positive control (PMA/IO), confirming that SEB was efficient in activating CD4+ T cells. However, this frequency decreased when anti-SEB was used at 5 μg/mL ($p < 0.01$). Similarly, using the SEB1741 at concentrations of 2, 5, and 10 μg/mL, the frequency of CD4+Ki-67+ T cells was also decreased ($p < 0.05$).

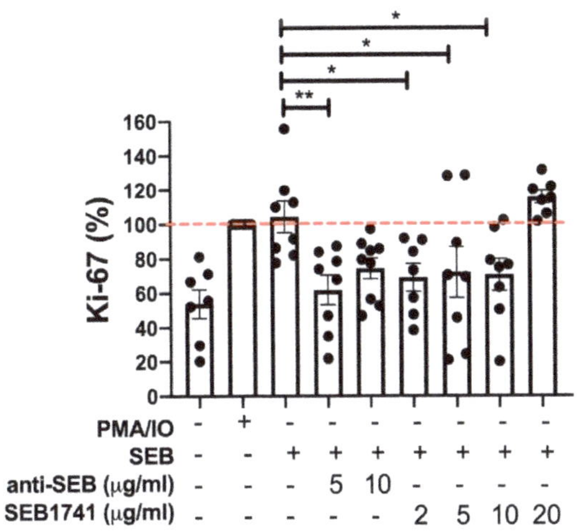

Figure 2. Staphylococcal enterotoxin B (SEB) increases the frequency of CD4+Ki-67+ T cells, but the SEB1741 aptamer inhibits this proliferation. The frequency of Ki-67 in human CD4+ T cells was assessed 48 h after stimulation with SEB, anti-SEB, or SEB1741 aptamer. PMA/IO was used as a positive control of activation, and it was normalised as 100% (dashed red line). Cell surface and intracellular staining were performed for the assessment using flow cytometry. n = six independent experiments, dots represent individual data, and bars represent mean value ± SEM. Differences between groups were analysed using the ANOVA and Dunnett's tests as post hoc analyses. * $p < 0.05$, ** $p < 0.01$.

It is essential to note that high concentrations of both anti-SEB (10 μg/mL) and the SEB1741 aptamer (10 μg/mL) were unable to block the proliferation of CD4+ T cells in response to SEB stimulus, suggesting that an excess of SEB blockers does not avoid the SAg function of SEB.

2.3. The SEB1741 Aptamer Decreases the CD4+ T-Cell Activation Induced by SEB

The expression of CD69 and CD25 on T cell surfaces is considered a classical marker of early and late activation, respectively [19]. Our data showed that SEB increased the frequency of CD4+CD69+ T cells after 48 h of culture, but not to the same level as PMA/IO (Figure 3a). Using the SEB1741 aptamer at 5 and 10 µg/mL was efficient in decreasing the frequency of the CD4+CD69+ T cells ($p < 0.01$ and $p < 0.05$, respectively). The anti-SEB also reduced the frequency of CD4+CD69+ T cells at 5 and 10 µg/mL ($p < 0.05$) (Figure 3a).

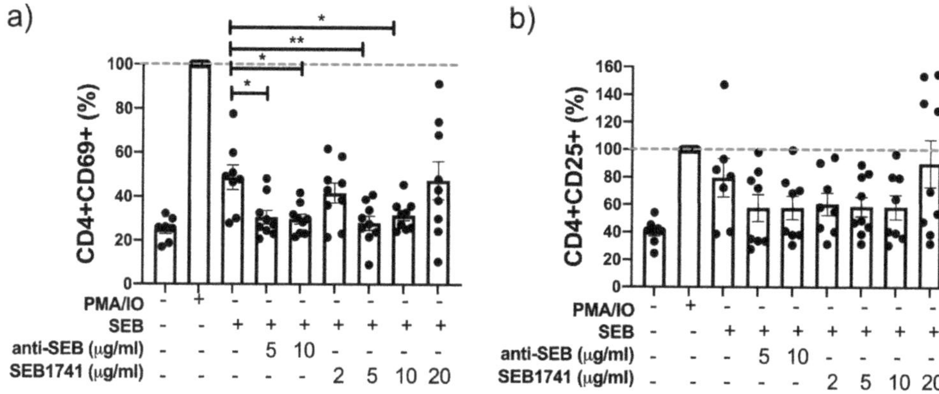

Figure 3. The frequency of CD4+CD69+ T cells, but not that of CD4+CD25+ T cells, is decreased by the SEB1741 aptamer. CD4+ T cells were recovered after 48 h of culture stimulated with staphylococcal enterotoxin B (SEB), anti-SEB, or SEB1741 aptamer, and PMA/IO was used as the activation control. Cell surface staining was performed and assessed using flow cytometry. The frequency of CD4+CD69+ (**a**) and CD4+CD25+ (**b**) was evaluated. n = six independent experiments, dots represent individual data, and bars represent mean value ± SEM. Differences between groups were analysed using the ANOVA and Dunnett's tests as post hoc analyses. * $p < 0.05$, ** $p < 0.01$.

Regarding the frequency of CD4+CD25+ T cells or late activation cells, our data showed that SEB increased the frequency of this subpopulation, similar that the frequency induced with PMA/IO after 48 h of culture; however, neither anti-SEB nor SEB1741 decreased the frequency of CD4+CD25+ T cells (Figure 3b) significantly.

Together, these results suggest that the SEB1741 aptamer efficiently prevented the interaction of SEB/CD4+ T cells because of a decrease in activated CD4+ T cells.

2.4. Low Concentration of the SEB1741 Aptamer Decreases the Release of the Proinflammatory Cytokines IL-2 and IFN-γ but Not That of TNF

To confirm the efficiency of the SEB1741 aptamer in preventing the activation of CD4+ T cells, we evaluated the profile of pro- and anti-inflammatory cytokines in the supernatant of cultured CD4+ T cells. IL-2 is a cytokine that is produced as a response to a correct activation of T cells. In fact, CD25, the marker of cellular activation, is the receptor to IL-2, whereas IFN-γ and TNF are two of the most common proinflammatory cytokines [20,21].

Our in vitro system showed that SEB induced higher levels of IL-2, IFN-γ, and TNF than the unstimulated cells. However, these levels were lower than those of the positive control (PMA/IO), suggesting that after 48 h of culture, SEB efficiently activated and induced the release of proinflammatory cytokines by CD4+ T cells (Figure 4).

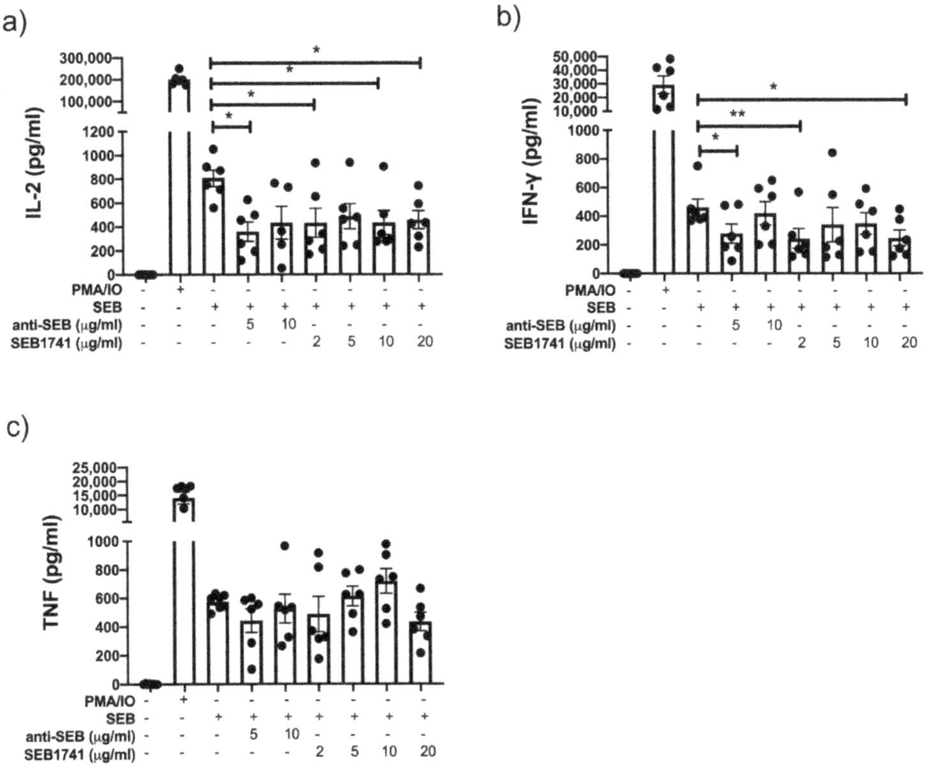

Figure 4. The SEB1741 aptamer efficiently decreases interleukin (IL)-2 and interferon-gamma (IFN)-γ production, but not tumour necrosis factor (TNF), in CD4+ T cells. IL-2 (**a**), IFN-γ (**b**), and TNF (**c**) levels were measured in supernatants collected from CD4+ T cells stimulated by staphylococcal enterotoxin B (SEB), anti-SEB, and SEB1741 aptamer by multiple cytokine assay. n = six independent experiments, dots represent individual data, and bars represent mean value ± SEM. Differences between groups were analysed using the ANOVA and Dunnett's tests as post hoc analyses. * $p < 0.05$, ** $p < 0.01$.

Thus, the SEB1741 aptamer at 2, 10, and 20 μg/mL decreased the release of IL-2 ($p < 0.05$), like 5 μg/mL of anti-SEB (Figure 4a). Furthermore, regarding IFN-γ levels, they were reduced using 2 and 20 μg/mL of SEB1741 ($p < 0.01$ and $p < 0.05$, respectively), whereas anti-SEB decreased IFN-γ levels at 5 μg/mL ($p < 0.05$) (Figure 4b). Finally, TNF was increased under SEB stimulus but was not downregulated with either anti-SEB or SEB1741 (Figure 4c).

Together, these results suggest that SEB efficiently induces the release of proinflammatory cytokines and that the SEB1741 aptamer efficiently decreases their release.

2.5. SEB Induces the Release of GM-CSF and IL-10 but Not That of Another Anti-Inflammatory Cytokine

The release of an anti-inflammatory profile was also evaluated in the supernatant of CD4+ T cells cultured under SEB stimulation.

Our data shows that PMA/IO is an optimal stimulus for inducing the release of a broad spectrum of cytokines. SEB induced higher levels of GM-CSF and IL-10 than in unstimulated cells, but neither anti-SEB nor SEB1741 inhibited this production (Figure 5a,b, respectively). Moreover, SEB did not induce the secretion of IL-4, IL-5, IL-12p75, or IL-13 (Figure 5c–f, respectively). Consequently, anti-SEB or SEB1741 did not show any effect.

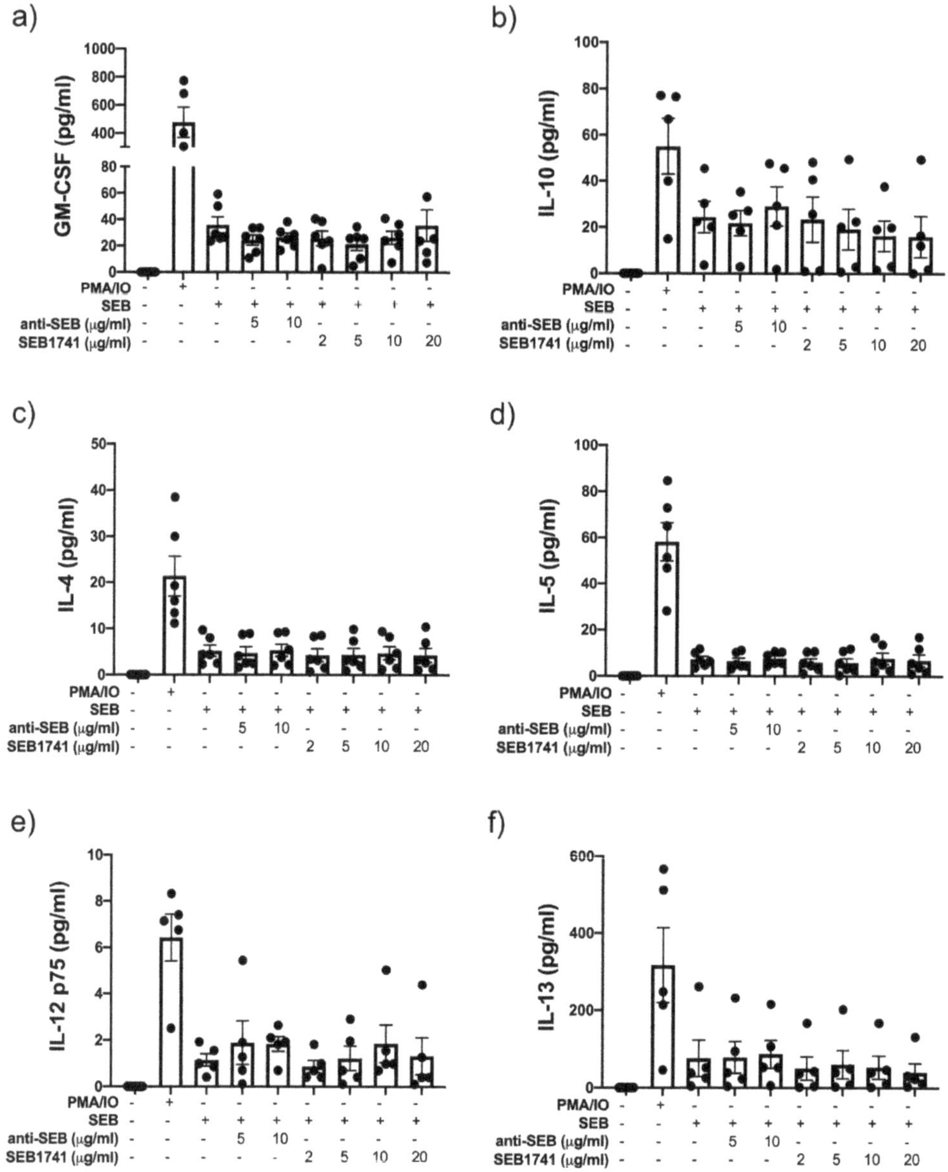

Figure 5. The addition of anti-staphylococcal enterotoxin B (SEB) and SEB1741 aptamers does not modify the production of GM-CSF (Granulocyte-macrophage colony-stimulating factor) interleukin (IL)-10, IL-4, IL-5, IL-12p75, and IL-13 in CD4+ T cells. The levels of GM-CSF (**a**), IL-10 (**b**), IL-4 (**c**), IL-5 (**d**), IL-12p75 (**e**), and IL-13 (**f**) were measured in supernatants collected from CD4+ T cells stimulated by SEB, anti-SEB and SEB1741 aptamer by multiple cytokine assay. n = six independent experiments, dots represent individual data, and bars represent mean value ± SEM.

3. Materials and Methods

3.1. Ethical Approval

All procedures performed in this study adhered to the principles stipulated in the Declaration of Helsinki. This study was approved by the Ethics Committee of the Military Central Hospital, Mexico City (protocol code: C.Inv.068). The healthy donors provided written informed consent for the use of their blood samples.

3.2. Peripheral Blood Mononuclear Cells (PBMCs)

Blood samples from six healthy donors (HD) were collected into EDTA tubes (BD Vacutainer, Franklin Lakes, NJ, USA). PBMCs were isolated within 1 h of blood collection and immediately used to isolate CD4+ T cells and develop cell cultures. PBMCs were isolated using standard Ficoll density gradient centrifugation (Lymphoprep Axis-Shield, Oslo, Norway), and cell viability was determined using trypan blue dye.

3.3. Isolation and Purification of CD4+ T Cells

CD4+ T cells were isolated by negative selection using the Human CD4+ T Cell Isolation Kit (Miltenyi Biotec, Germany) according to the manufacturer's instructions. As a result, the enrichment of the CD4+ T cell fraction was >96%, as determined by flow cytometry.

3.4. CD4+ T Cell Stimulation Assay with SEB

Isolated CD4+ T cells were plated in triplicate (2×10^5 cells/200 µL), and SEB was added (1.7 µg/mL, Oxoid, Basingstoke, England) as the stimulus. The optimal concentration of SEB was established according to a previous standardisation, where CD4+ T cells were cultured for 48 h at 37 °C in 5% CO_2 with SEB at 1.7 µg/mL and 8.5 µg/mL and the supernatant was recovered. TNF was quantified as an indirect marker of activation of CD4+ T cells.

Thus, the cell culture included one condition with SEB plus the monoclonal antibody (mAb) anti-SEB at 5 and 10 µg/mL (MyBioSource, San Diego, CA, USA) and another condition with the SEB1741 aptamer at concentrations 2, 5, 10, and 20 µg/mL (0.28 µM, 1.42 µM, 2.85 µM, and 5.7 µM, respectively). DNA aptamer (40 nt) was synthesised with a 5′-thiol group and HPLC (High-performance liquid chromatography) was purified by T4 Oligo® (Guanajuato, México) at a concentration of 100 µM. In addition, a negative control (no stimulation) and positive control stimulated with phorbol myristate acetate (PMA) 50 nM–ionomycin (IO) 2 ng/µL (Sigma Aldrich, St. Louis, MO, USA) were included.

The cell culture was maintained for 48 h at 37 °C in 5% CO_2. After incubation, the plate was centrifuged, and the supernatant was recovered and stored at −70 °C until it was used to measure the cytokine levels. However, the cells were prepared for flow cytometry.

3.5. Flow Cytometry

We used mAbs to evaluate the expression levels of CD4, CD25, CD69, and Ki-67 in cultured CD4+ T cells. To assess the expression of molecules on cell membranes, the cells were stained for 30 min at 4 °C in the dark. The cells were then fixed with 2% p-formaldehyde in phosphate-buffered saline (PBS, 0 mM sodium phosphate, 0.15 M sodium chloride, pH 7.2). The cells were washed to evaluate intracellular Ki-67 staining. Next, the cell pellet was suspended in a fixation/permeabilisation solution (eBioscience, Thermo Fisher, Waltham, MA, USA) at 4 °C, washed with permeabilisation buffer (eBioscience, Thermo Fisher, Waltham, MA, USA), and stained with Ki-67 for 30 min at 4 °C. Then, it was washed and analysed using flow cytometry. In addition, cells used for the Fluorescence Minus One (FMO) condition were stained and acquired in parallel to identify the background levels of staining.

Data were collected using a FACS Aria II (BD Biosciences, San Jose, CA, USA) and analysed using FlowJo v10.2 (FlowJo LLC, Inc., Ashland, OR, USA). For each case, 50,000 events are recorded. More details on the antibodies used are listed in Table A1.

3.6. Multiple Cytokine Assay in Supernatant Samples

A Bio-Plex Pro Human Cytokine Th1/Th2 Panel 9-Plex (Bio-Rad Laboratories, Richmond, CA, USA) was used. The level of Hu GM-CSF (Granulocyte-macrophage colony-stimulating factor), Hu IFN-γ, Hu IL-2, Hu IL-4, Hu IL-5, Hu IL-10, Hu IL-12(p70), Hu IL-13, and Hu TNF-a was measured in the supernatant of stimulated CD4+ T cell cultures, following the manufacturer's instructions. Data were acquired using a Bio-Plex 200 System and analysed using Bio120 Plex Manager 6.1 software.

3.7. Statistical Analysis

Data are presented as mean $\pm$ standard error of the mean (SEM). Multiple comparisons were performed with the one-way ANOVA test with Dunnett's multiple comparisons post hoc analysis. p-values < 0.05 were considered statistically significant (GraphPad Software V9, La Jolla, CA, USA).

4. Discussion

S. aureus is a highly ubiquitous pathogen with various virulence factors that have helped develop immune evasion strategies. Within the family of staphylococcal enterotoxins (SEs), in addition to SEB, toxic shock syndrome toxin 1 (TSST-1) is included, and it has been reported to be highly associated with SEB cases [22]. Staphylococcal toxic shock syndrome (TSS) has a mortality rate of 50–60%, depending on the severity of the clinical case [23].

The first epidemic of staphylococcal toxic shock syndrome was reported in 1980; subsequently, in the 1990s, the incidence of TSS associated with menstruation and the proportion of non-menstrual cases increased. Non-menstrual TSS has been reported mostly in association with any primary staphylococcal infection following surgery, mainly in cases where there has been disruption of mechanical barriers or placement of a medical device [24]. Many of these reasons prompted several studies to develop toxoid vaccines or peptides that mimic regions in the structure of superantigens and antibodies produced to block their activity [25].

SEB is an SAg produced by *S. aureus* that can interact with the active site of the major histocompatibility complex class II (MHC-II) to activate CD4+ T cells through the T-cell receptor (TCR) [3,26]. There is then massive activation of these T cells, which are a subset of cells that help coordinate the immune response and lead to the excessive release of proinflammatory cytokines [27]. This means that SAg binds outside the complementarity-determining regions (CDRs) of the TCR, causing the cross-linking of MHC-II/TCR. The magnitude of the hyperactivation status induced by an SAg could be dimensional, considering that in a normal adaptive immune response, only around 0.0001% of T cells are activated, whereas SAg activates up to 30% of the T-cell pool [28].

One of the consequences of massive T-cell activation is an increase in leukocyte recruitment, which could favour the survival of *S. aureus* within neutrophils and macrophages due to hyperactivation, eventually leading to T-cell anergy and cell death [29].

Currently, there are no approved vaccines or specific drugs for treating SEB-induced illnesses. Moreover, the use of mAbs has been suggested to block the action of SEB [30–33]. However, the production of anti-SEB mAbs has disadvantages such as high manufacturing costs, variable specificity, and difficulty in production or storage. Recently, the possibility of using a single-chain biparatopic bispecific antibody to target SEB and allosterically prevent TCR binding has been explored. However, this has been performed only as a computational analysis [33]. Antibodies against LXY8, Ig121, c19F1, and 20B1 have been used in experimental models [34–36]. Currently, they are not used in humans, indicating the need to focus on the development of therapeutic alternatives.

Thus, the search has not been limited to blocking the interaction between SEB and TCR. The use of anti-inflammatory drugs, such as sulfasalazine [37], or immunosuppressive drugs, such as rapamycin [38], could be useful in decreasing the excessive inflammatory status. Although they are alternatives to treat the symptoms, they do not prevent massive hyperactivation of T cells.

The development of aptamers has also been proposed to improve treatment schemes [15] and could be an optimal tool because aptamers have high specificity and sensitivity, similar to antibodies. Over the last 25 years, 18 nucleic acid-based therapeutics have been approved for the treatment of various diseases. Compared to mAbs, aptamers have advantages such as stability for long-term storage, simplicity of synthesis and function, and low immunogenicity [39,40].

In this context, previously, we reported the selection "in silico" of the SEB1741 aptamer using a strategy similar to "in vitro" SELEX, including the negative and positive rounds to select the best aptamer with the higher values of interaction with SEB and no other toxins [7]. This approach allowed us to find the SEB1741, and based on the "in silico" analysis, we proposed that it could inhibit the CD4+ T cell activation to regulate the inflammatory status induced by SEB. Here, we showed that the SEB1741 aptamer decreased the percentage of Ki-67 cells and IL-2, suggesting that these molecules inhibit T-cell activation.

Under diverse pathologies, regulating the excessive activation of CD4+ T cells is necessary to maintain homeostasis; therefore, targets have been studied to limit CD4+ T cell activation [41,42]. This study showed that CD4+ T cells treated with the SEB1741 aptamer showed decreased SEB-induced activation and the release of proinflammatory cytokines, specifically IL-2 and IFN-γ, but not TNF. This suggests that SEB1741 is efficient in blocking the IFN-γ-dependent inflammatory pathway. IFN-γ production is one of the most important characteristics of diseases with massive inflammation. Moreover, it has recently been reported that SAg induces excessive IFN-γ levels, which allows the replication of bacteria, such as *S. aureus*, within macrophages, promoting bacterial survival by manipulating the immune response [8].

TNF is a proinflammatory cytokine that activates cell death pathways, mainly via the TNF receptor 1. However, TNF was not decreased by either SEB1741 or anti-SEB, suggesting that this pathway was not inhibited, even when using the antibody. The regulation of TNF production induced by SEB stimuli probably requires the use of molecules, such as drugs, whose functions are performed inside the cell. Bisdemethoxycurcumin has been reported as a potential natural antibacterial agent to decrease the TNF induced by SEB or staphylococcal enterotoxin A (SEA) [43]. Similarly, black ginger has been proposed as a promising natural compound for reducing TNF production, even in methicillin-resistant *S. aureus* [44].

Evidence suggests that SEB blockers do not prevent TNF production; apparently, this pathway is regulated differently from IFN-γ production, at least by SEB stimulus. Thus, the search for therapies that block TNF production must be addressed simultaneously. Although cytotoxic cells were not evaluated in this study, it is a new open question because we do not know whether SEB1741 also inhibits the binding of SEB to CD8+ T cells or if it blocks only the interaction between SEB/CD4+ T cells.

Moreover, it is worth noting that an excess of aptamers can lose their ability to inhibit proliferation, similar to what is observed with an excess of anti-SEBs. Research has shown that an excess of antibodies or antigens can interfere with signalling for T cell activation and result in a lack of IL-2 production and phosphorylation of signalling molecules [45]. Thus, determining the optimal concentration of aptamers is critical to effectively assay the function of a blocking antibody.

Our study has some limitations. For instance, we did not experimentally evaluate other aptamers in parallel to confirm the high specificity of SEB1741. Nonetheless, as previously reported, the aptamer library we used to identify the best aptamers for SEB, cholera toxin, and botulinum toxin displayed high specificity to their respective toxins "in silico" and did not exhibit cross-reactivity [7]. Additionally, further studies should be conducted "in vivo" to evaluate inflammation and survival before suggesting the use of SEB1741 as an alternative treatment for toxic shock syndrome.

Despite the limitations in our study, these data strongly suggest that SEB1741 has the potential to be an optimal tool for preventing TSS through an excessive inflammatory response mediated by IFN-γ produced by CD4+ T cells. Like antibodies, aptamers bind their targets by folding them into a specific three-dimensional conformation dictated by the nucleic acid sequence. In addition to aptamers having a higher affinity, they display a lower Kd and thus bind better to the target molecule [46].

These features represent a combination of the best features of small molecules and antibodies, providing great potential for numerous therapeutic applications, including the inhibition of proinflammatory proteins. Additionally, from the perspective of the pharmaceutical industry, aptamer production could represent a lower cost than antibody production.

5. Conclusions

These results suggest that the SEB1741 aptamer has therapeutic potential as an alternative to antibodies and other protein inhibitors currently used to treat SAg-mediated diseases. Interestingly, this aptamer specifically blocks the exacerbated production of IFN-γ but not that of other proinflammatory cytokines, such as TNF. Moreover, further studies are needed to evaluate its efficacy and long-term safety in vivo.

Author Contributions: Conceptualisation, L.C.-G., K.M.-Q. and M.G.C.-E.; methodology, A.R., L.A.R.-L., B.T.-C. and A.S.-M.; formal analysis, A.R., L.C.-G. and A.E.-G.; investigation, A.R., L.A.R.-L., L.C.-G., A.E.-G., K.M.-Q. and M.G.C.-E.; writing—original draft preparation, A.R.; writing—review and editing, L.C.-G., A.E.-G., K.M.-Q. and M.G.C.-E. supervision: L.C.-G. and K.M.-Q. project administration: L.C.-G. and K.M.-Q. All authors have read and agreed to the published version of the manuscript.

Funding: This work was supported by the Research and Development Center of the Mexican Army and Air Force (C.I.D.E.F.A.M.) under the budget program A022 "Military research and development in coordination with public universities, public higher education institutions, and/or other public higher research centres."

Institutional Review Board Statement: This study was approved by the Ethics Committee of Military Central Hospital, Mexico City (protocol code C.Inv.068).

Informed Consent Statement: Informed consent was obtained from all participants involved in the study.

Data Availability Statement: The authors confirm that the raw data used to support the conclusions of this study are included in the manuscript. The corresponding author will provide more details upon request from any qualified researcher.

Conflicts of Interest: The authors declare no conflict of interest.

Appendix A

Table A1. Information on the antibodies used in this study.

Antibody	Company	Clone	Cat. Number
Brilliant Violet 421™ anti-human Ki-67	BioLegend	Ki67	B269687
PE/Cyanine7 anti-human CD4	BioLegend	OKT4	B159354
PE-A anti-human CD25	BD Pharmingen	BC96	3235775
Brilliant Violet 510™ anti-human CD69	BioLegend	FN50	B322621

References

1. Douglas-Louis, R.; Lou, M.; Lee, B.; Minejima, E.; Bubeck-Wardenburg, J.; Wong-Beringer, A. Prognostic significance of early platelet dynamics in *Staphylococcus aureus* bacteremia. *BMC Infect. Dis.* **2023**, *1*, 82. [CrossRef] [PubMed]
2. Howden, B.P.; Giulieri, S.G.; Lung, T.W.F.; Baines, S.L.; Sharkey, L.K.; Lee, J.Y.H.; Hachani, A.; Monk, I.R.; Stinear, T.P. *Staphylococcus aureus* host interactions and adaptation. *Nat. Rev. Microbiol.* **2023**, *1*, 1–16. [CrossRef]
3. Francis, D.; Bhairaddy, A.; Joy, A.; Hari, G.V.; Francis, A. Secretory proteins in the orchestration of microbial virulence: The curious case of *Staphylococcus aureus*. *Adv. Protein Chem. Struct. Biol.* **2023**, *133*, 271–350.

4. Lawrynowicz-Paciorek, M.; Kochman, M.; Piekarska, K.; Grochowska, A.; Windyga, B. The distribution of enterotoxin and enterotoxin-like genes in Staphylococcus aureus strains isolated from nasal carriers and food samples. *Int. J. Food Microbiol.* **2007**, *117*, 319–323. [CrossRef] [PubMed]
5. Jang, J.H.; Kim, S.; Kim, S.G.; Lee, J.; Lee, D.G.; Jang, J.; Jeong, Y.S.; Song, D.H.; Min, J.K.; Park, J.G.; et al. A Sensitive Immunodetection Assay Using Antibodies Specific to Staphylococcal Enterotoxin B Produced by Baculovirus Expression. *Biosensors* **2022**, *10*, 787. [CrossRef] [PubMed]
6. Purwanasari, H.N.; Permatasari, A.T.U.; Lestari, F.B.; Wasissa, M.; Zaini, K.; Salasia, S.I.O. Cellular immune response of *Staphylococcus aureus* enterotoxin B in Balb/c mice through intranasal infection. *Vet. World.* **2022**, *7*, 1765–1771. [CrossRef]
7. Escamilla-Gutiérrez, A.; Córdova-Espinoza, M.G.; Sánchez-Monciváis, A.; Tecuatzi-Cadena, B.; Regalado-García, A.G.; Medina-Quero, K. In silico selection of aptamers for bacterial toxins detection. *J. Biomol. Struct. Dyn.* **2022**, *1*, 1–10. [CrossRef]
8. Tuffs, S.W.; Goncheva, M.I.; Xu, S.X.; Craig, H.C.; Kasper, K.J.; Choi, J.; Flannagan, R.S.; Kerfoot, S.M.; Heinrichs, D.E.; McCormick, J.K. Superantigens promote *Staphylococcus aureus* bloodstream infection by eliciting pathogenic interferon-gamma production. *Proc. Natl. Acad. Sci. USA* **2022**, *8*, e2115987119. [CrossRef] [PubMed]
9. Noli Truant, S.; De Marzi, M.C.; Sarratea, M.B.; Antonoglou, M.B.; Meo, A.P.; Iannantuono López, L.V.; Fernández Lynch, M.J.; Todone, M.; Malchiodi, E.L.; Fernández, M.M. *egc* Superantigens Impair Monocytes/Macrophages Inducing Cell Death and Inefficient Activation. *Front. Immunol.* **2020**, *10*, 3008. [CrossRef]
10. Fraser, J.D.; Proft, T. The bacterial superantigen and superantigen-like proteins. *Immunol. Rev.* **2008**, *1*, 226–243. [CrossRef]
11. Mihara, H.; Uchida, K.; Watanabe, Y.; Nanjo, S.; Sakumura, M.; Motoo, I.; Ando, T.; Minemura, M.; Muhammad, J.S.; Yamamoto, H.; et al. Colonic TRPV4 overexpression is related to constipation severity. *BMC Gastroenterol.* **2023**, *23*, 13. [CrossRef]
12. Hu, H.; Liu, S.; Hon, K.; Psaltis, A.J.; Wormald, P.J.; Vreugde, S. Staphylococcal protein A modulates inflammation by inducing interferon signaling in human nasal epithelial cells. *Inflamm. Res.* **2023**, *2*, 251–262. [CrossRef]
13. Lowy, F.D. *Staphylococcus aureus* Infections. *N. Engl. J. Med.* **1998**, *8*, 520–532. [CrossRef] [PubMed]
14. Gottlieb, M.; Long, B.; Koyfman, A. The Evaluation and Management of Toxic Shock Syndrome in the Emergency Department: A Review of the Literature. *J. Emerg. Med.* **2018**, *6*, 807–814. [CrossRef] [PubMed]
15. Wang, K.; Gan, L.; Jiang, L.; Zhang, X.; Yang, X.; Chen, M.; Lan, X. Neutralization of Staphylococcal Enterotoxin B by an Aptamer Antagonist. *Antimicrob. Agents Chemother.* **2015**, *4*, 2072–2077. [CrossRef]
16. Mudili, V.; Makam, S.S.; Sundararaj, N.; Siddaiah, C.; Gupta, V.K.; Rao, P.V.L. Retraction Note: A novel IgY-Aptamer hybrid system for cost-effective detection of SEB and its evaluation on food and clinical samples. *Sci. Rep.* **2022**, *1*, 10939. [CrossRef] [PubMed]
17. Moltajaei, M.H.; Pourzare Mehrbani, S.; Motahari, P.; Rezapour, R. Clinicopathological and prognostic value of Ki-67 expression in oral malignant melanoma: A systematic review and meta-analysis. *J. Dent. Res. Dent. Clin. Dent. Prospects.* **2022**, 3, 140–146. [CrossRef]
18. Di Rosa, F.; Cossarizza, A.; Hayday, A.C. To Ki or Not to Ki: Re-Evaluating the Use and Potentials of Ki-67 for T Cell Analysis. *Front. Immunol.* **2021**, *12*, 653974. [CrossRef]
19. Chávez-Galán, L.; Illescas-Eugenio, J.; Alvarez-Sekely, M.; Baez-Saldaña, R.; Chávez, R.; Lascurain, R. Tuberculosis patients display a high proportion of $CD8^+$ T cells with a high cytotoxic potential. *Microbiol. Immunol.* **2019**, *8*, 316–327. [CrossRef]
20. Zhang, Y.; Su, J. Interleukin-2 family cytokines: An overview of genes, expression, signaling and functional roles in teleost. *Dev. Comp. Immunol.* **2023**, *141*, 104645. [CrossRef]
21. Villalba, N.; Ma, Y.; Gahan, S.A.; Joly-Amado, A.; Spence, S.; Yang, X.; Nash, K.; Yuan, S.Y. Lung infection by P. aeruginosa induces neuroinflammation and blood-brain barrier dysfunction in mice. *bioRxiv* **2023**, *1*, 31.
22. Spaulding, A.R.; Salgado-Pabón, W.; Kohler, P.L.; Horswill, A.R.; Leung, D.Y.M.; Schlievert, P.M. Staphylococcal and Streptococcal Superantigen Exotoxins. *Clin. Microbiol. Rev.* **2013**, *3*, 422–447. [CrossRef] [PubMed]
23. Bîrluţiu, V.; Crişţiu, O.; Baicu, M.; Bîrluţiu, R.M. The Management of Staphylococcal Toxic Shock Syndrome. A Case Report. *J. Crit. Care Med.* **2016**, *2*, 85–88. [CrossRef]
24. Chuang, Y.-Y.; Huang, Y.-C.; Lin, T.-Y. Toxic Shock Syndrome in Children. *Pediatr. Drugs.* **2005**, *1*, 11–25. [CrossRef] [PubMed]
25. Visvanathan, K.; Charles, A.; Bannan, J.; Pugach, P.; Kashfi, K.; Zabriskie, J.B. Inhibition of Bacterial Superantigens by Peptides and Antibodies. *Infect. Immun.* **2001**, *2*, 875–884. [CrossRef]
26. Shepherd, F.R.; Davies, K.; Miners, K.L.; Llewellyn-Lacey, S.; Kollnberger, S.; Redman, J.E.; Grant, M.M.; Ladell, K.; Price, D.A.; McLaren, J.E. The superantigens SpeC and TSST-1 specifically activate $TRBV12\text{-}3/12\text{-}4^+$ memory T cells. *Commun Biol.* **2023**, *1*, 78. [CrossRef]
27. Rha, M.S.; Kim, S.W.; Chang, D.Y.; Lee, J.K.; Kim, J.; Park, S.H.; Khalmuratova, R.; Lim, H.S.; Eun, K.M.; Hong, S.N.; et al. Superantigen-related TH2 $CD4^+$ T cells in nonasthmatic chronic rhinosinusitis with nasal polyps. *J. Allergy Clin. Immunol.* **2020**, *5*, 1378–1388. [CrossRef]
28. Deacy, A.M.; Gan, S.K.; Derrick, J.P. Superantigen Recognition and Interactions: Functions, Mechanisms and Applications. *Front. Immunol.* **2021**, *12*, 731845. [CrossRef]
29. Gresham, H.D.; Lowrance, J.H.; Caver, T.E.; Wilson, B.S.; Cheung, A.L.; Lindberg, F.P. Survival of Staphylococcus aureus inside neutrophils contributes to infection. *J. Immunol.* **2000**, *164*, 3713–3722. [CrossRef]
30. Larkin, E.A.; Stiles, B.G.; Ulrich, R.G. Inhibition of Toxic Shock by Human Monoclonal Antibodies against Staphylococcal Enterotoxin B. *PLoS ONE* **2010**, *10*, e13253.

31. Chen, G.; Karauzum, H.; Long, H.; Carranza, D.; Holtsberg, F.W.; Howell, K.A.; Abaandou, L.; Zhang, B.; Jarvik, N.; Ye, W.; et al. Potent Neutralization of Staphylococcal Enterotoxin B In Vivo by Antibodies that Block Binding to the T-Cell Receptor. *J. Mol. Biol.* **2019**, *431*, 4354–4367. [CrossRef] [PubMed]
32. Liu, Y.; Song, Z.; Ge, S.; Zhang, J.; Xu, L.; Yang, F.; Lu, D.; Luo, P.; Gu, J.; Zou, Q.; et al. Determining the immunological characteristics of a novel human monoclonal antibody developed against staphylococcal enterotoxin B. *Hum. Vaccines Immunother.* **2020**, *7*, 1708–1718. [CrossRef]
33. Bai, G.; Ge, Y.; Su, Y.; Chen, S.; Zeng, X.; Lu, H.; Ma, B. Computational Construction of a Single-Chain Bi-Paratopic Antibody Allosterically Inhibiting TCR-Staphylococcal Enterotoxin B Binding. *Front. Immunol.* **2021**, *12*, 732938. [CrossRef]
34. MacIntyre, J.L.; Varshney, A.K.; Wang, X.; Gatto, S.; Friedman, C.; Liu, Y.; Kerns, K.; Kovalenko, O.; Adkins, K.; Zollner, R.; et al. Optimization of experimental conditions for functional in vitro characterization of humanized antibodies specific for staphylococcal enterotoxin B. *Int. Immunopharmacol.* **2015**, *28*, 354–358. [CrossRef] [PubMed]
35. Verreault, D.; Ennis, J.; Whaley, K.; Killeen, S.Z.; Karauzum, H.; Aman, M.J.; Holtsberg, R.; Doyle-Meyers, L.; Didier, P.J.; Zeitlin, L.; et al. Effective Treatment of Staphylococcal Enterotoxin B Aerosol Intoxication in Rhesus Macaques by Using Two Parenterally Administered High-Affinity Monoclonal Antibodies. *Antimicrob. Agents. Chemother.* **2019**, *63*, e02049-18. [CrossRef]
36. Hu, N.; Qiao, C.; Wang, J.; Wang, Z.; Li, X.; Zhou, L.; Wu, J.; Zhang, D.; Feng, J.; Shen, B.; et al. Identification of a novel protective human monoclonal antibody, LXY8, that targets the key neutralizing epitopes of staphylococcal enterotoxin B. *Biochem. Biophys. Res. Commun.* **2021**, *549*, 120–127. [CrossRef] [PubMed]
37. Krakauer, T. Sulfasalazine Attenuates Staphylococcal Enterotoxin B-Induced Immune Responses. *Toxins* **2015**, *2*, 553–559. [CrossRef]
38. Krakauer, T.; Buckley, M. Intranasal Rapamycin Rescues Mice from Staphylococcal Enterotoxin B-Induced Shock. *Toxins* **2012**, *9*, 718–728. [CrossRef]
39. Fan, R.; Tao, X.; Zhai, X.; Zhu, Y.; Li, Y.; Chen, Y.; Dong, D.; Yang, S.; Lv, L. Application of aptamer-drug delivery system in the therapy of breast cancer. *Biomed. Pharmacother.* **2023**, *161*, 114444. [CrossRef]
40. Egli, M.; Manoharan, M. Chemistry, structure and function of approved oligonucleotide therapeutics. *Nucleic Acids Res.* **2023**, *51*, 2529–2573. [CrossRef]
41. Wu, M.; Gao, X.; Tang, Y.; Wu, W.; Zhou, J.; Shao, Y.; Hao, C.; Yang, Y.; Zhang, J. Cbl-b inhibited $CD4^+$ T cell activation by regulating the expression of miR-99a/miR-125b. *Int. Immunopharmacol.* **2023**, *115*, 109677. [CrossRef] [PubMed]
42. Park, J.; Son, M.J.; Ho, C.C.; Lee, S.H.; Kim, Y.; An, J.; Lee, S.K. Transcriptional inhibition of STAT1 functions in the nucleus alleviates Th1 and Th17 cell-mediated inflammatory diseases. *Front. Immunol.* **2022**, *13*, 1054472. [CrossRef] [PubMed]
43. Wang, S.; Kang, O.-H.; Kwon, D.-Y. Bisdemethoxycurcumin Reduces Methicillin-Resistant *Staphylococcus aureus* Expression of Virulence-Related Exoproteins and Inhibits the Biofilm Formation. *Toxins* **2021**, *11*, 804. [CrossRef] [PubMed]
44. Lee, Y.-S.; Kim, K.-W.; Yoon, D.; Kim, G.-S.; Kwon, D.-Y.; Kang, O.-H.; Lee, D. Comparison of Antivirulence Activities of Black Ginseng against Methicillin-Resistant *Staphylococcus aureus* According to the Number of Repeated Steaming and Drying Cycles. *Antibiotics* **2021**, *6*, 617. [CrossRef]
45. Goldman, S.J.; Uniyal, S.; Ferguson, L.M.; Golan, D.E.; Burakoff, S.J.; Kiener, P.A. Differential Activation of Phosphotyrosine Protein Phosphatase Activity in a Murine T Cell Hybridoma by Monoclonal Antibodies to CD45. *J. Biol. Chem.* **1992**, *267*, 6197–6204. [CrossRef] [PubMed]
46. Nimjee, S.M.; White, R.R.; Becker, R.C.; Sullenger, B.A. Aptamers as Therapeutics. *Annu. Rev. Pharmacol. Toxicol.* **2017**, *1*, 61–79. [CrossRef] [PubMed]

molecules

MDPI

Article

Highly Sensitive β-Lactoglobulin Fluorescent Aptamer Biosensors Based on Tungsten Disulfide Nanosheets and DNase I-Assisted Signal Amplification

Yuying Wang [1], Sisi Chen [1], Wanmei Chen [1], Jingjing Wang [1], Kun Li [2], Chengyi Hong [1,3,*], Kailong Zhang [2,4,5,*] and Quansheng Chen [1]

1 College of Ocean Food and Biological Engineering, Jimei University, Xiamen 361021, China
2 School of Life Sciences, Longyan University, Longyan 364012, China
3 Fujian Provincial Key Laboratory of Food Microbiology and Enzyme Engineering, Xiamen 361021, China
4 Fujian Provincial Key Laboratory for the Prevention and Control of Animal Infectious Diseases and Biotechnology, Longyan 364012, China
5 Fujian Province Universities Key Laboratory of Preventive Veterinary Medicine and Biotechnology, Longyan University, Longyan 364012, China
* Correspondence: cyhong@jmu.edu.cn (C.H.); klzhang@lyun.edu.cn (K.Z.)

Abstract: β-lactoglobulin (β-Lg) is a protein found in milk that can cause severe allergic reactions, including rash, vomiting, and diarrhea. Thus, it is crucial to develop a sensitive β-Lg detection method to protect people who are susceptible to allergies. Here, we introduce a novel and highly sensitive fluorescent aptamer biosensor for detecting β-Lg. First, a fluorescein-based dye (FAM)-labeled β-lactoglobulin aptamer (β-Lg aptamer) is adsorbed on the surface of tungsten disulfide (WS_2) nanosheets via van der Waals forces, resulting in fluorescence quenching. When β-Lg is present, the β-Lg aptamer selectively binds to β-Lg, causing a conformational change in the β-Lg aptamer and releasing it from the surface of WS_2 nanosheets, which restores the fluorescence signal. Simultaneously, DNase I in the system cleaves the aptamer bound to the target, producing a short oligonucleotide fragment and releasing β-Lg. The released β-Lg then binds to another β-Lg aptamer adsorbed on WS_2, initiating the next round of cleavage, resulting in significant amplification of the fluorescence signal. This method has a linear detection range of 1–100 ng mL^{-1}, and the limit of detection is 0.344 ng mL^{-1}. Furthermore, this approach has been successfully used for detecting β-Lg in milk samples with satisfactory results, providing new opportunities for food analysis and quality control.

Keywords: β-lactoglobulin; aptamer; WS_2 nanosheets; DNase I

check for updates

Citation: Wang, Y.; Chen, S.; Chen, W.; Wang, J.; Li, K.; Hong, C.; Zhang, K.; Chen, Q. Highly Sensitive β-Lactoglobulin Fluorescent Aptamer Biosensors Based on Tungsten Disulfide Nanosheets and DNase I-Assisted Signal Amplification. *Molecules* **2023**, *28*, 3502. https://doi.org/10.3390/molecules28083502

Academic Editors: Sai Wang, Wei Wu and Long Wu

Received: 23 February 2023
Revised: 31 March 2023
Accepted: 14 April 2023
Published: 16 April 2023

1. Introduction

Food allergy is an adverse reaction that occurs when the immune system responds to a specific food after repeated exposure, which can result in various skin and gastrointestinal problems and, in severe cases, can be life-threatening. Thus, food allergy has become a significant global food safety issue [1,2]. The Food and Agriculture Organization of the United Nations has identified cows' milk as one of the eight food allergens [3]. Cows' milk protein allergy (CMPA) is a common food allergy among infants and children, which can cause allergic rhinitis, asthma, eczema, diarrhea, gastrointestinal bleeding, and other symptoms [4]. CMPA is mainly caused by αs1-casein, α-lactalbumin (α-La), and β-lactoglobulin (β-Lg) as common allergens [5]. β-Lg, which exists mainly as a noncovalently linked dimer and contains 162 amino acid residues with a molecular weight of 18 kDa, is particularly significant. It accounts for 50% of whey protein and 12% of total protein in cows' milk, with an average concentration of 2–4 g/L [6]. Epidemiological studies suggest that 2–3% of infants and children experience allergic reactions to proteins in cows' milk, with around 82% of cow milk allergy patients being allergic to cow's milk β-Lg [7]. Furthermore, β-lactoglobulin is widely employed in food processing due to its high nutritional value, hypotensive, antioxidant, anti-microbial activity, and immunomodulatory effects, increasing the risk of exposure for patients with cow's milk allergy [8].

Numerous analytical methods have been developed to detect β-Lg, which can be classified into three main categories based on the following detection principles: polymerase chain reaction assays based on allergenic DNA, chromatographic separation method based on allergenic proteins, and immunoassay based on allergenic proteins. However, bovine milk typically lacks allergen-related DNA, so chromatographic and immunological methods are primarily used [9–11]. Due to the accuracy requirements of the instrument, these approaches often require rigorous pretreatment procedures, long analysis times, and special laboratory skills [12]. In contrast, immunoassays based on antibody-antigen recognition have significantly enhanced the analysis of trace substances, with traditional methods mainly using enzyme-linked immunosorbent assays [13]. As technology advances, nanomaterials such as magnetic nanobeads, colloidal gold, quantum dots, graphene, and carbon nanotubes can be used to increase detection rates and sensitivity [14]. Although this approach has high specificity and sensitivity, it is expensive, and maintaining antibody stability can be challenging. Furthermore, the quality differences in antibody batches and cross-contamination can readily result in false positive results when accurate quantification is required [15]. Recently, emerging technologies, including surface-enhanced Raman spectroscopy [16], electrochemical immunosensors [17], visualized microarrays [18], and transient isochronous electrophoresis [19], have been developed. However, the sample preparation process of these approaches can be cumbersome, and external environmental factors can easily influence the assay's results. Therefore, it is crucial to develop a rapid, accurate, and effective analytical approach for monitoring β-Lg levels in milk and dairy products to contribute to the standardization of food package labeling and protect public health.

Nucleic acid aptamers (Apt) are DNA or RNA oligonucleotides obtained via the systematic evolution of the exponentially enriched ligand approach [20]. They offer several advantages, such as low cost, good stability, specificity, and affinity compared with traditional antibodies. Furthermore, the primary advantage of using an aptamer is that there are no target limits. Although antibody-based assays are well established, they have limitations in recognizing a wide range of targets, including hazardous small compounds and nonimmunogenic targets. They are also slow in detecting minute variations in large molecular targets, such as proteins. Therefore, aptamers can be employed as outstanding biological recognition elements to develop various aptamer-based bio-sensors for detecting small molecules, proteins, and even cells [21–23]. Recently, several aptamer-based biosensors have been developed using different techniques such as fluorescence [24], chemiluminescence [25], electrochemistry [26], surface plasmon resonance [27], and colorimetry [28]. Among these techniques, fluorescent biosensors have gained significant attention due to their simplicity, rapidity, cost-effectiveness, and high sensitivity. Layered tungsten disulfide (WS_2) nanosheets are transition metal dichalcogenides with an S-W-S intercalation structure. They have a high specific surface area, excellent photothermal conversion efficiency, good biocompatibility, and superior electrical properties. Furthermore, WS_2 nanosheets have significantly higher fluorescence quenching efficiency compared with graphene oxide and can accurately identify single-stranded DNA (ssDNA) and double-stranded DNA (dsDNA), which is crucial for the sensitivity of quench- and recovery-based fluorescent aptamer biosensors [29]. Due to these advantages, WS_2 has been successfully employed for detecting various substances. For instance, a new type of fluorescent aptamer biosensor based on up-conversion nanoparticles and WS_2 nanosheets has been developed for detecting Escherichia coli [30]. Similarly, based on the high fluorescence quenching efficiency of WS_2 nanosheets, effective fluorescent biosensors have been developed for detecting zearalenone [31] and kanamycin [32]. Therefore, WS_2 nanosheets are an ideal material for constructing fluorescent aptamer biosensors to detect β-Lg.

In real food samples, the content of the target molecule is usually low, so a signal amplification strategy is necessary to enhance the sensitivity of the biosensor. Researchers have successfully applied various signal amplification methods in biosensors, including those employing catalytic and self-propagating cascade reactions. Rolling cycle amplification, catalytic hairpin assembly, strand displacement reaction, and hybridization chain reaction are also widely used by researchers [33–35]. Deoxyribonuclease I (DNase I) is frequently used as a nucleic acid endonuclease that can cleave ssDNA for aptamer biosensors [36]. DNase I can also cleave the aptamer/target complex in the detection system and release the target into the solution for recycling, thus providing a fluorescence amplification effect. For instance, a DNase I-assisted fluorescence signal amplification strategy was developed to reduce the quantitative limit of the reaction in Patulin [37] and prostate-specific antigen detection [38]. In this study, we developed a WS_2 nanosheet-aptamer fluorescent biosensor based on fluorescein-based dye (FAM)-labeled β-lactoglobulin aptamer (β-Lg aptamer) by combining the enzymatic cycle amplification reaction of DNase I and the selective binding ability as well as the fluorescence quenching effect of WS_2 nanosheets. In this sensing system, in the absence of the β-Lg, the β-Lg aptamer is adsorbed onto the WS_2 nanosheet due to the van der Waals forces between the DNA nucleobase and the basal plane of the WS_2 nanosheet [39,40], which results in a fluorescence resonance energy transfer phenomenon, leading to fluorescence quenching. However, in the presence of β-Lg, the β-Lg aptamer preferentially binds to β-Lg, resulting in a change in the aptamer's conformation and weakening the interaction between the β-Lg aptamer and WS_2 nanosheets. This releases the aptamer from the surface of WS_2 nanosheets and restores fluorescence. At this time, DNase I present in the sensing system digests the fluorescent probe bound to the target, releasing β-Lg and generating a short oligonucleotide fragment. The cleaved FAM-labeled oligonucleotide fragment does not adsorb into the WS_2 nanosheet due to its weak affinity with the WS_2 nanosheet and thus retains a strong fluorescent signal. Meanwhile, β-Lg continued to bind to another fluorescent probe adsorbed on WS_2 nanosheets and initiates the next round of cleavage after releasing the target β-Lg, resulting in significant amplification in the fluorescence signal. The change in fluorescence signal intensity detects the β-Lg content (Figure 1). This approach is highly specific and sensitive and has been demonstrated to be useful for practical sample detection.

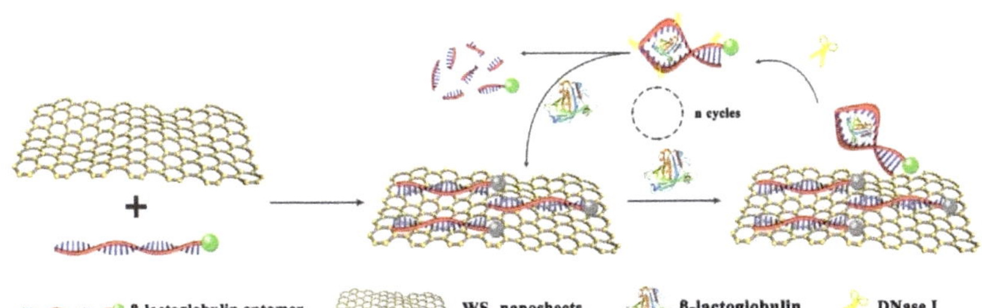

Figure 1. Working principle of the fluorescent aptamer biosensor for β-Lg detection.

2. Results

2.1. Feasibility Study of the Fluorescent Biosensor

To verify the feasibility of the enzymatic cycle amplification strategy, we conducted gel electrophoresis analyses using β-Lg and a β-Lg aptamer as models. After incubating the β-Lg aptamer with DNase I, it was completely cleaved (Figure 2a, lane 2). DNase I could also cleave the β-Lg aptamer in the presence of β-Lg (Figure 2a, lane 3). However, the β-Lg aptamer adsorbed on WS_2 nanosheets showed no evident enzymatic hydrolysis, indicating its resistance to enzymatic digestion (Figure 2a, lane 4). To further determine the feasibility of the fluorescent biosensor for β-Lg detection, Figure 2b illustrates the fluorescence emission spectra of the β-Lg aptamer under different conditions. When WS_2

nanosheets were introduced into the system, the aptamer was easily adsorbed on the WS_2 surface due to the van der Waals force between the strong β-Lg aptamer base and WS_2 nanosheets. Furthermore, WS_2 nanosheets have broad absorption ranging from 200 to 800 nm (Figure S1 in Supplementary Materials), and the β-Lg aptamer's fluorescence emission spectra overlapped well, resulting in FRET and a substantial decrease in fluorescence intensity. The high specificity between the β-Lg and β-Lg aptamer in the presence of β-Lg caused the aptamer to be released from the WS_2's surface and formed a β-Lg-aptamer complex, resulting in fluorescence recovery (blue line). To further enhance sensitivity, we employed an enzymatic cycle amplification approach. The β-Lg-aptamer complex was formed when the system contained the targeted β-Lg. Meanwhile, DNase I could cleave the β-Lg aptamer, releasing β-Lg and liberating the FAM fluorophore. Then, the released β-Lg could bind to another β-Lg aptamer, initiating the next round of cleavage and leading to significant fluorescence recovery compared to the system without DNase I. Due to this advantage, upon adding DNase I, fluorescence increased by approximately 3-fold, serving the purpose of signal amplification (pink line). These results confirm the feasibility of the proposed fluorescence amplification method.

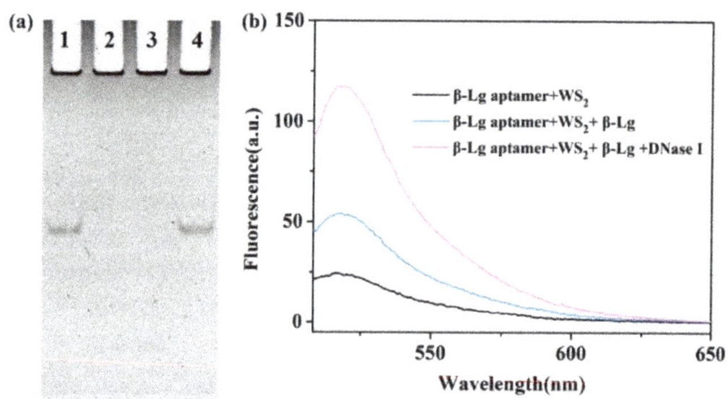

Figure 2. (**a**) Gel electrophoresis of the β-Lg aptamer. Lane 1: β-Lg aptamer; lane 2: β-Lg aptamer + DNase I; lane 3: β-Lg aptamer + β-Lg + DNase I; lane 4: β-Lg aptamer + WS_2 + DNase I. (**b**) Fluorescence spectra of β-Lg aptamer in different cases: (black line) β-Lg aptamer + WS_2; (blue line) β-Lg aptamer + WS_2 + β-Lg; (pink line) β-Lg aptamer + WS_2 + β-Lg + DNase I.

2.2. *Optimization of the Testing Conditions*

Several important parameters were optimized to obtain the best fluorescence signal. These included the concentration of WS_2 nanosheets, the quenching time of the method, the fluorescence recovery time of the system, and the concentration of DNase I. First, the concentration of WS_2 nanosheets was optimized to enhance the sensing performance. The fluorescence intensity of the β-Lg aptamer decreased substantially with the increasing concentration of WS_2 nanosheets, while the concentration of β-Lg aptamer remained constant (Figure 3a). The fluorescence quenching efficiency reached 96% when the concentration of WS_2 nanosheets increased to 750 $\mu g \cdot mL^{-1}$, demonstrating that WS_2 nanosheets can be an effective quencher. A further increase in the concentration of WS_2 nanosheets did not result in any significant changes in fluorescence intensity. Therefore, WS_2 nanosheets at a concentration of 750 $\mu g \cdot mL^{-1}$ were chosen for subsequent experiments to achieve the required sensitivity and selectivity of the assay. Subsequently, we analyzed the quenching time of the method as well as the fluorescence recovery time of the reaction system. Figure 3b illustrates the experimental results. The fluorescence intensity decreased rapidly and reached equilibrium in about 10 min when conducting WS_2 with the β-Lg aptamer (Figure 3c). Therefore, we considered 10 min as the optimal quenching time. With the addition of β-Lg and DNase I, the fluorescence intensity gradually increased over time, rapidly increasing within 40 min and then slowly increasing from 40 to 80 min. Accordingly, we selected

40 min as the optimal incubation time for β-Lg and DNase I. Finally, we optimized the concentration of DNase I by adding varying concentrations of DNase I (0, 0.01, 0.02, 0.03, 0.04, and 0.05 U·μL^{-1}) to the reaction system while keeping the other conditions consistent. The fluorescence intensity increased with increasing enzyme concentration and reached saturation at 0.04 U·μL^{-1} (Figure 3d). Therefore, we used this concentration of DNase I in subsequent assays.

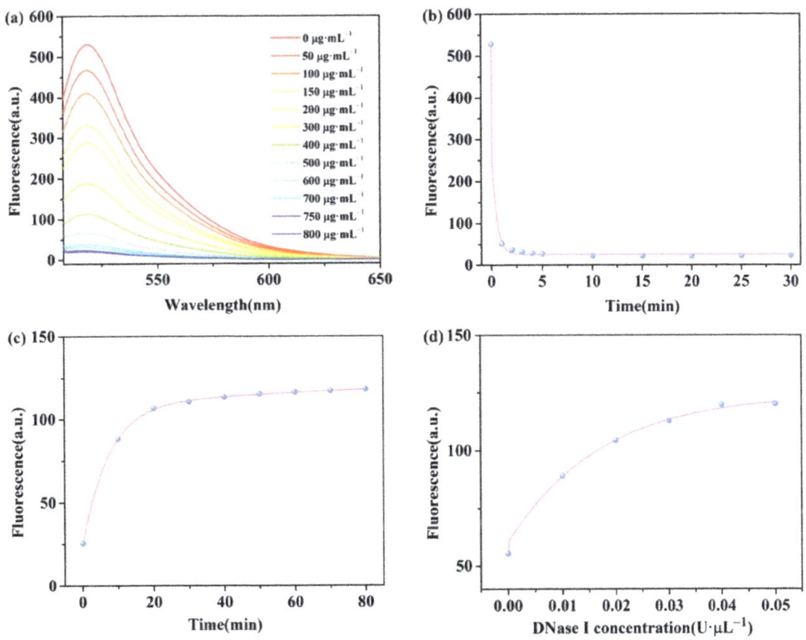

Figure 3. (**a**) Fluorescence spectra of β-Lg aptamer after adding different concentrations (from a to l: 0, 50, 100, 150, 200, 300, 400, 500, 600, 700, 750, and 800 μg·mL^{-1}) of WS_2 nanosheets. (**b**) Fluorescence intensity of β-Lg aptamer in response to various times. (**c**) Effect of recovery time on the fluorescence response of the β-Lg (10 μg·mL^{-1}) detection system. (**d**) Effect of enzyme concentration on the fluorescence response of the β-Lg (10 μg·mL^{-1}) detection system.

2.3. Sensitivity of the Aptamer Biosensor

We measured different concentrations of β-Lg under the optimal experimental conditions described above to analyze the sensitivity of the proposed fluorescent biosensor. We added a series of concentrations (0, 0.001, 0.01, 0.025, 0.05, 0.1, 1, 10, 20, 50, 75, 100, 150, and 200 μg·mL^{-1}) of β-Lg to the proposed detection system. The emission peaks were recorded at 518 nm. As shown in Figure 4a, the fluorescence intensity at 518 nm increased with an increase in β-Lg concentration from 0 to 200 μg·mL^{-1}, indicating the high dependence of the biosensor on the target concentration in detecting β-Lg. A good linear relationship was observed between the fluorescence intensity and the concentration of β-Lg within the range of 1–100 ng mL^{-1} (Figure 4b). The calibration function is $F = 581.5C + 24.6$ ($R^2 = 0.9969$), where C represents the concentration of β-Lg, and F represents the fluorescence intensity. Furthermore, we calculated the limit of detection of the method as 0.344 ng mL^{-1} using equation $3S_0/K$, where S_0 is the standard deviation of the blank test (n = 10), and K is the slope of the calibration curve. We also established the analytical performance for detecting β-Lg without DNase I (Figure 5). The fluorescence intensity at 518 nm increased with an increase in β-Lg concentrations from 0.1 to 150 μg·mL^{-1}. The detection limit was determined to be 35 ng mL^{-1}. The sensitivity of the proposed fluorescent biosensor was more than two orders of magnitude higher than that of unamplified fluorescent biosensor assays. A series of five repetitive measurements with the same concentrations were used to investigate the precision of the proposed method. The relative standard deviations for 1 ng mL^{-1} and 10 ng mL^{-1} of β-Lg were 2.44% and 4.43%, respectively, indicating good reproducibility of the assay. Currently, there are few studies detailing the use of DNA aptamers for β-Lg-detection-exploiting enzymes, and a comparison with other β-Lg assays is provided in Table 1. The approach demonstrated similar or better analytical performance compared with other β-Lg assays. The high sensitivity of the aptamer biosensor based on WS_2 nanosheets and DNase I indicated its potential for detecting β-Lg in food samples.

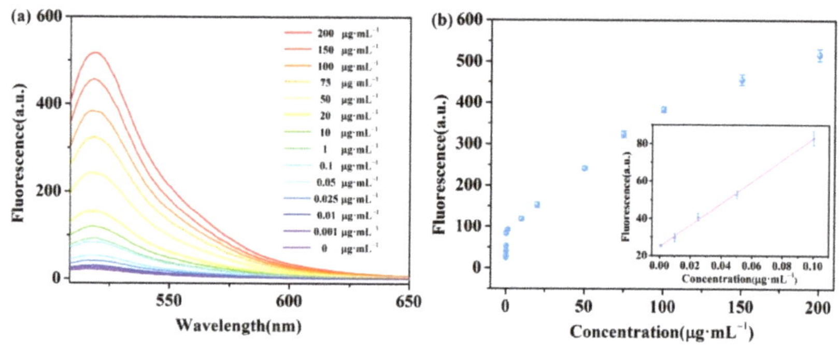

Figure 4. (**a**) Fluorescence emission spectra of β-Lg aptamer at different concentrations of β-Lg (from a to n: 0, 0.001, 0.01, 0.025, 0.05, 0.1, 1, 10, 20, 50, 75, 100, 150, and 200 μg·mL^{-1}). (**b**) The standard curve of β-Lg. Insert: linear fitting of fluorescence intensity versus target β-Lg concentration at 518 nm.

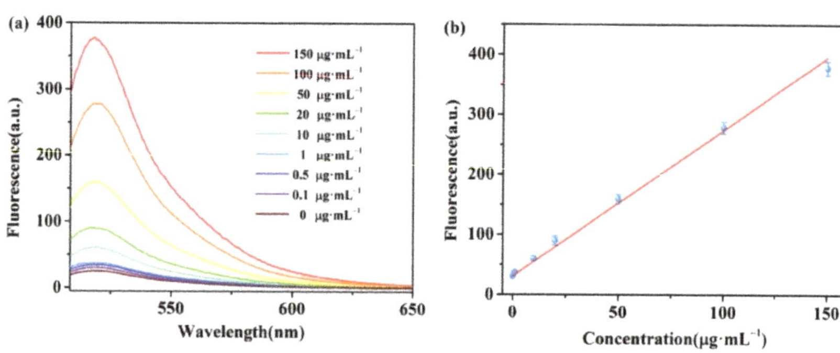

Figure 5. (**a**) Fluorescence emission spectra of β-Lg aptamer at various concentrations of β-Lg (from a to i: 0, 0.1, 0.5, 1, 10, 20, 50, 100, and 150 μg·mL^{-1}). (**b**) The standard curve of β-Lg. Insert: linear fitting of fluorescence intensity versus target β-Lg concentration at 518 nm.

Table 1. Comparison of different β-Lg assays.

Methods	Linear Range (μg·mL^{-1})	LOD (μg·mL^{-1})	References
QCM immunochip	0.5–1000	0.04	[8]
LC-MS/MS	0.48–31.25	0.2	[11]
UPLC	20–560	7	[12]
Sandwich ELISA	0.03125–8	1.96×10^{-3}	[14]
Fluorescence sensor	100–800	43	[15]
Surface Plasmon Resonance	0.49–1000	0.164	[41]
Electrochemical aptasensor	1×10^{-4}–0.01	9×10^{-5}	[42]
Fluorescence aptamer biosensor	1×10^{-3}–0.1	3.44×10^{-4}	This work

2.4. Specificity and Stability of the Aptamer Biosensor

Another important factor that affects the performance of an analytical assay is its specificity toward the targeted analyte. We evaluated the specificity of the method using two different approaches. First, we selected some other proteins, namely ovalbumin, γ-globulin, casein, and BSA (10 μg·mL^{-1}), as interfering proteins. The presence of these proteins did not significantly restore the fluorescence intensity of the β-Lg aptamer compared with β-Lg, as demonstrated in Figure 6. For the second approach, we used a negative control DNA sequence to adsorb onto the WS_2 nanosheet and evaluated the fluorescence change in the presence of different β-Lg concentrations (0.1, 1, 10, and 100 μg·mL^{-1}). Little fluorescence changes were observed in the presence of the negative control DNA sequence, independent of the concentration of β-Lg (Figure S2). These findings demonstrate that the designed fluorescent biosensor has high selectivity for β-Lg. To verify the stability of the proposed aptamer biosensor, we stored the β-Lg aptamer/WS_2 complex at 4 °C and measured its response to 10 μg·mL^{-1} of β-Lg every two days. The β-Lg aptamer could still recover its fluorescence signal in the β-Lg aptamer/WS_2 complex and retained 90.8% of its initial fluorescence signal after two weeks (Figure S3). This result demonstrates that the quenching efficiency of WS_2 nanosheets for the β-Lg aptamer was stable.

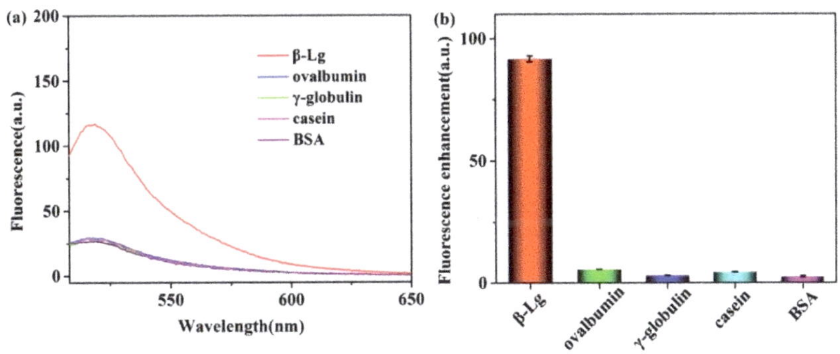

Figure 6. Selectivity of the method. (**a**) Fluorescence emission spectra of the system in the presence of different proteins. (**b**) Fluorescence changes of the system toward β-Lg and other interfering proteins.

2.5. Actual Sample Testing

Finally, to further examine the potential application of fluorescent biosensors in food samples, we used milk and infant formula samples for the assay. As indicated in Table 2, the testing results of β-Lg in milk samples obtained using the proposed fluorescent biosensors and HPLC were similar. The t-test values for the correlation were less than 2.78 (the t-test value was 2.78 at a 95% confidence level). Since no β-Lg was detected in the infant formula samples, the recovery rates of β-Lg were determined using a standard addition approach. We added known concentrations (1, 10, and 100 ng·mL^{-1}) of the β-Lg standard solution to the detection system. The recovery rates of β-Lg ranged from 94.3% to 99.5% for the infant formula samples, as shown in Table 3. These application's findings demonstrate that the established fluorescent biosensor has excellent accuracy and a high potential for quantifying β-Lg in real complex matrices.

Table 2. Determination of β-Lg in milk samples (n = 3) using different methods.

Milk Samples	Detected Concentration (mg·mL^{-1})		*t*-Test
	This Work	HPLC	
1	3.48 ± 0.16	3.39 ± 0.04	0.92
2	3.53 ± 0.17	3.62 ± 0.08	0.82
3	3.90 ± 0.09	4.07 ± 0.07	2.60

t_{table} = 2.78, p = 0.95, and v = 4.

Table 3. Determination of β-Lg in infant formula.

Samples	β-Lg Spiked ($ng \cdot mL^{-1}$)	β-Lg Found ($ng \cdot mL^{-1}$)	Recovery (%)	RSD (%)
Infant formula	-	Not detected	-	-
	1	0.978	97.8	3.8
	10	9.95	99.5	3.3
	100	94.31	94.3	5.4

3. Materials and Methods

3.1. Materials and Reagents

DNase I was purchased from New England Biolabs (Beijing, China), Ltd., and WS_2 nanosheets were obtained from Nanjing XFNANO Materials Tech Co., Ltd. (Nanjing, China). The β-Lg aptamer (5′-FAM-CGACGATCGGACCGCAGTACCCACAGCCCCAAC-ATCATGCCCATCCGTGTGTG-3′) [43] and the negative control DNA sequence (5′-FAM-CTAAGTCTGAAACATTACAGCTTGCTACACGAGAAGAGCCGCACGAGAACCCT) were synthesized by Shanghai Sangon Biotechnology Co., Ltd. (Shanghai, China). β-Lg, ovalbumin, and casein were purchased from Sigma Aldrich Chemical Co., Ltd. (St. Louis, MO, USA). The milk samples and infant formula samples were purchased from supermarkets in Xiamen, China. Other reagents were of analytical grade and were used without further purification. All solutions were prepared and diluted using ultrapure water (18.2 MΩ cm) produced by the Millipore Milli-Q system.

3.2. Apparatus

The fluorescence measurements were recorded using an LS 55 fluorescence spectrometer (PerkinElmer Ltd., Waltham, MA, USA). Ultraviolet–visible–near-infrared light (UV-Vis-NIR) absorption spectrums were recorded using a Lambda 265 UV-vis spectrophotometer (PerkinElmer Ltd., USA). The pH measurements were conducted using a digital PE 28 pH meter (Mettler Toledo, Greifensee, Switzerland).

3.3. Optimization of the Experimental Conditions

First, we optimized the concentration of WS_2 nanosheets. We added various concentrations of WS_2 nanosheets (0–800 $\mu g \cdot mL^{-1}$) to a fixed concentration of β-Lg aptamer (200 nM) and incubated it for 10 min at room temperature to determine the optimal quenching concentration. Next, we optimized the fluorescence quenching time (0, 1, 2, 3, 4, 5, 10, 15, 20, 25, and 30 min) and recovery time (0, 10, 20, 30, 40, 50, 60, 70, and 80 min) under the above optimal conditions based on the fluorescence emission intensity. Furthermore, to determine the optimized concentration of DNase I, we added different concentrations of DNase I (0, 0.01, 0.02, 0.03, 0.04, and 0.05 $U \cdot \mu L^{-1}$) to the β-Lg aptamer/WS_2 detection system and incubated it for 40 min at 37 °C. The excitation wavelength was 490 nm, and the fluorescent spectra were measured between 500 and 650 nm. The fluorescence emission intensity was recorded at 518 nm. The fluorescence quenching efficiency (Q_E) was computed using Equation (1):

$$Q_E = \frac{F_0 - F}{F_0} \tag{1}$$

where F_0 represents the fluorescence intensity at 518 nm in the absence of WS_2 nanosheets, and F_q represents the fluorescence intensity at 518 nm in the presence of WS_2 nanosheets.

3.4. Detection of β-Lg

In a typical fluorescence experiment, a mixture of 150 μL of the β-Lg aptamer (200 nM) and 150 μL of WS_2 nanosheets (750 $\mu g \cdot mL^{-1}$) was gently shaken for 10 min at room temperature. Then, β-Lg at various concentrations (0, 0.001, 0.01, 0.025, 0.05, 0.1, 1, 10, 20, 50, 75, 100, 150, and 200 $\mu g \cdot mL^{-1}$) in 141 μL of reaction buffer (50 mM Tris-HCl, 150 mM NaCl, 2 mM $MgCl_2$, and pH 7.5) and DNase I (2000 $U \cdot mL^{-1}$) in 9 μL were added to the above solution at 37 °C for 40 min to initiate signal amplification. The fluorescent spectra were measured using the above procedure.

3.5. Selectivity Experiment

For the assay of interfering proteins, the β-Lg aptamer/WS_2 detection system was mixed with 10 μg·mL^{-1} of β-Lg, ovalbumin, γ-globulin, casein, and BSA. The fluorescent spectra were measured according to the above procedure.

3.6. Stability and Reproducibility Experiment

To assess the stability of the biosensor, the β-Lg aptamer/WS_2 complex was stored at 4 °C for two weeks. Every two days, 300 μL of the complex solution was added to 141 μL of β-Lg solution (10 μg·mL^{-1}) and 9 μL of DNase I (2000 U·mL^{-1}). The resulting mixture solution was incubated at 37 °C for 40 min, and then the fluorescence spectra were measured. To demonstrate the reproducibility of the method, the fluorescence intensity of the biosensor was measured for the same concentration of β-Lg (10 μg·mL^{-1}) under optimal conditions on five consecutive occasions.

3.7. Detection of β-Lg in Actual Samples

The actual samples included milk and infant formula samples. These samples were pretreated according to previous reports [44]. For milk samples, the following steps were followed: The samples were heated at 40 °C for 30 min and then centrifuged at 8000 rpm for 20 min. After that, the samples were cooled for 15 min. The supernatant was collected, and the pH was adjusted to 4.6. The supernatant was further centrifuged at 8000 rpm for 20 min to precipitate casein and filtered through a 0.2 μm polycarbonate membrane. Then, the filtered solution was adjusted to pH 7.5 with 1 M NaOH. Finally, the pretreated milk samples were diluted to produce concentrations within the linear detection range. For infant formula samples, milk powder (8 g) was dissolved in 40 mL ultrapure water, and the remaining steps were the same as for the milk samples. β-Lg solutions with different concentrations (1, 10, and 100 ng·mL^{-1}) were diluted with the above solution, and detecting β-Lg in the actual samples was conducted according to the steps of the β-Lg assay.

4. Conclusions

In this study, we presented a highly sensitive β-Lg fluorescent aptamer biosensor based on WS_2 nanosheets and DNase I-assisted cyclic signal amplification approach. The sensitive detection performance of this biosensor can be easily achieved by combining the β-Lg aptamer/WS_2 complex, DNase I, and target β-Lg. This approach takes advantage of the strong quenching ability of WS_2 and the unique function of DNase I to release FAM-labeled short oligonucleotide fragments via a cyclic cleavage process, resulting in a significant amplification of the fluorescence signal. The sensing system demonstrates excellent selectivity for β-Lg, with a limit of detection of 0.344 ng·mL^{-1}, which is around two orders of magnitude higher than that of unamplified fluorescent biosensor assays. In summary, the proposed biosensor offers a simple and convenient alternative to standard approaches for food monitoring and demonstrates an ultrasensitive detection of β-Lg.

Supplementary Materials: The following supporting information can be downloaded at https://www.mdpi.com/article/10.3390/molecules28083502/s1. Figure S1. UV–Vis absorption spectra and TEM image of WS_2 nanosheets; Figure S2. The selectivity of the method using negative control DNA sequence; Figure S3. The stability of β-Lg aptamer/WS_2 complex for β-Lg detection.

Author Contributions: Conceptualization, C.H. and Q.C.; methodology, C.H.; software, Y.W.; validation, Y.W., S.C. and W.C.; formal analysis, C.H. and Y.W.; investigation, S.C. and J.W.; data curation, K.L.; writing—original draft preparation, Y.W. and K.Z.; writing—review and editing, C.H. and Q.C.; funding acquisition, C.H. and K.Z. All authors have read and agreed to the published version of the manuscript.

Funding: This work was funded by the Natural Science Foundation of Xiamen, China (3502Z20227061), the Natural Science Foundation of Fujian Province of China (2022J01803, 2020J05201), and the Longyan University Scientific Research Fund for the Doctoral Young Scholars (LB2020007).

Institutional Review Board Statement: Not applicable.

Informed Consent Statement: Not applicable.

Data Availability Statement: Not applicable.

Conflicts of Interest: The authors declare no conflict of interest.

Sample Availability: Samples of the compounds are available from the companies listed.

References

1. Renz, H.; Allen, K.J.; Sicherer, S.H.; Sampson, H.A.; Lack, G.; Beyer, K.; Oettgen, H.C. Food Allergy. *Nat. Rev. Dis. Prim.* **2018**, *4*, 17098. [CrossRef] [PubMed]
2. Sicherer, S.H.; Sampson, H.A. Food Allergy. *J. Allergy Clin. Immunol.* **2010**, *125*, S116–S125. [CrossRef] [PubMed]
3. McClain, S.; Bowman, C.; Fernández-Rivas, M.; Ladics, G.S.; Ree, R.V. Allergic Sensitization: Food-and Protein-Related Factors. *Clin. Transl. Allergy* **2014**, *4*, 11. [CrossRef]
4. Pecora, V.; Mennini, M.; Calandrelli, V.; Patriarca, G.; Valluzzi, R.; Fierro, V. How to Actively Treat Food Allergy. *Curr. Opin. Allergy Clin. Immunol.* **2018**, *18*, 248–257. [CrossRef]
5. Park, Y.W.; Haenlein, G.F. *Milk and Dairy Products in Human Nutrition: Production, Composition and Health*; John Wiley & Sons: Hoboken, NJ, USA, 2013.
6. Yin, J.Y.; Huo, J.S.; Xin, X.; Jing, S.; Huang, J. Study on the Simultaneously Quantitative Detection for β-Lactoglobulin and Lactoferrin of Cow Milk by Using Protein Chip Technique. *Biomed. Environ. Sci.* **2017**, *30*, 875–886. [PubMed]
7. Aoki, T.; Iskandar, S.; Yoshida, T.; Takahashi, K.; Hattori, M. Reduced Immunogenicity of β-Lactoglobulin by Conjugating with Chitosan. *Biosci. Biotechnol. Biochem.* **2006**, *70*, 2349–2356. [CrossRef]
8. Pan, M.; Hong, L.; Yang, J.; Xie, X.; Liu, K.; Wang, S. Fabrication and Evaluation of a Portable and Reproducible Quartz Crystal Microbalance Immunochip for Label-Free Detection of β-Lactoglobulin Allergen in Milk Products. *Food Sci. Hum. Wellness* **2022**, *11*, 1315–1321. [CrossRef]
9. Kürzl, C.; Wohlschläger, H.; Schiffer, S.; Kulozik, U. Concentration, Purification and Quantification of Milk Protein Residues Following Cleaning Processes Using a Combination of SPE and RP-HPLC. *MethodsX* **2022**, *9*, 101695. [CrossRef]
10. Li, L.; Wang, J.; Li, M.; Yang, Y.; Wang, Z.; Miao, J.; Zhao, Z.; Yang, J. Detection of the Adulteration of Camel Milk Powder with Cow Milk by Ultra-High Performance Liquid Chromatography (UPLC). *Int. Dairy J.* **2021**, *121*, 105117. [CrossRef]
11. Ji, J.; Zhu, P.; Pi, F.; Sun, C.; Sun, J.; Jia, M.; Ying, C.; Zhang, Y.; Sun, X. Development of a Liquid Chromatography-Tandem Mass Spectrometry Method for Simultaneous Detection of the Main Milk Allergens. *Food Control* **2017**, *74*, 79–88. [CrossRef]
12. Boitz, L.I.; Fiechter, G.; Seifried, R.K.; Mayer, H.K. A Novel Ultra-High Performance Liquid Chromatography Method for the Rapid Determination of β-Lactoglobulin as Heat Load Indicator in Commercial Milk Samples. *J. Chromatogr. A* **2015**, *1386*, 98–102. [CrossRef]
13. Khan, M.U.; Lin, H.; Hang, T.; Zhao, J.; Dasanayaka, B.P.; Zhang, J.; Ahmed, I.; Zhang, Z.; Jiang, Y.; Qazi, I.M. Development of a Sandwich Enzyme-Linked Immunosorbent Kit for Reliable Detection of Milk Allergens in Processed Food. *Anal. Biochem.* **2022**, *648*, 114667. [CrossRef]
14. He, S.; Li, X.; Gao, J.; Tong, P.; Chen, H. Development of Sandwich ELISA for Testing Bovine β-Lactoglobulin Allergenic Residues by Specific Polyclonal Antibody against Human IgE Binding Epitopes. *Food Chem.* **2017**, *227*, 33–40. [CrossRef] [PubMed]
15. Hong, L.; Pan, M.; Yang, X.; Xie, X.; Liu, K.; Yang, J.; Wang, S.; Wang, S. A UCMPs@ MIL-100 Based Thermo-Sensitive Molecularly Imprinted Fluorescence Sensor for Effective Detection of β-Lactoglobulin Allergen in Milk Products. *J. Nanobiotechnol.* **2022**, *20*, 51. [CrossRef]
16. Duan, N.; Yao, T.; Li, C.; Wang, Z.; Wu, S. Surface-Enhanced Raman Spectroscopy Relying on Bimetallic Au–Ag Nanourchins for the Detection of the Food Allergen β-Lactoglobulin. *Talanta* **2022**, *245*, 123445. [CrossRef] [PubMed]
17. Hong, S.P.; Mohd Naim, N.F.; Keasberry, N.A.; Ahmed, M.U. Electrochemical Detection of Β-Lactoglobulin Allergen Using Titanium Dioxide/Carbon Nanochips/Gold Nanocomposite-based Biosensor. *Electroanalysis* **2022**, *34*, 684–691. [CrossRef]
18. Li, Z.; Wen, F.; Li, Z.; Zheng, N.; Jiang, J.; Xu, D. Simultaneous Detection of α-Lactoalbumin, β-Lactoglobulin and Lactoferrin in Milk by Visualized Microarray. *BMC Biotechnol.* **2017**, *17*, 72. [CrossRef]
19. Horká, M.; Šalplachta, J.; Karásek, P.; Roth, M. Sensitive Identification of Milk Protein Allergens Using On-Line Combination of Transient Isotachophoresis/Micellar Electrokinetic Chromatography and Capillary Isoelectric Focusing in Fused Silica Capillary with Roughened Part. *Food Chem.* **2022**, *377*, 131986. [CrossRef] [PubMed]
20. Song, S.; Wang, L.; Li, J.; Fan, C.; Zhao, J. Aptamer-Based Biosensors. *TrAC Trends Anal. Chem.* **2008**, *27*, 108–117. [CrossRef]
21. Kim, Y.S.; Raston, N.H.A.; Gu, M.B. Aptamer-Based Nanobiosensors. *Biosens. Bioelectron.* **2016**, *76*, 2–19.
22. Liu, C.; Lu, C.; Tang, Z.; Chen, X.; Wang, G.; Sun, F. Aptamer-Functionalized Magnetic Nanoparticles for Simultaneous Fluorometric Determination of Oxytetracycline and Kanamycin. *Microchim. Acta* **2015**, *182*, 2567–2575. [CrossRef]

23. Jo, H.; Gu, H.; Jeon, W.; Youn, H.; Her, J.; Kim, S.-K.; Lee, J.; Shin, J.H.; Ban, C. Electrochemical Aptasensor of Cardiac Troponin I for the Early Diagnosis of Acute Myocardial Infarction. *Anal. Chem.* **2015**, *87*, 9869–9875. [CrossRef] [PubMed]
24. Shen, Y.; Wu, T.; Zhang, Y.; Ling, N.; Zheng, L.; Zhang, S.-L.; Sun, Y.; Wang, X.; Ye, Y. Engineering of a Dual-Recognition Ratiometric Fluorescent Nanosensor with a Remarkably Large Stokes Shift for Accurate Tracking of Pathogenic Bacteria at the Single-Cell Level. *Anal. Chem.* **2020**, *92*, 13396–13404. [CrossRef] [PubMed]
25. Zhang, H.; Wang, Z.; Wang, F.; Zhang, Y.; Wang, H.; Liu, Y. In Situ Formation of Gold Nanoparticles Decorated Ti3C2 MXenes Nanoprobe for Highly Sensitive Electrogenerated Chemiluminescence Detection of Exosomes and Their Surface Proteins. *Anal. Chem.* **2020**, *92*, 5546–5553. [CrossRef] [PubMed]
26. Dong, X.; Yan, X.; Li, M.; Liu, H.; Li, J.; Wang, L.; Wang, K.; Lu, X.; Wang, S.; He, B. Ultrasensitive Detection of Chloramphenicol Using Electrochemical Aptamer Sensor: A Mini Review. *Electrochem. Commun.* **2020**, *120*, 106835. [CrossRef]
27. Sun, L.; Wu, L.; Zhao, Q. Aptamer Based Surface Plasmon Resonance Sensor for Aflatoxin B1. *Microchim. Acta* **2017**, *184*, 2605–2610. [CrossRef]
28. Qi, Y.; Chen, Y.; Xiu, F.-R.; Hou, J. An Aptamer-Based Colorimetric Sensing of Acetamiprid in Environmental Samples: Convenience, Sensitivity and Practicability. *Sens. Actuators B Chem.* **2020**, *304*, 127359. [CrossRef]
29. Zuo, X.; Zhang, H.; Zhu, Q.; Wang, W.; Feng, J.; Chen, X. A Dual-Color Fluorescent Biosensing Platform Based on WS_2 Nanosheet for Detection of Hg^{2+} and Ag^{+}. *Biosens. Bioelectron.* **2016**, *85*, 464–470. [CrossRef]
30. Wang, P.; Wang, A.; Hassan, M.M.; Ouyang, Q.; Li, H.; Chen, Q. A Highly Sensitive Upconversion Nanoparticles-WS_2 Nanosheet Sensing Platform for Escherichia Coli Detection. *Sens. Actuators B Chem.* **2020**, *320*, 128434. [CrossRef]
31. Bi, X.; Li, L.; Luo, L.; Liu, X.; Li, J.; You, T. A Ratiometric Fluorescence Aptasensor Based on Photoinduced Electron Transfer from CdTe QDs to WS_2 NTs for the Sensitive Detection of Zearalenone in Cereal Crops. *Food Chem.* **2022**, *385*, 132657. [CrossRef]
32. Tang, Y.; Hu, Y.; Zhou, P.; Wang, C.; Tao, H.; Wu, Y. Colorimetric Detection of Kanamycin Residue in Foods Based on the Aptamer-Enhanced Peroxidase-Mimicking Activity of Layered WS_2 Nanosheets. *J. Agric. Food Chem.* **2021**, *69*, 2884–2893. [CrossRef]
33. Hong, C.; Zhang, X.; Ye, S.; Yang, H.; Huang, Z.; Yang, D.; Cai, R.; Tan, W. Aptamer-Pendant DNA Tetrahedron Nanostructure Probe for Ultrasensitive Detection of Tetracycline by Coupling Target-Triggered Rolling Circle Amplification. *ACS Appl. Mater. Interfaces* **2021**, *13*, 19695–19700. [CrossRef]
34. Zou, L.; Zhang, M.; Li, M.; Xiao, Z.; Ling, L. Hybridization Chain Reaction and DNAzyme-based Dual Signal Amplification Strategy for Sensitive Fluorescent Sensing of Aflatoxin B1 by using the Pivot of Triplex DNA. *Food Res. Int.* **2022**, *158*, 111538. [CrossRef]
35. Zhou, H.; Duan, S.; Huang, J.; He, F. An Ultrasensitive Electrochemical Biosensor for Pseudomonas Aeruginosaassay based on a Rolling Circle Amplification-Assisted Multipedal DNA Walker. *Chem. Commun.* **2020**, *56*, 6273–6276. [CrossRef]
36. Lu, C.; Li, J.; Lin, M.; Wang, Y.; Yang, H.; Chen, X.; Chen, G. Amplified Aptamer-based Assay through Catalytic Recycling of the Analyte. *Angew. Chem. Int. Ed.* **2010**, *122*, 8632–8635. [CrossRef]
37. Ma, L.; Guo, T.; Pan, S.; Zhang, Y. A Fluorometric Aptasensor for Patulin based on the Use of Magnetized Graphene Oxide and DNase I-Assisted Target Recycling Amplification. *Microchim. Acta* **2018**, *185*, 487. [CrossRef] [PubMed]
38. Fang, B.; Wang, C.; Li, C.; Wang, H.; Zhao, Y. Amplified using DNase I and Aptamer/Graphene Oxide for Sensing Prostate Specific Antigen in Human Serum. *Sens. Actuators B Chem.* **2017**, *244*, 928–933. [CrossRef]
39. Yuan, Y.; Li, R.; Liu, Z. Establishing water-soluble layered WS_2 nanosheet as a platform for biosensing. *Anal. Chem.* **2014**, *86*, 3610–3615. [CrossRef] [PubMed]
40. Mo, L.; Li, J.; Liu, Q.; Qiu, L.; Tan, W. Nucleic acid-functionalized transition metal nanosheets for biosensing applications. *Biosens. Bioelectron.* **2017**, *89*, 201–211. [CrossRef] [PubMed]
41. Ashley, J.; Aurelio, R.D.; Piekarska, M.; Temblay, J.; Pleasants, M.; Trinh, L.; Rodgers, T.L.; Tothill, I.E. Development of a β-Lactoglobulin Sensor Based on SPR for Milk Allergens Detection. *Biosensors* **2018**, *8*, 32. [CrossRef]
42. Amor-Gutiérrez, O.; Selvolini, G.; Fernández-Abedul, M.T.; de la Escosura-Muñiz, A.; Marrazza, G. Folding-Based Electrochemical Aptasensor for the Determination of β-Lactoglobulin on Poly-L-Lysine Modified Graphite Electrodes. *Sensors* **2020**, *20*, 2349. [CrossRef] [PubMed]
43. Eissa, S.; Zourob, M. In Vitro Selection of DNA Aptamers Targeting β-Lactoglobulin and Their Integration in Graphene-Based Biosensor for the Detection of Milk Allergen. *Biosens. Bioelectron.* **2017**, *91*, 169–174. [CrossRef] [PubMed]
44. Hong, C.; Wang, J.; Wang, Y.; Huang, Z.; Yang, H.; Yang, D.; Cai, R.; Tan, W. Fluorescence Detection of Milk Allergen β-Lactoglobulin Based on Aptamers and WS_2 Nanosheets. *J. Mater. Chem. B* **2022**, *10*, 6752–6757. [CrossRef] [PubMed]

molecules

Article

Structure-Guided Development of Bivalent Aptamers Blocking SARS-CoV-2 Infection

Md Shafiqur Rahman [1], Min Jung Han [1], Sang Won Kim [1], Seong Mu Kang [1], Bo Ri Kim [1], Heesun Kim [2], Chang Jun Lee [1], Jung Eun Noh [1], Hanseong Kim [1], Jie-Oh Lee [1,*] and Sung Key Jang [1,2,*]

1 Department of Life Sciences, POSTECH Biotech Center, Pohang University of Science and Technology, 77 Cheongam-ro, Nam-gu, Pohang-si 37673, Republic of Korea; shafiqmb@postech.ac.kr (M.S.R.); hmj@postech.ac.kr (M.J.H.); ksw5758@postech.ac.kr (S.W.K.)

2 Division of Integrative Bioscience & Biotechnology, POSTECH Biotech Center, Pohang University of Science and Technology, Nam-gu, Pohang-si 37673, Republic of Korea

* Correspondence: jieoh@postech.ac.kr (J.-O.L.); sungkey@postech.ac.kr (S.K.J.)

Abstract: Severe acute respiratory syndrome coronavirus 2 (SARS-CoV-2) has caused devastation to human society through its high virulence, infectivity, and genomic mutations, which reduced the efficacy of vaccines. Here, we report the development of aptamers that effectively interfere with SARS-CoV-2 infection by targeting its spike protein, which plays a pivotal role in host cell entry of the virus through interaction with the viral receptor angiotensin-converting enzyme 2 (ACE2). To develop highly effective aptamers and to understand their mechanism in inhibiting viral infection, we determined the three-dimensional (3D) structures of aptamer/receptor-binding domain (RBD) complexes using cryogenic electron microscopy (cryo-EM). Moreover, we developed bivalent aptamers targeting two distinct regions of the RBD in the spike protein that directly interact with ACE2. One aptamer interferes with the binding of ACE2 by blocking the ACE2-binding site in RBD, and the other aptamer allosterically inhibits ACE2 by binding to a distinct face of RBD. Using the 3D structures of aptamer–RBD complexes, we minimized and optimized these aptamers. By combining the optimized aptamers, we developed a bivalent aptamer that showed a stronger inhibitory effect on virus infection than the component aptamers. This study confirms that the structure-based aptamer-design approach has a high potential in developing antiviral drugs against SARS-CoV-2 and other viruses.

Keywords: SARS-CoV-2; aptamer–RBD complex; cryo-EM structure; Nap-dU; neutralization; viro-SELEX

check for updates

Citation: Rahman, M.S.; Han, M.J.; Kim, S.W.; Kang, S.M.; Kim, B.R.; Kim, H.; Lee, C.J.; Noh, J.E.; Kim, H.; Lee, J.-O.; et al. Structure-Guided Development of Bivalent Aptamers Blocking SARS-CoV-2 Infection. *Molecules* **2023**, *28*, 4645. https://doi.org/10.3390/molecules28124645

Academic Editors: Long Wu, Wei Wu and Sai Wang

Received: 22 May 2023
Revised: 4 June 2023
Accepted: 5 June 2023
Published: 8 June 2023

1. Introduction

In December 2019, the novel coronavirus, severe acute respiratory syndrome coronavirus 2 (SARS-CoV-2), was identified for the first time in Wuhan, China [1,2], and afterward caused a global pandemic. SARS-CoV-2 caused the coronavirus disease 2019 (COVID-19). Over 765 million people were infected by SARS-CoV-2, and more than 6.9 million deaths attributed to COVID-19 were confirmed up to 10 May 2023 [3]. SARS-CoV-2's spike protein is essential for the virus to enter the host cell through interaction with the viral receptor protein, angiotensin-converting enzyme 2 (ACE2) [4–6]. The spike protein consists of S1 and S2 subunits, and the S1 subunit contains an N-terminal domain (NTD) and a receptor-binding domain (RBD) [4,7]. The receptor-binding motif (RBM) of RBD is responsible for interaction with the helix in the N-terminal peptidase domain of ACE2 [5]. Owing to its essential role in virus infection, the trimeric spike protein of SARS-CoV-2 has become the major target for vaccine development, neutralizing antibodies (nAbs), drugs, and SARS-CoV-2 diagnosis [5,8].

Several vaccines have been developed to prevent infection of SARS-CoV-2, and they play a key role in preventing COVID-19 and/or reducing the severity of COVID-19 [9–12].

Moreover, a few drugs that have been found to be effective against COVID-19 originated from antiviral drug candidates for other viruses. For instance, Paxlovid, a protease inhibitor, and Molnupiravir, an RNA-dependent RNA polymerase inhibitor, are given to COVID-19 patients, even though their efficacy is limited [13,14]. Alternatively, nAbs targeting the spike have been used to treat patients with COVID-19 [15]. However, potential adverse effects of antibodies, such as antibody-dependent enhancement (ADE), represent a concern for antibody-based therapy [16,17]. The major concern for COVID-19 is the evolution of SARS-CoV-2. Numerous variants of SARS-CoV-2 have been discovered that escape from the immune response stimulated by vaccines and antiviral drugs [12,18–20]. Therefore, it is essential to conduct additional research for the development of better vaccines and antiviral medications to combat SARS-CoV-2 variants.

Aptamers are short, single-stranded oligonucleotides (DNA or RNA) that exhibit high affinity and specificity for a range of target molecules. Systematic evolution of ligands by exponential enrichment (SELEX) is a method used to generate aptamers [21]. Compared to antibodies, aptamers have a number of benefits. For example, aptamers can be produced chemically, are produced at low cost, exhibit low batch variability, can be readily modified, are smaller, and provoke a very low immune response [22]. Additionally, the use of modified nucleotide bases in lieu of conventional nucleotide bases has been shown to increase affinity and stability [23–26]. Therefore, it is anticipated that in the near future, aptamers will be used often in diagnostic and therapeutic applications. In the context of SARS-CoV-2, several studies have reported the development of aptamers against RBD of SARS-CoV-2 for diagnostic and therapeutic purposes [27–31]. Notably, one study developed an inhibitory aptamer independent of RBD [32]. Insights for the minimization and optimization of the aptamer might be gained from the structural information of an aptamer–protein complex, which would provide an understanding of the mechanism behind the interaction between the aptamer and its target protein [33]. However, to date, there is no structural analysis of an aptamer–SARS-CoV-2 spike complex showing the inhibition of SARS-CoV-2 infection, except for one aptamer binding to NTD of SARS-CoV-2 used for virus detection [34].

Here, we report the generation of aptamers that bind strongly to the spike protein of SARS-CoV-2 using viro-SELEX combined with conventional SELEX technology [35,36]. We determined the structure of aptamers bound to the RBD of SARS-CoV-2 using cryogenic electron microscopes (cryo-EM). Through the analysis of the aptamer/RBD complex structure, we revealed the molecular bases of the binding of aptamers to the target molecule. The inhibitory effects of aptamers on SARS-CoV-2 infection were investigated using pseudoviruses with envelope proteins containing the spike proteins of SARS-CoV-2 [31,37], mimicking the entry process of SARS-CoV-2. Two aptamers, namely AM032-0 and AM047-0, inhibited the infection of pseudoviruses. Furthermore, we optimized the aptamers to have minimum sizes and stable configurations based on the three-dimensional structure of the aptamer/RBD complex. Finally, we generated bivalent aptamers that comprised two optimized aptamers, AM032-4 and AM047-6, connected together by flexible linkers. A bivalent aptamer composed of the optimized aptamers showed a very strong inhibitory effect on the pseudovirus infection.

2. Results

2.1. Selection and Characterization of Aptamers Binding to Spike Proteins of SARS-CoV-2

To develop aptamers that interact strongly and specifically with the spike proteins of SARS-CoV-2, we performed viro-SELEX and conventional protein-SELEX [35,36] alternately using a modified nucleotide, i.e., 5-[N-(1-naphthylmethyl) carboxamide]-2′-deoxy uridine (Nap-dU), instead of the natural nucleotide deoxythymidine [23]. We used Nap-dU instead of deoxythymidine to increase the probability of acquiring aptamers with high affinities to the target molecule [23–26]. In the SELEX process, we used a purified trimeric spike protein of wild-type (Wuhan strain) SARS-CoV-2 (Supplementary Figure S1a) for protein-SELEX and a modified baculovirus (surrogate virus) containing wild-type spike proteins

(Supplementary Figure S1d) on the viral envelope for viro-SELEX [35,36] to enrich aptamers binding to the natural form of spikes present on SARS-CoV-2; this was in accordance with the scheme presented in Supplementary Figure S2a. Moreover, purified trimeric spike proteins of other variants (D614G, alpha, beta, and delta) of SARS-CoV-2 were used to investigate the binding of aptamers to the variants (Supplementary Figure S1a). The trimeric nature of spike proteins was further confirmed by size exclusion chromatography (SEC) and cryo-EM (Supplementary Figure S1e,h).

After carrying out SELEX, the protein-binding affinities of DNA pools from the 9th–12th rounds of SELEX were examined using filter binding assay [35]. For example, the DNA pool from 10th round showed strong binding affinities to various parts of the wild-type spike (WT-S) protein without showing binding affinity to spike proteins of other human betacoronaviruses HCoV HKU1 and HCoV OC43, which cause common cold (Supplementary Figure S2b). Among the DNA pools evaluated, we selected the DNA pool from the 11th round of SELEX, which showed the strongest binding affinity to the S1 subunit and WT-S proteins (K_D of ~0.45 nM and ~0.25 nM, respectively; Supplementary Figure S2c), for cloning plus sequencing by Sanger method and for direct sequencing by the next-generation sequencing (NGS) method. It is noteworthy to mention that many aptamers with various affinities to the spike protein were identified (Supplementary Figure S2f). Among the aptamers, we selected the two aptamers AM032 and AM047, which strongly bound to the RBD of the wild-type SARS-CoV-2 spike, for further analyses (Figure 1a,b). In addition, as shown in Figure 1c, aptamers AM032 and AM047 did not compete for binding to the RBD, indicating their different binding sites in the spike protein. Next, we synthesized derivatives of AM032 and AM047 with serial truncations from both 5′ and 3′ ends and measured their binding affinities to spike proteins. We selected an 80-mer derivative of AM032 and a 52-mer derivative of AM047 (designated as AM032-0 and AM0470, respectively), which had minimum sizes without losing binding affinities, for further studies. Both aptamers, i.e., AM032-0 and AM047-0, showed strong binding affinities to the spike proteins of SARS-CoV-2 variants in the K_D ranges of 3 nM–10.5 nM and 0.04 nM–0.4 nM, respectively (Figure 1d,e; Supplementary Figure S2d,e; and Table S1).

2.2. *Aptamers AM032-0 and AM047-0 Block SARS-CoV-2 Infection through Inhibition of the Interaction between RBD and ACE2*

Since aptamers AM032-0 and AM047-0 bind to the RBD of spike protein, they have the potential to inhibit SARS-CoV-2 infection by interfering with the interaction between spike protein and ACE2 proteins. We investigated their effects on the infectivity of pseudoviruses, which mimic the entry step of SARS-CoV-2 into a host cell as mediated by the spike proteins of SARS-CoV-2 on the surface of the pseudoviruses. Pseudoviruses containing the spike proteins of either D614G or delta variant and HEK293T cells expressing human ACE2 were used as the pseudoviruses and host cells, respectively. Firefly luciferase activity from the reporter gene integrated into the RNA of a pseudovirus was monitored to determine its infectivity. Pseudoviruses were cultured for 48 h in the presence of serially diluted aptamers, and the effects of aptamers on the infectivity of pseudoviruses were monitored by measuring firefly luciferase activities in the cell extracts. Both aptamers AM032-0 and AM047-0 inhibited infection of the pseudoviruses containing the spike proteins of D614G or delta variant in dose-dependent manners (Figure 1f,g). The half-maximal inhibitory concentrations (IC_{50}s) of AM032-0 and AM047-0 against the pseudovirus with D614G spike were 188 nM and 134 nM, respectively (Figure 1f). Additionally, the IC_{50}s of AM032-0 and AM047-0 against the pseudovirus with the delta spike were 197 nM and 141 nM, respectively (Figure 1g). The results indicate that both aptamers AM032-0 and AM047-0 interfered with the infection of SARS-CoV-2 variants and that the aptamer AM047-0, which had higher affinity to the spike protein than aptamer AM032-0, also had a higher inhibitory effect on SARS-CoV-2 infection. To understand the underlying mechanism of the interference of SARS-CoV-2 infection by the aptamers AM032-0 and AM047-0, we examined whether they inhibited the interaction between the spike and ACE2 proteins. Therefore, we performed enzyme-linked immunosorbent assays (ELISAs) measuring the interaction between ACE2 and the RBD of wild-type spike in the presence or absence of aptamers (Figure 1h). Both aptamers AM032-0 and AM047-0 strongly inhibited the interaction between ACE2 and the wild-type RBD. The half-maximal effective concentrations (EC_{50}s) of aptamers AM032-0 and AM047-0 were 41.0 nM and 17.4 nM, respectively (Figure 1h). Again, the binding affinities of aptamers were well correlated with the interfering actions of aptamers in the interaction between ACE2 and RBD. Taken together, these findings conclude that aptamers AM032-0 and AM047-0 inhibit SARS-CoV-2 infection by interfering with the interaction between the spike and ACE2 proteins.

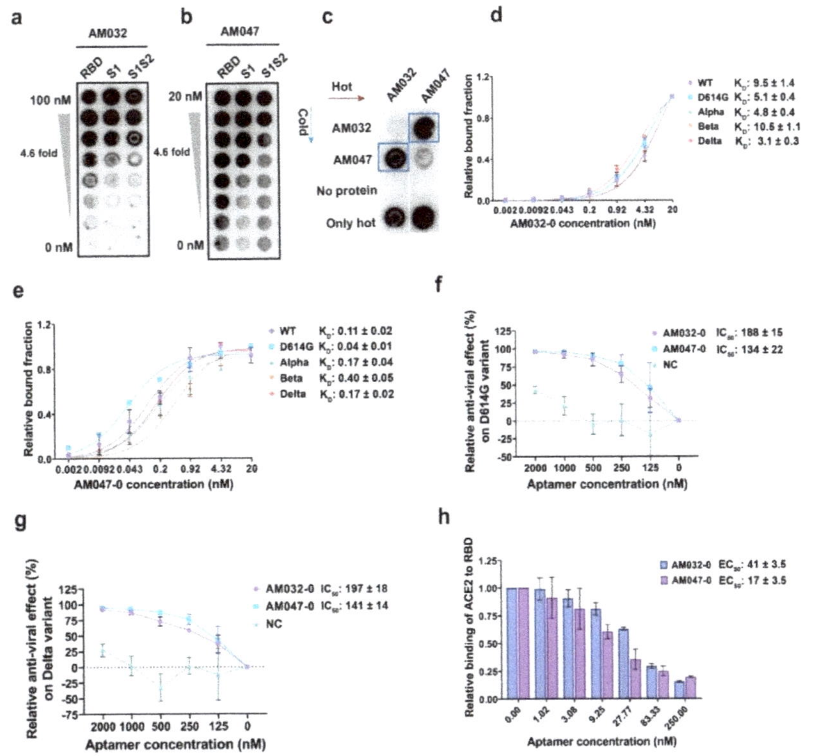

Figure 1. Development and characterization of aptamers. (**a**,**b**) Filter binding assays showing the binding of aptamers AM032 and AM047 to the various parts of a SARS-CoV-2 spike protein. (**c**) Protein-binding competition assay against S1S2 protein. [P^{32}]-labeled (hot) aptamers and unlabeled (cold) aptamers were used as probes and competitors, respectively. (**d**,**e**) Binding affinities of AM032-0 (**d**) and AM047-0 (**e**) to the spike proteins of SARS-CoV-2 variants. The binding affinities were determined by filter binding assays (see Section 4). (**f**,**g**) Inhibitory effects of aptamers on the infection of SARS-CoV-2 pseudoviruses containing spike proteins of D614G (**f**) and delta (**g**) variants. (**h**) Inhibitory effects of aptamers on the interaction between ACE2 and RBD. Enzyme-linked immunosorbent assays (ELISAs) were performed to monitor the inhibitory effects. NC, negative control. The values presented for K_D, IC_{50}, and EC_{50} were measured in nM.

2.3. Determination of the Three-Dimensional Structure of RBD–Aptamer Complex by Cryo-EM

To understand the detailed molecular mechanism of the aptamer binding to the spike protein and the inhibition of SARS-CoV-2 infection, we determined the three-dimensional structure of the RBD–aptamer complex using cryo-EM. We used a previously published Fab domain of an antibody, namely REGN10987 [38], against the RBD of the spike to increase the overall size of the RBD–aptamer complex and make it suitable for cryo-EM analysis (Supplementary Figure S3a). Both aptamers AM032-0 and AM047-0 were applied to the RBD–antibody complex to acquire a complex containing the aptamers: RBD–Fab–AM032-0–AM047-0. From 21,340 movie frames and from post-processing the cryo-EM movies in cryoSPARC [39], we obtained the cryo-EM map of the protein–aptamer complex at an overall resolution of 3.32 Å (Supplementary Figure S3b–f and Table S2). The initial atomic model for the protein part was constructed by fitting the previously published structures of RBD (PDB code 7KGJ) [40] and Fab (PDB code 6XDG) [38] to the map. The aptamer structures were built de novo using Coot [41] (see Materials and Methods for detail). The cryo-EM map corresponding to the constant region of Fab and the 5′ and 3′ end areas of the aptamer AM032-0 were not clearly visible, presumably due to high structural flexibility, and were therefore omitted from the refined structure (Figure 2a,b and Supplementary Figure S3g). The RBD structure of the complex is superimposable with a root mean square (r.m.s.) deviation of 1.41 Å and 1.48 Å (193 Cα pairs) with PDB IDs 6M0J [5] and 7KGJ [40], respectively, demonstrating that no significant conformational changes in the RBD molecule were induced by the aptamer binding. The complex structure illustrates that the aptamers AM032-0 and AM047-0 bind to two different sites in RBD (Figure 2a,b). AM032-0 stands astride the ACE2-binding surface of RBD, spanning residues from G476 to Y505 and distant residues R403 and F456. AM047-0 binds to an area composed of residues from Y365 to N388 that is distal from the ACE2-binding site (Figure 2c). The bound aptamers adopt regular double-helical structures at the areas close to their 5′ and 3′ ends but form complicated loop structures near the RBD–aptamer interface areas (Figure 2a).

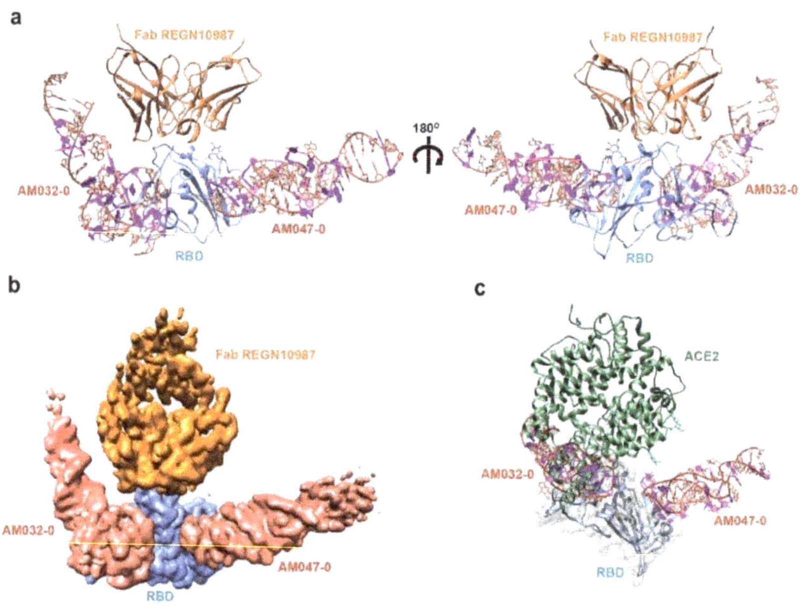

Figure 2. Cryo-EM structure of the RBD–Fab–AM032-0–AM047-0 complex. (**a**) Structure of the RBD–Fab–AM032-0–AM047-0 complex. Front view (left) and back view (right). (**b**) Cryo-EM density map of the protein–aptamer complex. (**c**) Structure of the RBD–aptamer complex superimposed with that of the RBD–ACE2 complex (PDB ID: 6M0J). Nap-dU residues of the aptamers are shown in purple.

Aptamer AM032-0 is composed of 80 nucleotides with 11 modified Nap-dU nucleotides (Figure 3a). The three-dimensional structure of AM032-0 has one stem and three turns, i.e., T1, T2, and T3 (Figure 3b; see Supplementary Note for structural details). This unusual three-turn structure is stabilized by a hydrophobic core composed of seven Nap-dU residues at positions 33, 34, 39, 44, 48, 51, and 53 (Figure 3b,c). Mutagenesis of some of these nucleotides individually with deoxythymidine abolishes the binding of the aptamer to RBD (Figure 3e). Among the three turns, the T1 plays the main role in RBD binding. Bases 32, 34, 35, 36, and 43 make intimate polar interactions with side chains of N487, Q493, R403, and Y505 (Figure 3c and Supplementary Table S3). The aromatic amino acids F456, F486, and Y489 of RBD make hydrophobic interactions with C32, Nap-dU33, Nap-dU34, C43, Nap-dU44, G45, and Nap-dU53 (Figure 3c and Supplementary Table S3). Mutations of these aromatic residues disrupt aptamer binding (Figure 3d).

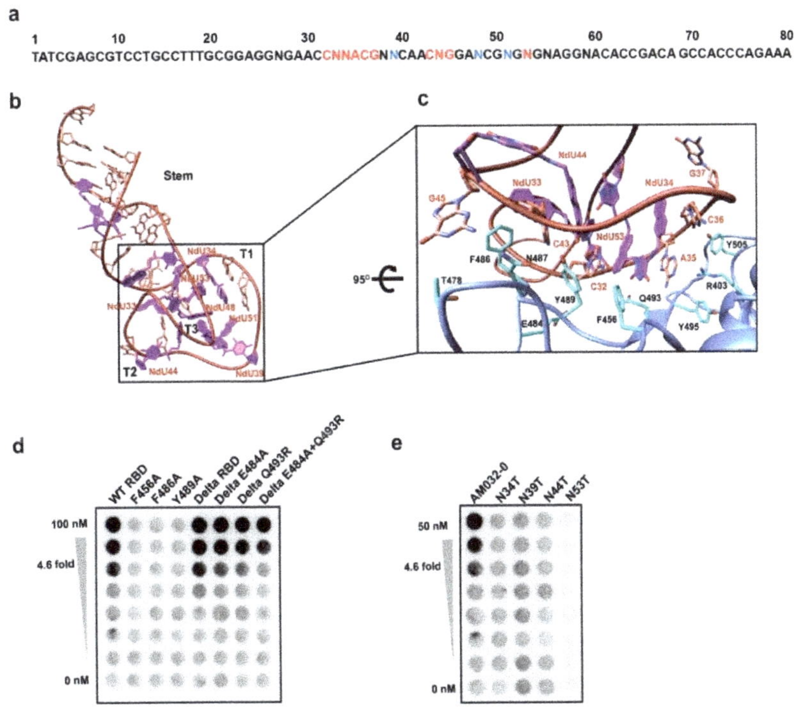

Figure 3. Structure of AM032-0 bound to the RBD of SARS-CoV-2. (**a**) Sequence of the aptamer AM032-0. The bases required for RBD binding and aptamer folding are written in red and blue, respectively. (**b**) The overall structure of aptamer AM032-0. For clarity, some nucleotides are not shown. Ndu represents Nap-dU. (**c**) Close-up view of the binding interface between AM032-0 and RBD. AM032-0 and RBD are colored in red and light blue, respectively. Nap-dU nucleotides (NdU) and amino acids critical for binding are shown in purple and cyan, respectively. (**d**) Effects of mutations in RBD on binding to aptamers. Mutations in RBDs of wild type (lanes 2–4) and delta variant (lanes 6–8) were introduced, and the effects of the mutations on binding to aptamer AM032-0 were monitored by filter binding assay. (**e**) Effects of nucleotide substitution of aptamer AM032-0 on binding to wild-type RBD. N, Nap-dU = 5-[N-(1-naphthylmethyl) carboxamide]-2′-deoxyuridine; A, 2′-deoxyadenosine; T, 2′-deoxythymidine; G, 2′-deoxyguanosine; C, 2′-deoxycytidine.

The aptamer AM047-0 comprises 52 nucleotides with 11 Nap-modified nucleotides (Figure 4a). The refined structure of AM047-0 in complex with RBD has two stems, S1 and S2; two turns, T1 and T2; and two loops, L1 and L2 (Figure 4b; see Supplementary Note for structural details). The L1 and L2 loops of AM047-0 are responsible for binding to RBD protein. The aptamer-binding site consists of beta-strand β2, alpha helix α4, and a loop preceding the β2 strand of RBD (Figure 4c). Both hydrophobic and polar interactions mediated by several residues are involved in the interaction with RBD (Figure 4c and Supplementary Table S3). The Nap-dU residues at positions 30 and 35 of the L2 loop and positions 17 and 20 of the L1 loop form a hydrophobic surface in the aptamer that is critical in the interaction with the hydrophobic surface formed by Y365, L368, Y369, F374, F377, P384, and L387 residues of the spike protein (Figure 4c and Supplementary Table S3). This core hydrophobic interaction is supported by hydrophilic interactions in the periphery. The G19 residue of the L1 loop makes a hydrogen bond with S366 residue of RBD. The G31 and G33 residues of the L2 loop make a polar interaction with T376 and K378 residues of the spike (Figure 4c and Supplementary Table S3). Consistent with these observations, the replacement of Nap-dU35 with dT abolished the binding of aptamer AM047-0 with RBD (Figure 4e), suggesting its central role in binding and stabilizing the interface structure of AM047-0. This hypothesis was also supported by the binding assay showing that the reverting of the amino acid mutations in omicron BA.4/5 (P373 and F375), which binds weakly to AM047-0, to the wild-type (S373 and S375) restores the binding to AM047-0 (Figure 4d and Supplementary Table S4).

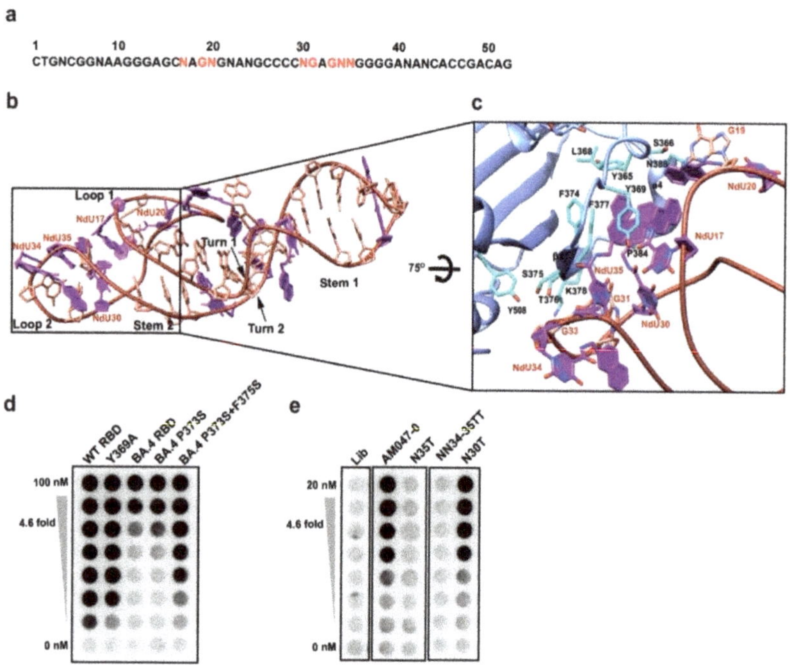

Figure 4. Structure of AM047-0 bound to the RBD of SARS-CoV-2. (**a**) Sequence of the aptamer AM047-0. The bases required for RBD binding are written in red. (**b**) The overall structure of AM047-0. (**c**) Close-up view of the binding interface between AM047-0 (red) and RBD (light blue). Nap-dUs are shown in purple. (**d**) Effects of mutations in RBD on binding to aptamers. Mutations in RBDs of wild type (lane 2) and BA.4/5 variant (lanes 4–5; BA.4/5 denoted as BA.4) were introduced, and the effects of the mutations on binding to aptamer AM047-0 were monitored by filter binding assay. (**e**) Effects of nucleotide substitution of aptamer AM047-0 aptamer on binding to wild-type RBD. N, Nap-dU = 5-[N-(1-naphthylmethyl) carboxamide]-2′-deoxyuridine; A, 2′-deoxyadenosine; T, 2′-deoxythymidine; G, 2′-deoxyguanosine; C, 2′-deoxycytidine; Lib, aptamer library.

2.4. Aptamers AM032-0 and AM047-0 Block the Association of ACE2 to RBD through Distinct Mechanisms

As shown in Figure 1f–h, both AM032-0 and AM047-0 aptamers compete with ACE2 for binding to the spike protein and inhibit the infection of SARS-CoV-2 pseudovirus.

The inhibition of the ACE2–RBD interaction by AM032-0 is easily comprehensible since ACE2 and AM032-0 occupy an overlapped region on RBD, as shown in Figure 2c. Additionally, several amino acid residues in RBD (F456, F486, Y489, and Q493) participate in the interactions with both ACE2 [5] and AM032-0 (Figures 2c and 3c). Therefore, ACE2 and AM032-0 cannot bind to the spike protein at the same time since the binding of one molecule (either ACE2 or AM032-0) would not allow the binding of the other molecule to the spike (Figure 1h). However, aptamer AM047-0 binds to RBD at a distinct region from the ACE2-binding site, and AM047-0 and ACE2 may not directly bump into each other even if both of the molecules are associated with RBD, as shown in Figure 2c. A possible explanation for the inhibitory effect of AM047-0 is described in Section 3.

2.5. Optimization of Aptamers AM032-0 and AM047-0

Optimization of aptamers (an increase of binding affinity, minimization of size, minimization of modified nucleotide Nap-dU, reduction of susceptibility to nucleases, and stabilization of structure) is required for use in diagnostic or therapeutic purposes. Through the structural analysis of the aptamers AM032-0 and AM047-0, we recognized that the 5′ and 3′ ends of the aptamers form partially double-stranded configurations that stabilize the overall structures of the aptamers (Figure 2a). On the other hand, the central parts of the aptamers AM032-0 and AM047-0 form complex structures participating in the protein bindings (Figures 3b,c and 4b,c). We designed several aptamers to confirm the structures and to optimize aptamers (Supplementary Figure S4a,b). In the optimization process, we maintained the central regions of the aptamers in order to not disturb the structures required for protein binding. On the other hand, we removed the peripheral regions as much as possible to minimize the aptamer size and changed the sequences to develop strong stem structures with many G-C base pairings (Supplementary Figure S4a,b) to stabilize the aptamer structures [31,42].

Various oligonucleotides related to AM032-0 and AM047-0 were synthesized, and their binding affinities to wild-type RBD were monitored by filter binding assays (Supplementary Figure S4c,d). Aptamers AM032-4 and AM047-6 (Supplementary Figures S4a,b and S5h,i) were the finally optimized aptamers that had strong binding affinities to the spike proteins of SARS-CoV-2 variants (Figure 5a,b; Supplementary Figure S4e,f; and Supplementary Table S1). To evaluate the antiviral activities of the optimized aptamers, we analyzed the effects of AM032-4 and AM047-6 on the RBD–ACE2 interaction and their inhibitory effects on the infection of the SARS-CoV-2 delta pseudovirus (Figure 5c,d). AM032-4 and AM047-6 inhibited RBD–ACE2 interaction with EC_{50} values of 8.3 nM and 5.8 nM, respectively. Notably, AM032-4 and AM047-6 demonstrated approximately five and three times stronger inhibitory effects compared with the original aptamers, AM032-0 and AM047-0 (Figures 1h and 5d). However, the inhibitory effects of AM032-4 and AM047-6 on pseudovirus infection were similar to those of AM032-0 and AM047-0 (Figures 1g and 5c).

2.6. Structure of RBD–Fab–AM032-4–AM047-6 Complex

We determined the structures of the optimized aptamers complexed with RBD associated with Fab since the optimized aptamers were likely to show higher resolution due to their compact structures around the termini with strengthened double-stranded stems. Indeed, we obtained the cryo-EM map of the protein–aptamer complex at a resolution of 3.43 Å, with an improved resolution in the binding interfaces and less flexibility at the termini of aptamers (Figure 5e–g; Supplementary Figure S5a–g; and Table S2). As expected, these aptamers have similar structures at the binding interfaces compared to AM032-0 and AM047-0 (Figure 5h,i). Major differences were observed at the 5′ and 3′ ends, where truncations and changes in bases were newly introduced in the optimized aptamers (Figure 5g).

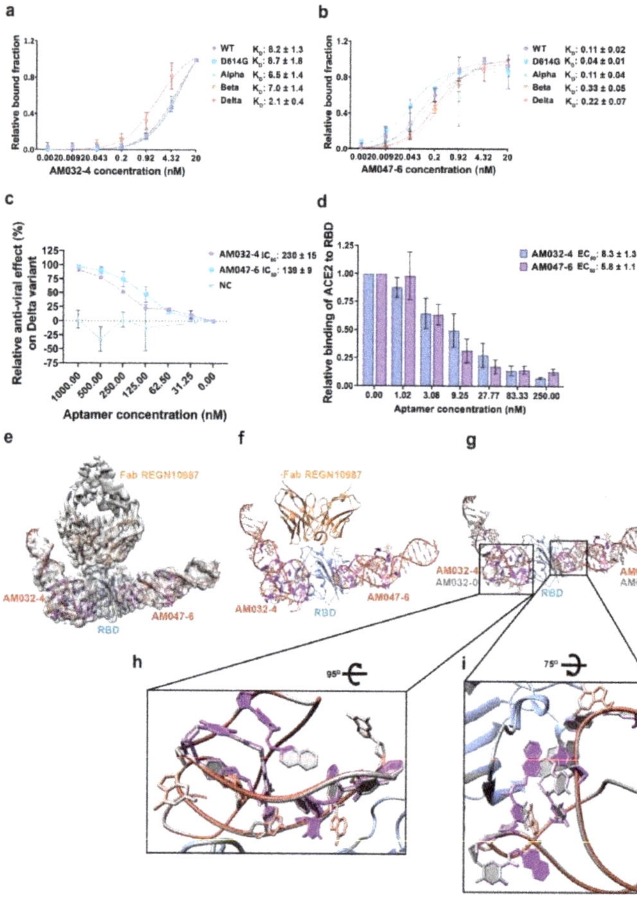

Figure 5. Characterization of aptamers AM032-4 and AM047-6 and cryo-EM structure of RBD–Fab–AM032-4–AM047-6 complex. (**a**,**b**) Binding affinities to the spike proteins of SARS-CoV-2 variants. (**c**) Inhibitory effects of aptamers on the infection of SARS-CoV-2 pseudovirus containing the spike protein of the delta variant. (**d**) Inhibitory effects of aptamers on the interaction between ACE2 and RBD. ELISAs were performed to monitor the inhibitory effects. (**e**) Cryo-EM density map of the protein–aptamer complex (front view) is superimposed with the refined model. (**f**) Structure of the RBD–Fab–AM032-4–AM047-6 complex. (**g**) The structure of the RBD–Fab–AM032-4–AM047-6 complex superimposed with that of the RBD–Fab–AM032-0–AM047-0 complex (gray). Nap-dUs of AM032-4 and AM047-6 are shown in purple. (**h**,**i**) Structures of the binding interfaces of AM032-4 (**h**) and AM047-6 (**i**) are superimposed with that of the RBD-Fab-AM032-0-AM047-0 complex (gray). NC, negative control. The values presented for K_D, IC_{50}, and EC_{50} were measured in nM.

2.7. Generation of a Bivalent Aptamer That Strongly Inhibits SARS-CoV-2 Pseudovirus Infection

It is possible to increase the activities of aptamers by connecting two or more aptamers together through a linker(s), which would increase the avidity of aptamers to a target molecule [43,44]. We designed and synthesized two bivalent aptamers, i.e., AM-B1 and AM-B2, using the optimized aptamers AM032-4 and AM047-6. Aptamers AM032-4 and AM047-6 were connected by a long flexible linker [7× or 8× hexaethylene glycol (HEG = Spacer 18)] to obtain bivalent aptamers AM-B1 and AM-B2 based on the three-dimensional structure of the AM032-4–RBD–AM047-6 complex (Figure 6a–c). The 5′ end of AM032-4 was connected with the 3′ end of AM047-6 using a linker of 7× HEG to obtain AM-B1, and the 5′ end of AM047 was connected with the 3′ end of AM032-4 using a linker of 8× HEG to obtain AM-B2 (Figure 6a). AM-B1 and AM-B2 strongly interacted with the spike proteins of several SARS-CoV-2 variants with similar K_D values to each other (Figure 6d,e and Supplementary Figure S6a,b). Interestingly, AM-B1 showed stronger inhibitory effects on ACE2 binding (EC_{50} = 15.6 nM) and on pseudovirus infection (IC_{50} = 47.0 nM) than AM-B2 (EC_{50} = 125.9 nM, IC_{50} = 159 nM) (Figure 6f,g). It should be noted that the inhibitory effect of AM-B1 (IC_{50} = 47.0 nM) is higher than those of individual aptamers AM032-4 and AM047-6 (IC_{50} = 230 nM and 139 nM, respectively). The results indicate that a properly arranged bivalent aptamer (e.g., AM-B1) can have stronger inhibitory activity than individual aptamers, but a poorly arranged aptamer (e.g., AM-B2) shows similar inhibitory activity to the stronger monomeric aptamer (e.g., AM047-6).

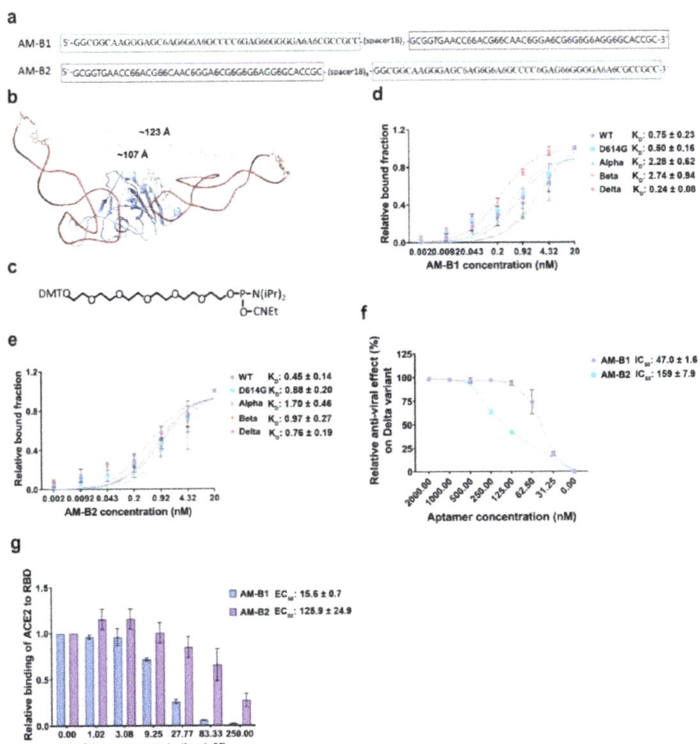

Figure 6. A bivalent aptamer shows strongly augmented antiviral activity. (**a**) Sequences of bivalent aptamers. Aptamers AM032-4 and AM047-6, shown in purple and blue boxes, respectively, are connected together by flexible linkers composed of 7× or 8× spacer 18s (HEGs). (**b**) Distances between the 5′ and 3′ ends of the aptamers AM032-4 and AM047-6 are depicted in the complex structure. Distance information helped in designing bivalent aptamers. (**c**) Chemical structure of phosphoramidite 18 linker that generates spacer18 linker by the polymerization reaction. (**d**,**e**) Determination of binding affinities to the spike proteins of SARS-CoV-2 variants. (**f**) Inhibitory effects of aptamers on the infection of SARS-CoV-2 pseudovirus of the delta variant. (**g**) Inhibitory effects of aptamers on the interaction between ACE2 and RBD. ELISAs were performed to monitor the inhibitory effects. The values presented for K_D, IC_{50}, and EC_{50} were measured in nM.

3. Discussion

Herein, we could develop several aptamers having modified nucleotides (Nap-dUs) with high affinities for the spike protein of SARS-CoV-2. Our study focused on the aptamers AM032 and AM047 and their derivatives that bind to the RBD of spike protein. Interestingly, these aptamers showed high affinities (K_D = 0.04 nM~10.5 nM) not only to the wild-type spike protein but also to the spike proteins of alpha, beta, and delta variants. This indicates that developing aptamers with a broad spectrum of target molecules is possible by using an appropriate selection process.

To investigate the feasibility of developing aptamers with anti-SARS-CoV-2 activities via a structure-based aptamer-designing approach, we determined the three-dimensional structure of the aptamer/RBD complex using cryo-EM technology. We used RBD instead of a trimeric spike protein due to the flexible nature of RBD relative to the rest of the spike protein [4], which reduces the resolution of the RBD region of the spike in the structural analysis. The three-dimensional structure of the aptamer/RBD complex (resolution up to ~3.43 Å in Supplementary Figure S5f) provided us with the information required for understanding the mechanism of aptamer structure formation and of protein–aptamer interaction. Outstanding features of the aptamer/RBD complex are as follows: (1) The naphthyl group of Nap-dU plays a key role in maintaining the loop structures of both aptamers AM032-0 and AM047-0 that enable specific interactions with RBD (Figures 3 and 4 and Supplementary Table S3). The unusual three-turn structure in aptamer AM032-0 is stabilized by a hydrophobic core composed of seven naphthyl residues, and the L1 and L2 loops in aptamer AM047-0 are maintained by a hydrophobic core composed of four naphthyl residues stretched out from L1 and L2 loops, which provide two naphthyl residues each. (2) The hydrophobic and specific polar interactions between the aptamers and the spike protein are the major forces for the strong and specific associations among the molecules. (3) Base pairing and/or base stacking that resides outside of the loop regions, which participate in aptamer–protein interaction, stabilize the overall structure of the aptamers. These features indicate that the usage of a modified nucleotide (Nap-dU, in this case) can dramatically increase the diversity of shapes of the aptamers by forming hydrophobic cores composed of naphthyl groups. The conformational diversity and the direct participation of the naphthyl group in binding to a target protein seem to increase the possibility of finding aptamers with high affinities, which was experimentally proven previously without knowing the structures [23–26].

The molecular bases of the antiviral effects of aptamers were revealed by the three-dimensional structure of the aptamer–RBD complex. A large part of the interface of RBD that participates in the interaction with aptamer AM032-0 (F456 and G476~Y505 on RBD) overlaps with that participating in the interaction with ACE2 (G446~Y505 on RBD) [5]. The competition between AM032-0 and ACE2 for the same binding sites on the RBD is the inhibitory mechanism of AM032-0 for the ACE2 binding to the RBD. On the other hand, the inhibitory effect of AM047-0 on the RBD-ACE2 interaction was not attributable to the occupation of the same binding site in RBD by AM047-0 since the aptamer AM047-0 localized to a completely distinct region from the ACE2-binding area (Figure 2c). It is likely that the binding of AM047-0 indirectly inhibits ACE2 binding through the conformational change of RBD.

Our structure-based and optimized aptamers AM032-4 and AM047-6 showed stronger or similar affinities to spike variants compared with the original aptamers (Figures 1 and 5). Moreover, the existence of more stable stem structures due to the incorporation of G-C base pairs at the termini of the optimized aptamers AM032-4 and AM047-6 compared with the original aptamers AM032-0 and AM047-0 was indicated by the higher resolution of the cryo-EM map around the aptamers' termini of the protein–aptamer complex (Supplementary Figures S3g and S5g). A strong, double-stranded stem structure at the termini of oligonucleotides protects an aptamer from degradation by exonucleases, which are the major nucleases that destroy DNAs or RNAs in the plasma [31,42]. These aptamers can be even further stabilized by replacing the nucleotides forming the stems with locked nucleic acids (LNA) or 2′-O-methyl nucleotides, which are more resistant to nuclease-mediated degradation and form more stable stem structures with a higher melting temperature [45–47].

Based on the three-dimensional structure of the RBD–Fab–AM032-4–AM047-6 complex, we designed and synthesized bivalent aptamers containing both AM032-4 and AM047-6 connected by linkers with different lengths. One of the bivalent aptamers (AM-B1) showed an approximately 3-fold higher inhibitory effect than AM047-6 on the infection of pseudovirus SARS-CoV-2 delta. The other bivalent aptamer (AM-B2) showed the same inhibitory effect as AM047-6 on the infection of the delta pseudovirus. These results demonstrate the feasibility of developing highly effective bivalent aptamers based on the structures of aptamers binding to two different regions of a target protein.

While we were carrying out this study, SARS-CoV-2 evolved rapidly to evade host immunity acquired by vaccine treatment and/or by the viral infection of early SARS-CoV-2 variants [12,18–20]. In addition, the variants with higher infectivity through their stronger binding affinity to the viral receptor ACE2 outnumbered the earlier variants by their faster-spreading capability [48]. As a consequence, the omicron variants of SARS-CoV-2 have become the most prevalent variants to date [49]. The mutations in SARS-CoV-2 variants are accumulated mostly in the spike protein since it plays a pivotal role in the interaction with ACE2 and is targeted by neutralizing antibodies generated by acquired immunity [50]. The aptamers AM032-0 and AM047-0 and their derivatives have high affinities to the spike proteins of various variants up to the delta variant (Figures 1 and 5), even though these variants contain many mutations in the spike gene that nullify the binding of many neutralizing antibodies [51]. This indicates that these aptamers have a broad spectrum of binding to various spike proteins compared with many monoclonal antibodies. However, unfortunately, these aptamers have reduced binding affinities to the spike of omicron variants that have 15 or more mutations within the RBD region that nullify the binding of many neutralizing antibodies [52] (Supplementary Figure S6e and Supplementary Table S1; AM032-4 and AM047-4 have K_D values of 26.7 and 19.9 nM, respectively).

Several mutations exist in the aptamer-binding interfaces in the RBD of omicron variants. For instance, eight (S477N, T478K, E484A, Q493R, G496S, Q498R, N501Y, and Y505H) [53] and seven (S477N, T478K, E484A, F486V, Q498R, N501Y, and Y505H) [54] mutations exist in the AM032-binding interfaces of omicron B.1.1.529 and BA.4/5, respectively (Supplementary Figure S6c and Table S4). A slight movement (1.5 to 2.4 Å) of aromatic amino acids in the loop of omicron variants composing the AM032-binding interface and several mutations in amino acids that participate in aptamer–RBD interaction seem to attribute to the reduced affinity of the omicron spike to AM032 (Supplementary Figure S6c). Moreover, we confirmed the importance of E484, F486, and Q493 residues in aptamer binding by testing the effect of mutations in RBD (Figure 3d). Three (S371L, S373P, and S375F) and four (S371L, S373P, S375F, and T376A) mutations exist in the AM047-binding interfaces of omicron B.1.1.529 and BA.4/5, respectively (Supplementary Table S2). The presence of these mutations causes conformational differences in the AM047-binding interface of omicron, as shown in Supplementary Figure S6d, thereby explaining the reduced binding of omicron to AM047. Moreover, we confirmed the importance of residues 373 and 375 of the spike in the interaction with AM047 by reverting these mutations to the wild type (P373S and F375S) and testing the affinity of the reverted RBD for AM047. The reversion mutations of omicron BA.4/5 restored the binding affinity to AM047-0 almost completely (Figure 4d).

4. Materials and Methods

4.1. Cloning, Expression, and Purification of Proteins

In order to synthesize the spike proteins of SARS-CoV-2, we used a baculovirus gene expression system [35,36]. The DNA fragment encoding the ectodomain of the spike protein (designated as WT-S, residues 1–1208) of SARS-CoV-2 containing two proline substitutions at 986–987 amino acid positions, a furin cleavage site mutation (RRAR to GSAS at position of 682–685 residues), and a C-terminal trimerization motif [4] was amplified by PCR using a plasmid kindly provided by Prof. Jason S. McLellan of University of Texas at Austin as a template. A TEV protease cleavage site was newly included during PCR using a proper primer. The synthesized DNA fragment was cloned into a p42 vector (a modified pFastBac) containing 6xHis and a Flag-tag [35,36]. To create a plasmid encoding D614G mutation in the spike gene, site-directed mutagenesis (SDM) was performed. Other DNA fragments encoding the spike genes of alpha, beta, delta, and omicron variants with two proline substitutions and the furin cleavage site mutations were synthesized by Gene Universal and cloned into the p42 vector, as described for WT-S construction. In order to produce plasmids encoding human ACE2 protein (residues 19-615) and the polypeptide corresponding to the RBD of the spike (319−541 aa) with an N-terminal honeybee melittin (HBM) signal sequence [55], C-terminal 6xHis-tag, and an Fc-tag, DNA fragments corresponding to the genes were amplified by PCR using the plasmids kindly provided by Prof. Jason S. McLellan as templates. The synthesized DNA fragments were cloned into the p42 vector as described above.

Recombinant proteins were produced using a baculovirus-expression system as previously described [35,36]. Sf9 insect cells (2×10^6 cells/mL) were inoculated with baculoviruses expressing recombinant proteins and cultivated for three days, and then, culture media were collected. The culture media were centrifuged at 6000 rpm for 10 min to remove cell debris and then filtered through a 0.45 μm syringe filter (Sartorius). The filtrate was passed through a His-tag purification resin (Roche) for protein binding and then washed with wash buffer (50 mM Tris-HCl (pH 8.0), 200 mM NaCl, and 20 mM imidazole), and then, the proteins were eluted in elution buffer (50 mM Tris-HCl (pH 8.0), 200 mM NaCl, and 200–300 mM imidazole). Eluted fractions were pooled, concentrated, and injected into a Superose® 6 Increase 10/300 GL column (GE Healthcare) for size-exclusion chromatography (SEC) in a buffer of 20 mM Tris-HCl (pH 8.0) and 200 mM NaCl. Finally, SEC fractions were pooled, concentrated, and stored at −70 °C for further experiments. RBD protein was purified (Supplementary Figure S1b,f) by the same method with a final buffer of 25 mM Tris-HCl (pH 7.5) and 150 mM NaCl in SEC (Superdex 200 increase 10/300 GL, GE healthcare). Other RBD mutants of the wild-type (Y369A, F456A, F486A, and Y489A), delta (E484A, Q493R, and E484A + Q493R), and omicron BA.4/5 (P373S and P373S + F375S) varieties were generated by SDM and purified by the same method as wild-type RBD. Fc-tagged ACE2 proteins were purified (Supplementary Figure S1c,g) by passing the filtrate through protein A resin, washed with wash buffer (50 mM Tris-HCl (pH 8.0) and 200 mM NaCl), and eluted with buffer (100 mM glycine (pH 3.0) and 150 mM NaCl). The pH of the eluent was raised to ~pH 8.0 by adding 1M Tris-HCl (pH 9.0), and then, the buffer was exchanged for 50 mM Tris-HCl (pH 8.0) and 200 mM NaCl. Buffer-exchanged protein was subsequently purified by SEC (Supplementary Figure S1g).

For the expression and purification of Fab (PDB ID: 6XDG) [38], genes for the heavy and light chains of Fab were synthesized by Twist Bioscience. A signal sequence was added at the N-terminal ends of both chains, and an alpha-tag [56] with a thrombin cleavage site and an 8x His-tag were added at the C-terminal ends of the heavy and light chains, respectively. Genes were cloned into the pAcGP67 vector for the generation of recombinant baculovirus [57]. Hi5 insect cells were co-infected with baculoviruses expressing heavy and light chains. Upon clarification of cells by centrifugation, the secreted Fab complex was purified by Ni-affinity column (Roche) with wash buffer (20 mM Tris-HCl (pH 8.0), 200 mM NaCl, and 1 mM PMSF) and gradient elution buffer (20 mM Tris-HCl (pH 8.0), 200 mM NaCl, and 100-500 mM imidazole). Elution fractions were pooled and purified by SEC (Superdex 200), and peak fractions were concentrated and stored at −80 °C.

4.2. Preparation of Surrogate Virus

Surrogate baculovirus preparation was performed as described previously [36]. Briefly, the ectodomain (polypeptide spanning the outer part of the viral envelope: 1-1208 residues) of the SARS-CoV-2 spike conjugated with the foldon domain of the T4 phage [4] was cloned in front of the G-stem of the vesicular stomatitis virus (VSV) with a C-terminal Flag-tag as a single open-reading frame in a p42 vector. The VSV G-stem was used as an anchor on the viral envelope to display the spike protein on the surface of the baculovirus as described previously [58]. The surrogate baculovirus was generated using a baculovirus-expression system. At three days after infection, the surrogate viruses in Sf9 culture media were collected by centrifugation (500× *g* for 5 min) and cleared by a 0.2 μm syringe filter. The viruses in the media were precipitated by ultra-centrifugation (100,000× *g* for 2 h) in a 20% sucrose gradient in a sucrose gradient buffer (5 mM Tris-HCl (pH 8.0), 100 mM NaCl, and 1 mM EDTA). The surrogate virus pellet was resuspended in cold PBS buffer (137 mM NaCl, 2.7 mM KCl, 10 mM Na_2HPO_4, and 1.8 mM KH_2PO_4) with 2.5% glycerol and stored at −70 °C.

4.3. Systematic Evolution of Ligands by Exponential Enrichment (SELEX)

The SELEX procedures were performed as described by Kwon et al. and Narayan et al., with a few modifications [35,36]. The SELEX process is schematically illustrated in Supplementary Figure S2a, depicting positive and negative selection processes. Protein-based SELEX (left) and virus-based SELEX (right) were executed alternatively. For the enrichment of aptamers specifically interacting with the spike protein of SARS-CoV-2, positive and negative selection processes were included. For positive selection, monomeric spike protein mS (Sino Biological Inc., Wayne, PA, USA), purified spike protein of SARS-CoV-2 (WT-S), and a surrogate virus containing the spike of SARS-CoV-2 (SARS-CoV-2 surrogate virus) were used in the viro-SELEX. For negative selection, a mixture of spike proteins (purchased from Sino Biological Inc.) of human betacoronaviruses (HCoV HKU1 and HCoV OC43) were used in the SELEX (Supplementary Figure S2a). Among the 12 rounds of SELEX, the seventh round was performed with a SARS-CoV-2 surrogate virus (in positive selection) and baculovirus (in negative selection). The ssDNA library contains 90-mer DNAs composed of a randomized nucleic acid sequence (40-mer) in the center flanked by 25-mer constant regions at both ends. A modified nucleotide, i.e., 5-[N-(1-naphthylmethyl) carboxamide]-2′-deoxy uridine (Nap-dU), was used instead of natural nucleotide deoxythymidine for the random sequence of the library. The sequence pattern of library DNAs is similar to 5′-ATATATATCGAGCGTCCTGCCTTTG-40N-CACCGACAGCCACCCAGAAAAAAAA-3′, where 40N represents 40 random sequences including a modified nucleotide (Nap-dU).

To perform aptamer refolding, an aptamer library mixture (1 nmole) dissolved in S buffer (40 mM HEPES (pH 7.5), 102 mM NaCl, 5 mM KCl, 5 mM $MgCl_2$, and 0.05% Tween-20) was heated at 95 °C for 5 min, followed by slow cooling to 37 °C at a rate of 0.1 °C/s. To remove unwanted magnetic beads and His-tag binders, aptamer library DNAs were pre-incubated with 6xHis-tag-associated Talon beads (Invitrogen, Carlsbad, CA, USA). The bead-bound DNAs were separated by applying a magnetic field, and the remaining solution with unbound DNAs was incubated with purified proteins (50 pmol) for 60 min at 37 °C. Next, the solution was incubated with Talon beads for 10 min with agitation, and the beads were collected using a magnetic field and washed three times with S buffer. The target-protein-bound aptamers were eluted from the beads using a 2 mM NaOH solution. The eluted DNAs were then amplified by PCR using a 5′-primer (5′-ATATATATCGAGCGTCCTGCCTTTG-3′) and a biotin-conjugated 3′-primer [5′-(2x) Biotin-TTTTTTTTCTGGGTGGCTGTCGGTG-3′]. After the PCR, biotin-conjugated DNAs were trapped on Myone SA beads (Invitrogen), and 20 mM NaOH elution buffer was used to elute the positive-sense DNAs. In order to synthesize positive-sense aptamers containing Nap-dU, the negative-sense DNAs trapped on the beads were used as templates, and the beads were incubated in a polymerization reaction buffer comprising DNA polymerase, dATP, dGTP, dCTP, and Nap-dUTP. The synthesized aptamers were used for the next round of SELEX. The SELEX process was executed in the absence or presence of dextran sulfate ($DxSO_4$). A decreasing amount of spike protein (down to 0.005 pmol) and an increasing amount of $DxSO_4$ (up to 10 μM) were added in the binding step and/or after the binding of protein with a DNA pool. The ssDNAs, enriched in the 11th round of SELEX, were amplified by PCR using the 5′- and 3′-primers and purified using a PCR purification kit (Qiagen, Germany). The purified DNAs were cloned and sequenced (Solgent, South Korea). Alternatively, the amplified DNAs were sequenced using a next-generation sequencing (NGS) method.

4.4. SARS-CoV-2 Spike Aptamer Synthesis

The aptamers were synthesized by Aptamer Sciences Inc. as described previously [35,36].

4.5. Measurement of the K_D Values of Aptamers

A filter binding method as described by Gold et al. (2010) was used to determine the equilibrium dissociation constant (K_D) between an aptamer and a protein [26]. In short, [^{32}P] radioisotope and [^{32}P]-ATP (Perkinelmer, Waltham, MA, USA) were used to label the 5′ ends of aptamers using T4 polynucleotide kinase (Takara Bio Inc.). The radiolabeled aptamers were diluted to 20,000 cpm in 200 μL of S buffer and then heated at 95 °C for 3 min and cooled slowly to 37 °C at a rate of 0.1 °C/s. The radiolabeled aptamers were incubated with various amounts of the target proteins (20 nM to 2.1 pM with a 4.6-fold dilution) at 37 °C for 30 min. Then, 1/30 of them were spotted on a non-charged nylon membrane. Zorbax resin (Agilent, Santa Clara, CA, USA) in H_2O (5.5 μL, 400 mg/mL) was added to the aptamer–protein solution and mixed with a thermomixer for 1 min at 1300 rpm. The target protein and radiolabeled aptamer mixer were then applied to a polyvinylidene difluoride (PVDF) filter plate and washed, and the radioactivity remaining on the filter plate and nylon membrane was measured with a phosphor-imager (Amersham Typhoon 5 Biomolecular Imager, GE healthcare, Japan). The radioactivity in each slot was normalized to that of the negative control ([^{32}P]-labeled library DNAs) slot. The K_D values of aptamers were calculated by plotting the data in one site-specific binding model using GraphPad Prism software, version 9.5.

4.6. Protein-Binding Competition Assay

Competition between aptamers for binding to SARS-CoV-2 spike proteins was monitored as described by Kwon et al. [35]. [^{32}P]-labeled (hot) aptamers (2000 cpm) were incubated with SARS-CoV-2 spike protein (mS) in the presence of unlabeled (cold) aptamers (25 pmole) as competitors at 37 °C for 30 min, and then, aptamer–protein complexes were captured by Zorbax resin. The amount of radio-labeled aptamers on the Zorbax resin was measured with a phosphor-imager (Amersham Typhoon 5 Biomolecular Imager).

4.7. Enzyme Linked Immunosorbent Assay (ELISA)

The aptamer activity in blocking the interaction between RBD and ACE2 was measured by competitive ELISA. In short, 100 μL of RBD protein (1 μg/mL) was pre-coated on Maxi binding ELISA plates (SPL Life Sciences) by incubating overnight at 4 °C. The next day, the plate was blocked with 5% BSA in PBST buffer (PBS buffer with 0.05% Tween-20) for 1 h at 37 °C and then washed three times with PBST buffer (200 μL/well). Aptamers were heated at 95 °C for 5 min and slowly cooled to 37 °C at a rate of 0.1 °C/s. Aptamers were serially diluted in PBST with 2 mM $MgCl_2$ (50 μL) and mixed with 50 ng of ACE2 (50 μL). This reaction mixture (100 μL) was applied to each well of the ELISA plate and incubated for 15 min at 37 °C. The wells were washed three times with PBST buffer, and then, HRP-conjugated anti-human IgG1-HRP antibody (Sigma-Aldrich, St. Louis, MO, USA) at a dilution of 1:40,000 was added to the well and incubated at 37 °C for 1 h. After washing, the color-development reaction was carried out by adding 100 μL of 3,3′,5,5′-tetramethylbenzidine (TMB) substrates (Sigma-Aldrich, T0440), and the reaction was stopped by adding 100 μL of 1N H_2SO_4. Optical density at 450 nm (OD_{450nm}) was read in a microplate reader (Tecan, Infinite M200pro, Männedorf, Switzerland). For all the experiments, triplicate wells were used.

4.8. Measurement of Antiviral Activities of Aptamers

Pseudoviruses based on an HIV backbone containing SARS-CoV-2 spike proteins of D614G and delta variants with a deletion (21 aa) at the C-termini were used to measure the antiviral effects of aptamers [31,37]. Briefly, for the production of pseudoviral particles, HEK293T cells were transfected with five plasmids: (1) pHAGE-CMV-Luc2-IRES-ZsGreen-W (NR-52516, BEI resources), (2) pHDM-Hgpm2 (NR-52517, BEI resources), (3) pHDM-tat1b (NR-52518, BEI resources), (4) pRC-CMV-Rev1b (NR-52519, BEI resources), and (5) pHDM-SARS-CoV-2 D614G (NR-53765, BEI resources) or a derivative of NR-53765 containing the SARS-CoV-2 delta variant gene. At 18–24 h after transfection, the media were changed, and the cells were further cultivated for 40 h. The media were harvested, filtered through a 0.45 μm pore-sized filter, and stored at −70 °C as a pseudovirus stock. For antiviral activity assays, the media containing pseudoviral particles were incubated with different amounts of aptamers (1000 nM, 500 nM, 250 nM, 125 nM, 62.5 nM, 31.2 nM, and 0 nM) at 37 °C for 60 min and then applied to 293T-ACE2 cells that were cultivated for 12 h after seeding of ~2×10^4 cells/well on a 96-well plate. After 6 h of incubation, the media were changed with fresh DMEM media with 10% FBS, and the cells were further cultivated for 48 h. The media were discarded, and 30 μL of passive lysis buffer (Promega, Madison, WI, USA) was applied to each well, and the plate was incubated for 5 min. Firefly luciferase substrate (E1501, Promega, Madison, WI, USA) was added to each well, and the luciferase activity in each well was measured for 1 s using a Tecan Infinite M200 Pro reader.

4.9. Cryo-EM Grid Preparation

RBD and Fab proteins were mixed at a 1:1.3 molar ratio, incubated on ice for one hour, and then fractionated by size-exclusion chromatography (Supplementary Figure S3a). Two aptamers (1.2-fold molar excess of each aptamer to RBD) were added to the RBD–Fab complex protein (1 mg/mL). The mixture was supplemented with 0.15% amphipol A8-35 (Anatrace, Maumee, OH, USA) and 1 mM $MgCl_2$ and incubated for 1 h on ice. Detergent Cymal-6 (Anatrace) of 0.003% was added to the sample before the vitrification. The protein–aptamer complex (2.5 μL) was applied to freshly prepared glow-discharged Quantifoil R1.2/1.3 300 or 400 mesh copper grids (EMS, Hatfield, PA, USA). The grids were blotted for 5 s at 4 °C with 100% humidity and plunge-frozen in liquid ethane using Vitrobot Mark IV (FEI). Cryo-EM data were collected using a 300 kV Titan Krios Electron Microscope equipped with a Gatan K3 BioQuantum direct electron detector in a fast mode with a magnification of 105,000×. For the complex of RBD–Fab–AM032-0–AM047-0, data were collected with an accumulated dose of 50 electrons/Å^2 and defocus range of −0.8 to −2.0 μm. In the case of the RBD–Fab–AM032-4–AM047-6 complex, data were collected with an accumulated dose of 60 electrons/Å^2 and defocus range of −0.7 to −1.6 μm.

4.10. Data Processing for RBD–Fab–AM032-0–AM047-0 Complex

The data processing workflow for the RBD–Fab–AM032-0–AM047-0 aptamer complex is summarized in Supplementary Figure S3. Briefly, 11,743 movies (dataset 1) of dose-fractionated image stacks were imported into cryoSPARC v3.3.2 [39], and then patch motion correction was used to align and dose weight the images. CTF parameters were calculated by means of patch CTF estimation. Micrographs with low quality were removed. The remaining 10,880 micrographs were used for data processing. The protein–aptamer particles were picked and extracted using the Topaz method [59]. An additional dataset was collected to increase the number of particles. Full-frame motion correction was used for aligning and dose weighting the second dataset, and then, patch CTF estimation was used to calculate the CTF parameters. The remaining 8720 micrographs were used for data processing after eliminating the low-quality micrographs. Particles from the two data sets were combined and underwent an additional round of 2D classification, which selected 1,837,985 particles. The particles were re-extracted without binning, duplicate particles were removed, and they were once again subjected to 2D classification. In total, 1,085,102 particles were selected from this round of 2D classes and used for two rounds of heterogeneous refinement. After heterogeneous refinement, 387,167 particles were subjected to non-uniform refinement, which yielded a map with a global resolution of 3.82 Å. Global CTF refinement and non-uniform refinement resolution improved the map to 3.8 Å. The non-uniform refined model was reconstructed in 3DFlex refinement [60] in cryoSPARC v4.1.0 [39], and the final resolution was improved to 3.32 Å for the overall map.

4.11. Data Processing for RBD–Fab–AM032-4–AM047-6 Complex

The data processing workflow for the RBD–Fab–AM032-4–AM047-6 aptamer complex is summarized in Supplementary Figure S5. CryoSPARC v4.1.0 [39] was used to import the dose-fractionated image stacks from 10,018 movies. Images were aligned and dose-weighted using patch motion correction, and CTF parameters were calculated using patch CTF estimation. Micrographs with low quality were removed. The remaining 9480 micrographs were used for data processing. The protein–aptamer particles were picked and extracted using the Topaz method. Selected particles from 2D classes were subjected to heterogeneous refinement. After heterogeneous refinement, 230,681 particles showing the complex were refined using non-uniform (NU) refinement. Particles from the NU refinement were further corrected for motion using local motion correction, global CTF refinement, and local CTF refinement, and then, 212,355 particles were refined again using non-uniform refinement that resulted in a 3.29 Å resolution map. The map from the NU refinement was next subjected to 3DFlex reconstruction [60], which improved the map to 3.43 Å resolution.

4.12. Model Building and Refinement

The auto-DRRAFTER program [61] was used to obtain the initial models of the aptamers using conventional nucleotides (see Supplementary Notes for detail). Some of the dU nucleotides in the initial model were replaced with Nap-dU using Coot [41]. Initial structures for the RBD and Fab parts were built using previously reported structures obtained from PDB IDs 7KGJ [40] and 6XDG [38], respectively, and were fitted to the cryo-EM density map using the Chimera [62] program. The fitted models of RBD, Fab, and aptamers were combined and adjusted again with the cryo-EM density map in Chimera [62]. Next, multiple rounds of real space refinement and manual modeling building were performed using Phenix [63] and Coot [41] programs to prepare the final model (Supplementary Figure S3h). The details of the refinement properties of each complex are also presented in Supplementary Table S2. The figures related to structure were built in Chimera [62].

4.13. Accession Numbers

The cryo-EM density maps and models of the RBD–Fab–AM032-0–AM047-0 complex (EMDB: EMD-35930, PDB code: 8J1Q) and of the RBD–Fab–AM032-4–AM047-6 complex (EMDB: EMD-35945, PDB code: 8J26) have been deposited to the EMDB and PDB.

5. Conclusions

This study clearly demonstrates that the acquisition of aptamers with high affinities to the viral envelope protein, i.e., the spike protein, is possible using a modified nucleotide and a carefully designed SELEX process. We revealed the molecular bases of aptamers' activities (protein binding and inhibition of virus infection) through the cryo-EM structure of the aptamer/RBD complex, and we also showed the possibility of aptamer optimization using a structure-based aptamer-designing approach. It may also be possible to improve the aptamers to have higher affinities to a target molecule by modifying the aptamers based on the structure of the aptamer–target complex in the future. Finally, we demonstrated that the development of bivalent aptamers, which have higher activity than individual aptamers, is possible by analyzing the structure of the aptamer–target complex.

Supplementary Materials: The following supporting information can be downloaded at: https://www.mdpi.com/article/10.3390/molecules28124645/s1, Figure S1: Protein expression and purification, Figure S2: Schematic diagram of SELEX process and characterization of aptamers, Figure S3: Workflow of data processing for determining the structure of RBD–Fab–AM032-0–AM047-0 complex, Figure S4: Optimization of aptamers AM032-0 and AM047-0, Figure S5: Workflow of data processing for determining the structure of RBD–Fab–AM032-4–AM047-6 complex, Figure S6: Characterization of bivalent aptamers AM-B1 and AM-B2; Table S1: Binding affinities of the aptamers, Table S2: Summary of the cryo-EM data collection and refinement, Table S3: Nucleotides in aptamers and amino acids residues at the interacting interfaces with a distance cut-off of 4 Å, Table S4: Mutations in RBD observed at the binding surface of aptamers AM032-0 and AM047-0. The reference [64–70] is in the Supplementary Materials.

Author Contributions: Conceptualization, M.S.R., J.-O.L. and S.K.J.; methodology, M.S.R., M.J.H., S.W.K., S.M.K., B.R.K., H.K. (Heesun Kim), C.J.L., J.E.N. and H.K. (Hanseong Kim); software, M.S.R., B.R.K., C.J.L. and H.K. (Hanseong Kim); validation, M.S.R., J.-O.L. and S.K.J.; formal analysis, M.S.R.; investigation, M.S.R. and S.K.J.; resources, S.K.J.; data curation, M.S.R., M.J.H. and S.W.K.; writing—original draft preparation, M.S.R.; writing—review and editing M.S.R., J.-O.L. and S.K.J.; visualization, M.S.R.; supervision, J.-O.L. and S.K.J.; project administration, S.K.J.; funding acquisition, S.K.J. All authors have read and agreed to the published version of the manuscript.

Funding: This research was funded by the Bio & Medical Technology Development Program of the National Research Foundation (NRF) funded by the Korean government (MSIT) (NRF-2019M3E5D6063871).

Institutional Review Board Statement: Not applicable.

Informed Consent Statement: Not applicable.

Data Availability Statement: Data will be provided upon request.

Acknowledgments: We thank Jason S. McLellan (University of Texas at Austin) for kindly providing plasmids for the SARS CoV-2 spike, RBD, and ACE2. We thank the team of cryo-EM facilities at Bio Open Innovation Center, POSTECH Biotech Center, and Institute of Membrane Proteins, POSTECH, for supporting the data collection.

Conflicts of Interest: The authors declare no conflict of interest.

Sample Availability: Not available.

References

1. Zhu, N.; Zhang, D.; Wang, W.; Li, X.; Yang, B.; Song, J.; Zhao, X.; Huang, B.; Shi, W.; Lu, R.; et al. A Novel Coronavirus from Patients with Pneumonia in China, 2019. *N. Engl. J. Med.* **2020**, *382*, 727–733. [CrossRef] [PubMed]
2. Zhou, P.; Yang, X.L.; Wang, X.G.; Hu, B.; Zhang, L.; Zhang, W.; Si, H.R.; Zhu, Y.; Li, B.; Huang, C.L.; et al. A pneumonia outbreak associated with a new coronavirus of probable bat origin. *Nature* **2020**, *579*, 270–273. [CrossRef] [PubMed]
3. Coronavirus Disease 2019 (COVID-19). WHO. Available online: https://covid19.who.int/ (accessed on 10 May 2023).
4. Wrapp, D.; Wang, N.; Corbett, K.S.; Goldsmith, J.A.; Hsieh, C.L.; Abiona, O.; Graham, B.S.; McLellan, J.S. Cryo-EM structure of the 2019-nCoV spike in the prefusion conformation. *Science* **2020**, *367*, 1260–1263. [CrossRef] [PubMed]
5. Lan, J.; Ge, J.; Yu, J.; Shan, S.; Zhou, H.; Fan, S.; Zhang, Q.; Shi, X.; Wang, Q.; Zhang, L.; et al. Structure of the SARS-CoV-2 spike receptor-binding domain bound to the ACE2 receptor. *Nature* **2020**, *581*, 215–220. [CrossRef] [PubMed]
6. Hoffmann, M.; Kleine-Weber, H.; Schroeder, S.; Kruger, N.; Herrler, T.; Erichsen, S.; Schiergens, T.S.; Herrler, G.; Wu, N.H.; Nitsche, A.; et al. SARS-CoV-2 Cell Entry Depends on ACE2 and TMPRSS2 and Is Blocked by a Clinically Proven Protease Inhibitor. *Cell* **2020**, *181*, 271–280.e278. [CrossRef] [PubMed]
7. Chan, J.F.; Kok, K.H.; Zhu, Z.; Chu, H.; To, K.K.; Yuan, S.; Yuen, K.Y. Genomic characterization of the 2019 novel human-pathogenic coronavirus isolated from a patient with atypical pneumonia after visiting Wuhan. *Emerg. Microbes Infect.* **2020**, *9*, 221–236. [CrossRef]
8. Piccoli, L.; Park, Y.J.; Tortorici, M.A.; Czudnochowski, N.; Walls, A.C.; Beltramello, M.; Silacci-Fregni, C.; Pinto, D.; Rosen, L.E.; Bowen, J.E.; et al. Mapping Neutralizing and Immunodominant Sites on the SARS-CoV-2 Spike Receptor-Binding Domain by Structure-Guided High-Resolution Serology. *Cell* **2020**, *183*, 1024–1042.e102. [CrossRef]
9. Polack, F.P.; Thomas, S.J.; Kitchin, N.; Absalon, J.; Gurtman, A.; Lockhart, S.; Perez, J.L.; Perez Marc, G.; Moreira, E.D.; Zerbini, C.; et al. Safety and Efficacy of the BNT162b2 mRNA COVID-19 Vaccine. *N. Engl. J. Med.* **2020**, *383*, 2603–2615. [CrossRef]
10. Baden, L.R.; El Sahly, H.M.; Essink, B.; Kotloff, K.; Frey, S.; Novak, R.; Diemert, D.; Spector, S.A.; Rouphael, N.; Creech, C.B.; et al. Efficacy and Safety of the mRNA-1273 SARS-CoV-2 Vaccine. *N. Engl. J. Med.* **2021**, *384*, 403–416. [CrossRef]

11. Su, S. Learning from the past: Development of safe and effective COVID-19 vaccines. *Nat. Rev. Microbiol.* **2021**, *19*, 211–219. [CrossRef]
12. Hassine, I.H. COVID-19 vaccines and variants of concern: A review. *Rev. Med. Virol.* **2022**, *32*, e2313. [CrossRef] [PubMed]
13. Bernal, A.J.; da Silva, M.M.G.; Musungaie, D.B.; Kovalchuk, E.; Gonzalez, A.; Delos Reyes, V.; Martin-Quiros, A.; Caraco, Y.; Williams-Diaz, A.; Brown, M.L.; et al. Molnupiravir for Oral Treatment of COVID-19 in Nonhospitalized Patients. *N. Engl. J. Med.* **2022**, *386*, 509–520. [CrossRef] [PubMed]
14. Najjar-Debbiny, R.; Gronich, N.; Weber, G.; Khoury, J.; Amar, M.; Stein, N.; Goldstein, L.H.; Saliba, W. Effectiveness of Paxlovid in Reducing Severe Coronavirus Disease 2019 and Mortality in High-Risk Patients. *Clin. Infect. Dis.* **2022**, *76*, 1158–1159. [CrossRef] [PubMed]
15. Barnes, C.O.; Jette, C.A.; Abernathy, M.E.; Dam, K.A.; Esswein, S.R.; Gristick, H.B.; Malyutin, A.G.; Sharaf, N.G.; Huey-Tubman, K.E.; Lee, Y.E.; et al. SARS-CoV-2 neutralizing antibody structures inform therapeutic strategies. *Nature* **2020**, *588*, 682–687. [CrossRef] [PubMed]
16. Lee, W.S.; Wheatley, A.K.; Kent, S.J.; DeKosky, B.J. Antibody-dependent enhancement and SARS-CoV-2 vaccines and therapies. *Nat. Microbiol.* **2020**, *5*, 1185–1191. [CrossRef]
17. Dejnirattisai, W.; Jumnainsong, A.; Onsirisakul, N.; Fitton, P.; Vasanawathana, S.; Limpitikul, W.; Puttikhunt, C.; Edwards, C.; Duangchinda, T.; Supasa, S.; et al. Cross-Reacting Antibodies Enhance Dengue Virus Infection in Humans. *Science* **2010**, *328*, 745–748. [CrossRef] [PubMed]
18. Aleem, A.; Akbar Samad, A.B.; Slenker, A.K. Emerging Variants of SARS-CoV-2 And Novel Therapeutics Against Coronavirus (COVID-19). In *StatPearls*; StatPearls Publishing: Treasure Island, FL, USA, 2022.
19. Hoffmann, M.; Arora, P.; Gross, R.; Seidel, A.; Hornich, B.F.; Hahn, A.S.; Kruger, N.; Graichen, L.; Hofmann-Winkler, H.; Kempf, A.; et al. SARS-CoV-2 variants B.1.351 and P.1 escape from neutralizing antibodies. *Cell* **2021**, *184*, 2384–2393. [CrossRef]
20. Korber, B.; Fischer, W.M.; Gnanakaran, S.; Yoon, H.; Theiler, J.; Abfalterer, W.; Hengartner, N.; Giorgi, E.E.; Bhattacharya, T.; Foley, B.; et al. Tracking Changes in SARS-CoV-2 Spike: Evidence that D614G Increases Infectivity of the COVID-19 Virus. *Cell* **2020**, *182*, 812–827. [CrossRef]
21. Tuerk, C.; Gold, L. Systematic Evolution of Ligands by Exponential Enrichment—Rna Ligands to Bacteriophage-T4 DNA-Polymerase. *Science* **1990**, *249*, 505–510. [CrossRef]
22. Keefe, A.D.; Pai, S.; Ellington, A. Aptamers as therapeutics. *Nat. Rev. Drug. Discov.* **2010**, *9*, 537–550. [CrossRef]
23. Xu, X.; Zhang, C.; Denton, D.T.; O'Connell, D.; Drolet, D.W.; Geisbrecht, B.V. Inhibition of the Complement Alternative Pathway by Chemically Modified DNA Aptamers That Bind with Picomolar Affinity to Factor B. *J. Immunol.* **2021**, *206*, 861–873. [CrossRef] [PubMed]
24. Vaught, J.D.; Bock, C.; Carter, J.; Fitzwater, T.; Otis, M.; Schneider, D.; Rolando, J.; Waugh, S.; Wilcox, S.K.; Eaton, B.E. Expanding the Chemistry of DNA for in Vitro Selection. *J. Am. Chem. Soc.* **2010**, *132*, 4141–4151. [CrossRef]
25. Gelinas, A.D.; Tan, T.K.; Liu, S.; Jaramillo, J.G.; Chadwick, J.; Harding, A.C.; Zhang, C.; Ream, B.E.; Chase, C.N.; Otis, M.R.; et al. Broadly neutralizing aptamers to SARS-CoV-2: A diverse panel of modified DNA antiviral agents. *Mol. Ther. Nucleic Acids* **2023**, *31*, 370–382. [CrossRef]
26. Gold, L.; Ayers, D.; Bertino, J.; Bock, C.; Bock, A.; Brody, E.N.; Carter, J.; Dalby, A.B.; Eaton, B.E.; Fitzwater, T.; et al. Aptamer-Based Multiplexed Proteomic Technology for Biomarker Discovery. *PLoS ONE* **2010**, *5*, e15004. [CrossRef]
27. Sun, M.; Wu, Z.J.; Zhang, J.L.; Chen, M.Y.; Lu, Y.; Yang, C.Y.; Song, Y.L. Spherical neutralizing aptamer suppresses SARS-CoV-2 Omicron escape. *Nano Today* **2022**, *44*, e101499. [CrossRef] [PubMed]
28. Sun, M.; Liu, S.W.; Wei, X.Y.; Wan, S.; Huang, M.J.; Song, T.; Lu, Y.; Weng, X.N.; Lin, Z.; Chen, H.L.; et al. Aptamer Blocking Strategy Inhibits SARS-CoV-2 Virus Infection. *Angew. Chem. Int. Ed.* **2021**, *60*, 10266–10272. [CrossRef] [PubMed]
29. Song, Y.; Song, J.; Wei, X.; Huang, M.; Sun, M.; Zhu, L.; Lin, B.; Shen, H.; Zhu, Z.; Yang, C. Discovery of Aptamers Targeting the Receptor-Binding Domain of the SARS-CoV-2 Spike Glycoprotein. *Anal. Chem.* **2020**, *92*, 9895–9900. [CrossRef]
30. Liu, X.H.; Wang, Y.L.; Wu, J.; Qi, J.J.; Zeng, Z.H.; Wan, Q.Y.; Chen, Z.H.; Manandhar, P.; Cavener, V.S.; Boyle, N.R.; et al. Neutralizing Aptamers Block S/RBD-ACE2 Interactions and Prevent Host Cell Infection. *Angew. Chem. Int. Ed.* **2021**, *60*, 10273–10278. [CrossRef]
31. Valero, J.; Civit, L.; Dupont, D.M.; Selnihhin, D.; Reinert, L.S.; Idorn, M.; Israels, B.A.; Bednarz, A.M.; Bus, C.; Asbach, B.; et al. A serum-stable RNA aptamer specific for SARS-CoV-2 neutralizes viral entry. *Proc. Natl. Acad. Sci. USA* **2021**, *118*, e2112942118. [CrossRef]
32. Schmitz, A.; Weber, A.; Bayin, M.; Breuers, S.; Fieberg, V.; Famulok, M.; Mayer, G. A SARS-CoV-2 Spike Binding DNA Aptamer that Inhibits Pseudovirus Infection by an RBD-Independent Mechanism. *Angew. Chem. Int. Ed.* **2021**, *60*, 10279–10285. [CrossRef]
33. Zhang, N.; Chen, Z.; Liu, D.; Jiang, H.; Zhang, Z.K.; Lu, A.; Zhang, B.T.; Yu, Y.; Zhang, G. Structural Biology for the Molecular Insight between Aptamers and Target Proteins. *Int. J. Mol. Sci.* **2021**, *22*, 4093. [CrossRef] [PubMed]
34. Kacherovsky, N.; Yang, L.F.; Dang, H.V.; Cheng, E.L.; Cardle, I.I.; Walls, A.C.; McCallum, M.; Sellers, D.L.; DiMaio, F.; Salipante, S.J.; et al. Discovery and Characterization of Spike N-Terminal Domain-Binding Aptamers for Rapid SARS-CoV-2 Detection. *Angew. Chem. Int. Ed.* **2021**, *60*, 21211–21215. [CrossRef]
35. Kwon, J.; Narayan, C.; Kim, C.; Han, M.J.; Kim, M.; Jang, S.K. Development of a Subtype-Specific Diagnostic System for Influenza Virus H3N2 Using a Novel Virus-Based Systematic Evolution of Ligands by Exponential Enrichment (Viro-SELEX). *J. Biomed. Nanotechnol.* **2019**, *15*, 1609–1621. [CrossRef] [PubMed]

36. Narayan, C.; Kwon, J.; Kim, C.; Kim, S.J.; Jang, S.K. Virus-based SELEX (viro-SELEX) allows development of aptamers targeting knotty proteins. *Analyst* **2020**, *145*, 1473–1482. [CrossRef] [PubMed]
37. Crawford, K.H.D.; Eguia, R.; Dingens, A.S.; Loes, A.N.; Malone, K.D.; Wolf, C.R.; Chu, H.L.Y.; Tortorici, M.A.; Veesler, D.; Murphy, M.; et al. Protocol and Reagents for Pseudotyping Lentiviral Particles with SARS-CoV-2 Spike Protein for Neutralization Assays. *Viruses* **2020**, *12*, 513. [CrossRef]
38. Hansen, J.; Baum, A.; Pascal, K.E.; Russo, V.; Giordano, S.; Wloga, E.; Fulton, B.O.; Yan, Y.; Koon, K.; Patel, K.; et al. Studies in humanized mice and convalescent humans yield a SARS-CoV-2 antibody cocktail. *Science* **2020**, *369*, 1010–1014. [CrossRef]
39. Punjani, A.; Rubinstein, J.L.; Fleet, D.J.; Brubaker, M.A. cryoSPARC: Algorithms for rapid unsupervised cryo-EM structure determination. *Nat. Methods* **2017**, *14*, 290–296. [CrossRef]
40. Ahmad, J.; Jiang, J.; Boyd, L.F.; Zeher, A.; Huang, R.; Xia, D.; Natarajan, K.; Margulies, D.H. Structures of synthetic nanobody-SARS-CoV-2 receptor-binding domain complexes reveal distinct sites of interaction. *J. Biol. Chem.* **2021**, *297*, 101202. [CrossRef]
41. Emsley, P.; Cowtan, K. Coot: Model-building tools for molecular graphics. *Acta Crystallogr. D* **2004**, *60*, 2126–2132. [CrossRef]
42. Xu, G.H.; Zhao, J.J.; Yu, H.; Wang, C.; Huang, Y.Y.; Zhao, Q.; Zhou, X.; Li, C.G.; Liu, M.L. Structural Insights into the Mechanism of High-Affinity Binding of Ochratoxin A by a DNA Aptamer. *J. Am. Chem. Soc.* **2022**, *144*, 7731–7740. [CrossRef]
43. Riccardi, C.; Napolitano, E.; Musumeci, D.; Montesarchio, D. Dimeric and Multimeric DNA Aptamers for Highly Effective Protein Recognition. *Molecules* **2020**, *25*, 5227. [CrossRef] [PubMed]
44. Tian, L.; Heyduk, T. Bivalent Ligands with Long Nanometer-Scale Flexible Linkers. *Biochemistry* **2009**, *48*, 264–275. [CrossRef] [PubMed]
45. Odeh, F.; Nsairat, H.; Alshaer, W.; Ismail, M.A.; Esawi, E.; Qaqish, B.; Al Bawab, A.; Ismail, S.I. Aptamers Chemistry: Chemical Modifications and Conjugation Strategies. *Molecules* **2020**, *25*, 3. [CrossRef] [PubMed]
46. Obika, S.; Nanbu, D.; Hari, Y.; Morio, K.-I.; In, Y.; Ishida, T.; Imanishi, T. Synthesis of 2′-O,4′-C-methyleneuridine and -cytidine. Novel bicyclic nucleosides having a fixed C_3, -endo sugar puckering. *Tetrahedron Lett.* **1997**, *38*, 8735–8738. [CrossRef]
47. Ni, S.J.; Yao, H.Z.; Wang, L.L.; Lu, J.; Jiang, F.; Lu, A.P.; Zhang, G. Chemical Modifications of Nucleic Acid Aptamers for Therapeutic Purposes. *Int. J. Mol. Sci.* **2017**, *18*, 1683. [CrossRef] [PubMed]
48. Zhu, R.; Canena, D.; Sikora, M.; Klausberger, M.; Seferovic, H.; Mehdipour, A.R.; Hain, L.; Laurent, E.; Monteil, V.; Wirnsberger, G.; et al. Force-tuned avidity of spike variant-ACE2 interactions viewed on the single-molecule level. *Nat. Commun.* **2022**, *13*, 7926. [CrossRef] [PubMed]
49. Iketani, S.; Liu, L.H.; Guo, Y.C.; Liu, L.Y.; Chan, J.F.W.; Huang, Y.M.; Wang, M.; Luo, Y.; Yu, J.; Chu, H.; et al. Antibody evasion properties of SARS-CoV-2 Omicron sublineages. *Nature* **2022**, *604*, 553–556. [CrossRef]
50. Magazine, N.; Zhang, T.Y.; Wu, Y.Y.; McGee, M.C.; Veggiani, G.; Huang, W.S. Mutations and Evolution of the SARS-CoV-2 Spike Protein. *Viruses* **2022**, *14*, 640. [CrossRef]
51. McCallum, M.; Walls, A.C.; Sprouse, K.R.; Bowen, J.E.; Rosen, L.; Dang, H.; DeMarco, A.; Franko, N.; Tilles, S.W.; Logue, J.; et al. Molecular basis of immune evasion by the Delta and Kappa SARS-CoV-2 variants. *Science* **2021**, *374*, 1621–1626. [CrossRef]
52. Cao, Y.L.; Wang, J.; Jian, F.C.; Xiao, T.H.; Song, W.L.; Yisimayi, A.; Huang, W.J.; Li, Q.Q.; Wang, P.; An, R.; et al. Omicron escapes the majority of existing SARS-CoV-2 neutralizing antibodies. *Nature* **2022**, *602*, 657–663. [CrossRef]
53. Chakraborty, C.; Bhattacharya, M.; Sharma, A.R.; Mallik, B. Omicron (B.1.1.529)—A new heavily mutated variant: Mapped location and probable properties of its mutations with an emphasis on S-glycoprotein. *Int. J. Biol. Macromol.* **2022**, *219*, 980–997. [CrossRef] [PubMed]
54. Tegally, H.; Moir, M.; Everatt, J.; Giovanetti, M.; Scheepers, C.; Wilkinson, E.; Subramoney, K.; Makatini, Z.; Moyo, S.; Amoako, D.G.; et al. Emergence of SARS-CoV-2 Omicron lineages BA.4 and BA.5 in South Africa. *Nat. Med.* **2022**, *28*, 1785–1790. [CrossRef] [PubMed]
55. Lai, X.L.; Soler-Lopez, M.; Wichers, H.J.; Dijkstra, B.W. Large-Scale Recombinant Expression and Purification of Human Tyrosinase Suitable for Structural Studies. *PLoS ONE* **2016**, *11*, e016169. [CrossRef] [PubMed]
56. Gotzke, H.; Kilisch, M.; Martinez-Carranza, M.; Sograte-Idrissi, S.; Rajavel, A.; Schlichthaerle, T.; Engels, N.; Jungmann, R.; Stenmark, P.; Opazo, F.; et al. The ALFA-tag is a highly versatile tool for nanobody-based bioscience applications. *Nat. Commun.* **2019**, *10*, 4403. [CrossRef]
57. Lim, C.S.; Jang, Y.H.; Lee, G.Y.; Han, G.M.; Jeong, H.J.; Kim, J.W.; Lee, J.O. TLR3 forms a highly organized cluster when bound to a poly(I:C) RNA ligand. *Nat. Commun.* **2022**, *13*, 6876. [CrossRef]
58. Rodriguez-Hernandez, A.P.; Martinez-Flores, D.; Cruz-Resendiz, A.; Padilla-Flores, T.; Gonzalez-Flores, R.; Estrada, K.; Sampieri, A.; Camacho-Zarco, A.R.; Vaca, L. Baculovirus Display of Peptides and Proteins for Medical Applications. *Viruses* **2023**, *15*, 411. [CrossRef]
59. Bepler, T.; Morin, A.; Rapp, M.; Brasch, J.; Shapiro, L.; Noble, A.J.; Berger, B. Positive-unlabeled convolutional neural networks for particle picking in cryo-electron micrographs. *Nat. Methods* **2019**, *16*, 1153–1160. [CrossRef]
60. Punjani, A.; Fleet, D. 3D Flexible Refinement: Structure and Motion of Flexible Proteins from Cryo-EM. *Microsc. Microanal.* **2022**, *288*, 1218. [CrossRef]
61. Kappel, K.; Zhang, K.; Su, Z.; Watkins, A.M.; Kladwang, W.; Li, S.; Pintilie, G.; Topkar, V.V.; Rangan, R.; Zheludev, I.N.; et al. Accelerated cryo-EM-guided determination of three-dimensional RNA-only structures. *Nat. Methods* **2020**, *17*, 699–707. [CrossRef]
62. Pettersen, E.F.; Goddard, T.D.; Huang, C.C.; Couch, G.S.; Greenblatt, D.M.; Meng, E.C.; Ferrin, T.E. UCSF chimera—A visualization system for exploratory research and analysis. *J. Comput. Chem.* **2004**, *25*, 1605–1612. [CrossRef]

63. Liebschner, D.; Afonine, P.V.; Baker, M.L.; Bunkoczi, G.; Chen, V.B.; Croll, T.I.; Hintze, B.; Hung, L.W.; Jain, S.; McCoy, A.J.; et al. Macromolecular structure determination using X-rays, neutrons and electrons: Recent developments in Phenix. *Acta Crystallogr. D Struct. Biol.* **2019**, *75*, 861–877. [CrossRef] [PubMed]
64. Dhawan, M.; Sharma, A.; Priyanka; Thakur, N.; Rajkhowa, T.K.; Choudhary, O.P. Delta variant (B.1.617.2) of SARS-CoV-2: Mutations, impact, challenges and possible solutions. *Hum. Vacc. Immunother.* **2022**, *18*, 2068883. [CrossRef] [PubMed]
65. Li, S.X.; Olson, W.K.; Lu, X.J. Web 3DNA 2.0 for the analysis, visualization, and modeling of 3D nucleic acid structures. *Nucleic Acids Res.* **2019**, *47*, W26–W34. [CrossRef] [PubMed]
66. Mills, J.E.J.; Dean, P.M. Three-dimensional hydrogen-bond geometry and probability information from a crystal survey. *J. Comput. Aid. Mol. Des.* **1996**, *10*, 607–622. [CrossRef] [PubMed]
67. Saenger, W. *Principles of Nucliec Acid Structure*; Springer: New York, NY, USA, 1984; pp. 120–121.
68. Wang, D.L.; Zhang, S.Y.; Li, L.A.; Liu, X.; Mei, K.R.; Wang, X.Q. Structural insights into the assembly and activation of IL-1 beta with its receptors. *Nat. Immunol.* **2010**, *11*, 905–911. [CrossRef]
69. Bao, L.; Zhang, X.; Jin, L.; Tan, Z.J. Flexibility of nucleic acids: From DNA to RNA. *Chinese Phys. B* **2016**, *25*. [CrossRef]
70. Zuker, M. Mfold web server for nucleic acid folding and hybridization prediction. *Nucleic Acids Res.* **2003**, *31*, 3406–3415. [CrossRef]

molecules

Article

Electrochemiluminescence Aptasensor with Dual Signal Amplification by Silica Nanochannel-Based Confinement Effect on Nanocatalyst and Efficient Emitter Enrichment for Highly Sensitive Detection of C-Reactive Protein

Ning Ma [1,2], Shuai Xu [3], Weidong Wu [1,2,*] and Jiyang Liu [3,*]

1 Shanxi Bethune Hospital, Shanxi Academy of Medical Sciences, Tongji Shanxi Hospital, Third Hospital of Shanxi Medical University, Taiyuan 030032, China; mnsxicu@163.com
2 Tongji Hospital, Tongji Medical College, Huazhong University of Science and Technology, Wuhan 430030, China
3 School of Chemistry and Chemical Engineering, Zhejiang Sci-Tech University, Hangzhou 310018, China; 202120104178@mails.zstu.edu.cn
* Correspondence: wuweidong@sxmu.edu.cn (W.W.); liujy@zstu.edu.cn (J.L.)

Abstract: The rapid and sensitive detection of the important biomarker C-reactive protein (CRP) is of great significance for monitoring inflammation and tissue damage. In this work, an electrochemiluminescence (ECL) aptasensor was fabricated based on dual signal amplification for the sensitive detection of CRP in serum samples. The sensor was constructed by modifying a silica nanochannel array film (SNF) on a cost-effective indium tin oxide (ITO) electrode using the Stöber solution growth method. Gold nanoparticles (AuNPs) were grown in situ within the nanochannels using a simple electrodeposition method as a nanocatalyst to enhance the active electrode area as well as the ECL signal. The negatively charged nanochannels also significantly enriched the positively charged ECL emitters, further amplifying the signal. The recognition aptamer was covalently immobilized on the outer surface of SNF after modification with epoxy groups, constructing the aptasensor. In the presence of CRP, the formation of complexes on the recognitive interface led to a decrease in the diffusion of ECL emitters and co-reactants to the supporting electrode, resulting in a reduction in the ECL signal. Based on this mechanism, ECL detection of CRP was achieved with a linear range of 10 pg/mL to 1 μg/mL and a low limit of detection (7.4 pg/mL). The ECL aptasensor developed in this study offers advantages such as simple fabrication and high sensitivity, making promising applications in biomarker detection.

Keywords: electrochemiluminescence aptasensor; dual dignal amplification; silica nanochannel array film; confinement; C-reactive protein

Citation: Ma, N.; Xu, S.; Wu, W.; Liu, J. Electrochemiluminescence Aptasensor with Dual Signal Amplification by Silica Nanochannel-Based Confinement Effect on Nanocatalyst and Efficient Emitter Enrichment for Highly Sensitive Detection of C-Reactive Protein. *Molecules* **2023**, *28*, 7664. https://doi.org/10.3390/molecules28227664

Academic Editors: Sai Wang, Wei Wu, Long Wu and Alejandro Baeza

Received: 20 September 2023
Revised: 10 November 2023
Accepted: 17 November 2023
Published: 19 November 2023

1. Introduction

C-reactive protein (CRP) is an acute phase protein induced by cytokines, primarily synthesized in the liver and regulated by interleukin [1,2]. In healthy individuals, CRP levels are very low (<1 μg/mL), but they rapidly increase during infections, tissue damage, or inflammation. Specifically, serum CRP concentration can rise to more than 500 μg/mL at about 6 h and peak at about 48 h, and the half-life is about 19 h [3]. For instance, sepsis is a severe infectious disease [4]. In patients with sepsis, the levels of CRP are typically elevated. This is because the infection stimulates the immune system to produce an inflammatory response, leading to an increase in CRP synthesis. Therefore, CRP is commonly used as an important indicator, assisting in the diagnosis and treatment monitoring of sepsis. In addition, CRP has been widely used as a non-specific biomarker for inflammation and tissue damage, such as cardiovascular diseases, in clinical medicine [5–7]. The American Heart Association (AHA) and the Centers for Disease Control and Prevention (CDC) use serum CRP levels to assess the risk level of cardiovascular diseases. Therefore, the rapid and reliable determination of CRP is of great significance.

Given the value of CRP as a pathological diagnostic indicator in clinical applications, CRP detection has been a research hotspot in the fields of medicine and bioanalytical chemistry. Immunoassay based on the principle of antigen-antibody binding is currently the mainstream technique for CRP clinical monitoring [8–10]. However, antibodies themselves have certain limitations, such as complex preparation processes, large batch-to-batch variability, and susceptibility to variability. In recent years, an aptamer-based CRP analysis has attractive attentions. Aptamers are typically short single-stranded oligonucleotide (DNA or RNA) sequences, ranging in length from 25 to 80 nucleotides [11]. In addition to high binding affinity and specificity, aptamers have advantages over traditional antibodies, including ease of modification, good stability, low manufacturing costs, non-toxicity, minimal batch-to-batch variation, reversible folding, and lack of immunogenicity [12]. They also have a wide range of target molecules, including small molecules, peptides, proteins, cells, and tissues. These characteristics make aptamers highly selective bio-recognition elements that have attracted wide attention in constructing various types of biosensors [13–15]. During the binding between the recognitive aptamers and target analytes, detection based on aptasensors can be achieved through monitoring optical signals such as fluorescence [16] and surface plasmon resonance (SPR) [17], or electrochemical signals like impedance spectroscopy (EIS), current and voltage changes [18,19], or electrochemiluminescence (ECL) signals [20,21]. Among these, ECL aptaensors offer advantages such as low background noise and high sensitivity. The development of convenient ECL aptasensors for the efficient detection of CRP is highly desirable.

Introducing a catalyst is an effective method to enhance the ECL signal. Commonly used catalysts include metal nanoparticles, enzymes, or organic molecules [22–25]. These catalysts can undergo specific reactions with an ECL emitter or co-reactor, resulting in stronger luminescent signals. In addition, catalysts can alter the kinetics of ECL reaction, making it more efficient. For instance, gold nanoparticles (AuNPs) possess a large surface area and abundant surface-active sites, promoting electron transfer processes [26] and accelerating the excitation/emission process. It has also been proven that the confinement of metal nanoparticles catalysts within nanospaces provides an effective approach for developing catalysts with high performance [27–29]. This is attributed to the fact that confining metal nanoparticles within nanospaces could enhance catalytic activity, selectivity, controllability, and stability. Specifically, precise control over the size and structure of metal nanoparticles catalysts can be achieved by adjusting parameters such as the size, shape, and chemical environment of the nanospaces [30–32]. For instance, confining metal nanoparticles within nanospaces provides a smaller size, higher specific surface area and abundant active sites. This facilitates improved contact and reaction rates between the catalyst and reactants, thereby enhancing catalytic activity. At the same time, restricting the diffusion and movement of reactants within nanospaces allows for control over the local environment of the reaction, leading to enhanced selectivity of the reaction. In addition, nanospace confinement can form a protective layer, reducing the interaction between the metal catalyst and the external environment or reactants, and minimizing the deactivation of active sites [33–36]. This extends the lifetime of the nanocatalyst and enhances its stability in complex reaction systems. Therefore, the utilization of metal nanoparticles confined within nanospaces holds promise for constructing highly sensitive ECL sensing platforms.

The loading and dispersion of metal nanoparticles is also a major difficulty. Porous materials have extensive applications in fields such as separation, catalysis, energy, and sensing due to their high specific surface area, permeability, tunability, and strength stability [37–39]. Silica nanochannel array film (SNF) is a thin nanofilm with nanoscale channel-like structures, offering tunability, high selectivity, and stability [40–43]. On the one hand, the pore size and film thickness of SNF can be adjusted by controlling the preparation conditions. The uniformly distributed nanochannels enable selective filtration, separation, or transport of molecules or ions of specific sizes or shapes [44–47]. This high selectivity gives them great potential in selective enrichment, separation, and filtration applications. For instance, the ionization of silicon hydroxyl groups (pK_a~2–3) on the SNF surface results in a negatively charged surface that can significantly enrich positively charged ions [48–51]. Furthermore, the high specific surface area of the densely packed ultrasmall nanochannels enhances this enrichment effect [52]. On the other hand, SNF exhibits excellent chemical and thermal stability due to their robust silica structure. Thus, SNF exhibits great potential in constructing high-performance ECL sensing platforms.

In this study, an electrochemiluminescence (ECL) aptasensor platform was constructed by combining the confinement effect of silica nanochannel array film (SNF) on nanocatalysts and the efficient enrichment of ECL emitters, which can be applied for the highly sensitive determination of CRP. Using a cost-effective and readily available ITO electrode as the supporting electrode, SNF was grown on ITO surface through the Stöber solution growth method. Confined gold nanoparticles (AuNPs) were in situ electrodeposited within the nanochannels as nanocatalysts. After the outer surface of SNF were derivatized by the epoxide group, the recognitive interface could be obtained after covalent immobilization of amino-modified aptamers followed by blocking the non-specific sites. The ECL signal could be enhanced combined with the electrostatic attraction of NH_2-SNF towards the ECL emitter, $Ru(bpy)_3^{2+}$, and the increased electrode active area as well as a catalytical process by AuNPs. Because the formed aptamer-CRP complex hinders the diffusion of the ECL emitter and co-reactant into the nanochannel array as well as the underlying electrode, it leads to a decrease in the ECL signal. Based on this principle, the ECL determination of CRP can be achieved. The constructed aptasensor also exhibits the advantages of a simple preparation method, high sensitivity, stability, and selectivity.

2. Results and Discussion

2.1. Construction of Aptasensor and CRP Detection Based on Dual Signal Amplification

The construction of the aptamer-based sensor and ECL determination of CRP based on dual signal amplification are shown in Figure 1. Firstly, a layer of SNF is grown on the patterned ITO electrode surface using the Stöber solution growth method. The growth mechanism involves the self-assembly of siloxane precursors under the surfactant micelle (SM) template, resulting in SM within the nanochannels of the obtained electrode (SM@SNF/ITO). Subsequently, the electrode is epoxy-functionalized. Since the nanochannels are sealed by SM, the reaction occurs selectively on the outer surface of SNF, leading to the electrode being abbreviated as SM@O-SNF/ITO. The sealed SM prevents an epoxy reaction inside the nanochannels, avoiding a blockage of the nanochannels. After removing SM, open nanochannels are obtained (O-SNF/ITO).

By placing the O-SNF/ITO electrode in a chloroauric acid solution for in situ electrodeposition, gold nanoparticles are deposited inside the nanochannels [34], resulting in the AuNPs@O-SNF/ITO electrode. Then, amino-functionalized aptamers undergo a covalent binding with the epoxy groups on the external surface of the nanochannels, forming the recognition and capture interface for CRP. After blocking the non-specific active site with BSA, the aptasensor, BSA/Apt/AuNPs@O-SNF/ITO, is constructed. The main role of BSA is to prevent non-specific binding through blocking the remaining space on the electrode surface after fixing the recognitive ligands. When the target substance, CRP, is present, it binds to the aptamer on the electrode surface. The formed large-sized complex obstructs the diffusion of the ECL probe $Ru(bpy)_3^{2+}$ and the co-reactant tri-n-propylamine (TPrA) through the nanochannels to reach the underlying electrode, resulting in a decrease in ECL signal. Based on this mechanism, ECL detection of CRP can be achieved.

Figure 1 also illustrates the dual signal amplification during ECL detection of CRP. On one hand, SNF features a vertically oriented array of silica nanochannels on the supporting electrode. The inner wall of nanochannel contains a large number of silanol groups with low pK_a (2~3), which ionize in common buffer systems, forming negatively charged sites that significantly enrich the positively charged ECL probe $Ru(bpy)_3^{2+}$ through electrostatic attraction, leading to a significantly enhanced ECL signal [53]. On the other hand, AuNPs deposited inside the nanochannels serve as nanocatalysts to enhance the signal of the $Ru(bpy)_3^{2+}$/TPrA system [28]. When CRP binds to the aptamer to form a complex on the outer surface of SNF (e.g., the entrance of nanochannel), the steric hindrance effect prevents the ECL emitter $Ru(bpy)_3^{2+}$ and the co-reactive agent TPrA from reaching the electrode surface, affecting the mass transfer and electron transfer of the probe. Thus, the ECL signal decreases. Based on this mechanism, CRP can be detected.

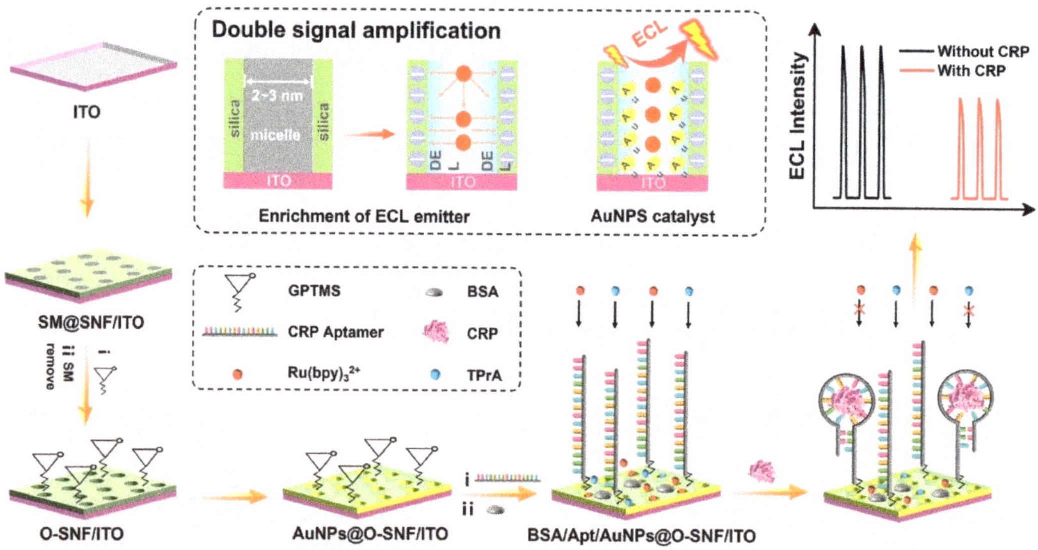

Figure 1. Illustration of the fabrication of the aptasensor and ECL detection of CRP based on dual signal amplification.

2.2. Characterization of the SNF/ITO and AuNPs@SNF/ITO Electrodes

The electrochemical properties of the SNF film grown on the electrode surface were characterized. The electrochemical signals of three redox probes with various charges including (neutral, Fc-MeOH; negatively charged $Fe(CN)_6^{3-}$, positively charged $Ru(NH_3)_6^{3+}$) were used to investigate the integrity and charge-selective permeability of the SNF film (Figure 2a–c). When SM@O-SNF/ITO was placed in a FcMeOH probe solution, SM@O-SNF/ITO exhibited an extraction effect on FcMeOH due to the hydrophobic interactions between FcMeOH and SM [40]. As shown in Figure 2a, the SM@O-SNF/ITO electrode showed a pair of oxidation-reduction peaks, with the peak potential shifting towards the positive direction compared to that obtained on an ITO electrode. Since the oxidation product of FcMeOH carries a positive charge and can be enriched by the negative-charged nanochannels, the reduction peak current of FcMeOH on O-SNF/ITO is greater than the oxidation peak current [40]. Due to the presence of hydrophobic SM inside the nanochannels, SM@O-SNF/ITO shows no signal in the solutions of $Fe(CN)_6^{3-}$ (Figure 2b) and $Ru(NH_3)_6^{3+}$ (Figure 2c) probes, only exhibiting a charging current. This indicates the complete coverage of SNF on the ITO surface with SM-blocked nanochannels. After the removal of SM, the electrode displays significant Faradaic current in the charged redox probes. Compared to ITO, SNF/ITO exhibits apparent suppression of Faradaic current in the anionic $Fe(CN)_6^{3-}$ solution. On the contrary, an enhancement current is observed in the case of cationic $Ru(NH_3)_6^{3+}$ probes. This is because the SNF surface and nanochannels contain numerous silanol groups, causing a negative charge in the detection medium. This negative charge can repel anions while attracting cations through electrostatic interactions, demonstrating significant charge-selective permeability. The electrochemical impedance spectroscopy (EIS) plots of the different electrode are shown in Figure 2d. As shown, SM@SNF/ITO electrode displays a significantly high electron transfer resistance (*R*et) of 7057 Ω, which results from hindered electron transfer between the probe and the electrode due to the sealing of nanochannels by SM. When SM was removed, *R*et obtained on SNF/ITO remarkably decreased (336 Ω) because of the open nanochannels, which is also higher than that of ITO (29 Ω) owing to the silica structure. These results also prove the successful modification of SNF on the ITO electrode.

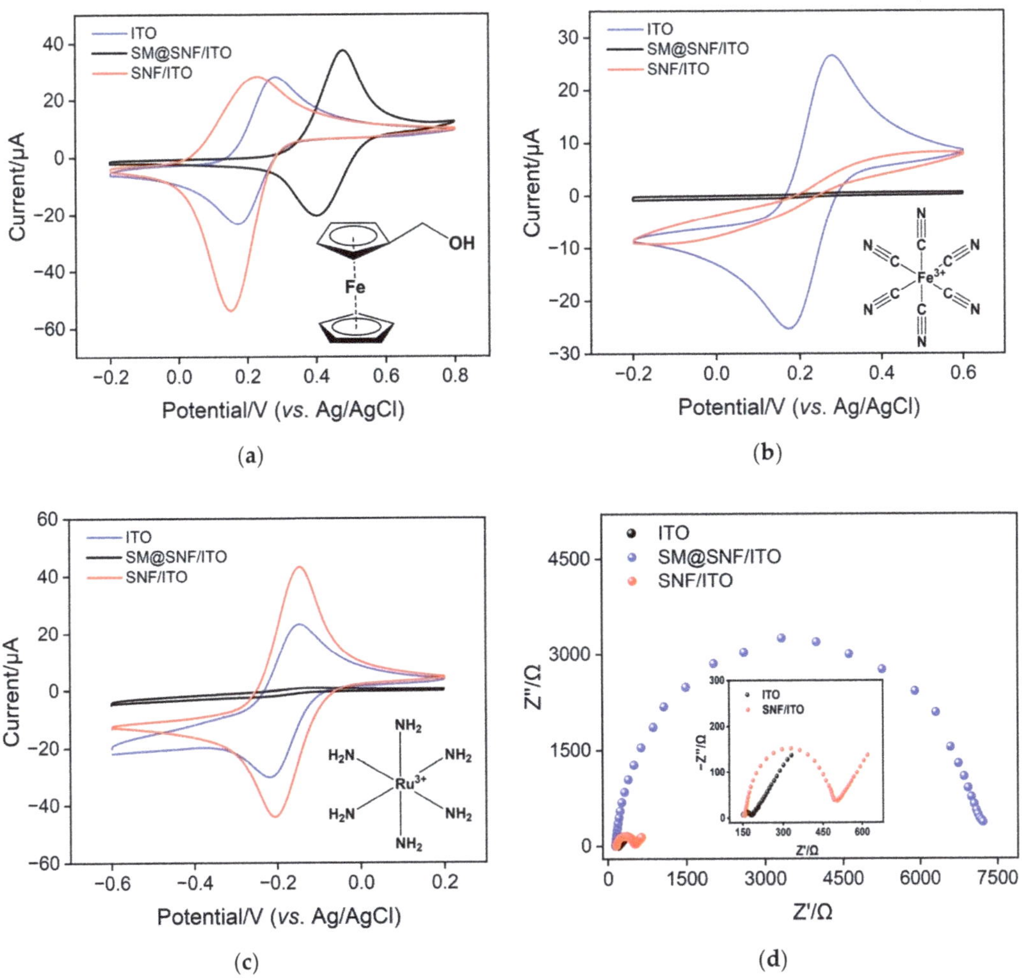

Figure 2. CV curves obtained on ITO, SM@SNF/ITO, and SNF/ITO in 50 mM KHP containing 0.5 mM (**a**) Fc-MeOH, (**b**) $Fe(CN)_6^{3-}$ and (**c**) $Ru(NH_3)_6^{3+}$. (**d**) EIS plots of SM@SNF/ITO and SNF/ITO in a 0.1 M KCl solution containing 2.5 mM $Fe(CN)_6^{3-/4-}$. The scan rate in (a), (b) and (c) was 0.05 V/s.

The morphology of SNF-modified ITO electrodes was characterized using scanning electron microscopy (SEM). As commonly known, the ITO electrode, also known as ITO conductive glass, is formed by depositing an ITO layer on the surface of glass. In this work, ITO electrodes were used as the support electrode and further modified with SNF. Figure 3a gives the SEM image of the cross-section of the SNF/ITO electrode. As shown, ITO/SNF displays a three-layer structure including the glass substrate, the ITO layer, and the SNF layer from bottom to top. Since ITO electrodes cannot be characterized by transmission electron microscopy (TEM), SNF was carefully scraped off using a scalpel, followed by ultrasonication, and then drop-cast onto a copper grid for TEM characterization. Figure 3b depicts the cross-sectional TEM images of two separate layers of SNF. Each SNF is composed of mutually parallel nanochannels with a thickness of approximately 100 nm. Figure 3c displays the top-view TEM image of SNF. As observed, SNF has densely packed nanochannels, as each bright spot represents a nanochannel. The average diameter of the nanochannels is 2.7 nm. Within the observed range, SNF shows no cracks.

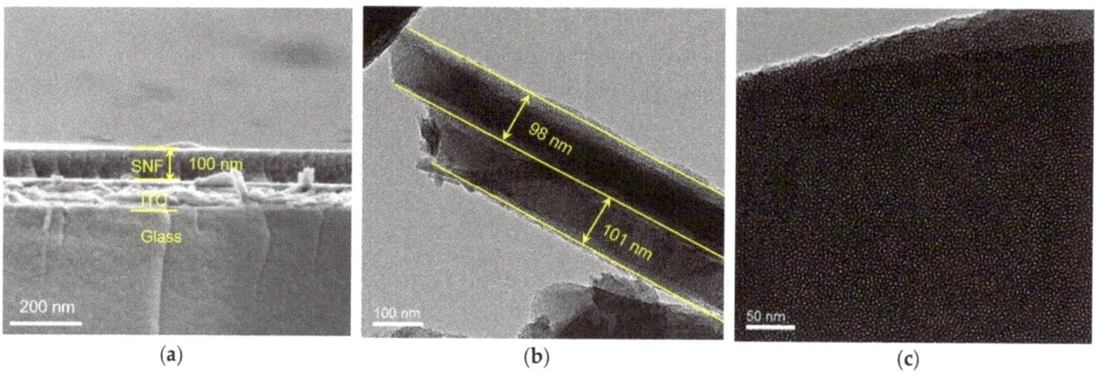

Figure 3. (**a**) SEM image of the cross-section of SNF/ITO. The cross-sectional (**b**) and top-view (**c**) TEM images of SNF.

To confirm the successful deposition and localization of AuNPs, the SNF/ITO electrode before and after AuNPs' deposition (AuNPs@SNF/ITO) was investigated using cyclic voltammetry (CV) and X-ray photoelectron spectroscopy (XPS). The CV curves of both electrodes are shown in Figure 4a. The SNF/ITO electrode exhibits virtually no signal in 0.1 M of sulfuric acid, while AuNPs@SNF/ITO shows characteristic oxidation-reduction peaks of gold at 1.18 V and 0.96 V, respectively. In comparison with SNF/ITO, characteristic peaks of Au_{4f} appear in the XPS spectrum for AuNPs@SNF/ITO, further confirming the successful modification of AuNPs (Figure 4b). Figure 4c,d show the CV curves of SNF/ITO and AuNPs@SNF/ITO electrodes in the absence or presence of different probe solutions. It can be found that the charging current and peak current of the AuNPs@SNF/ITO electrode are higher than those of the SNF/ITO electrode for both positive and negative probes. This indicates that the accessible electrode area is improved after the modification of AuNPs.

Due to the ultrasmall nanochannels of SNF, the size of in situ grown AuNPs is relatively small. The surface morphology of the obtained electrode was determined by dissolving the SNF membrane. Figure 5a,b show SEM and elemental mapping images of the electrode surface after removing SNF using sodium hydroxide solution. From the images, it can be observed that the nanoparticles appear irregular and spiky, and they tend to aggregate (Figure 5a). The elemental mapping images confirm the Au element (Figure 5b). As SNF is dissolved, the small AuNPs undergo aggregation, leading to the formation of larger-sized gold materials. SEM images of SNF/ITO before and after AuNPs deposition have been investigated. SNF/ITO surfaces remain smooth and flat before (Figure 5c) and after (Figure 5d) modification of AuNPs. Thus, AuNPs exposed outside the nanochannels are not observed. After combining the signal peak for the gold element in XPS characterization, which has a testing depth of only a few to several tens of nanometers, and the gold element mapping result after dissolving the SNF film, it is speculated that the deposited AuNPs are on the electrode surface or within the nanochannels.

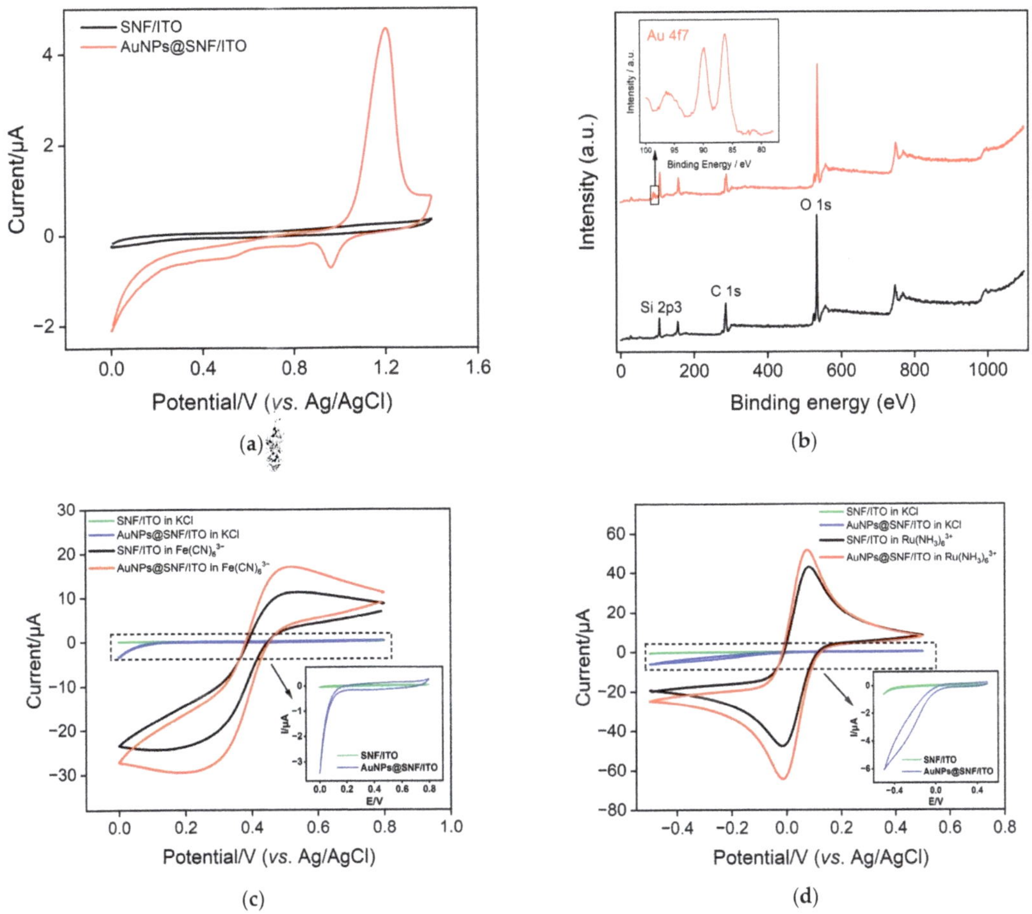

Figure 4. (**a**) CV curves of SNF/ITO and AuNPs@SNF/ITO in 0.1 M H_2SO_4 solution. (**b**) XPS profiles of SNF/ITO (black) and AuNPs@SNF/ITO (red). (**c**) CV curves of SNF/ITO and AuNPs@SNF/ITO in KCl solutions in absence or presence of 1 mM (**c**) $Fe(CN)_6^{3-}$ and (**d**) $Ru(NH_3)_6^{3+}$. Inset is the amplified figure of curves without probe.

2.3. Dual Signal Amplification Based on Enrichment of ECL Emitter and AuNPs' Nanocatalyst

Figure 6a presents the CV curves obtained on ITO, SNF/ITO, and AuNPs@SNF/ITO in a $Ru(bpy)_3^{2+}$/TPrA solution. No oxidation peak for gold around 1.2 V was observed. Additionally, the CV curve of AuNPs@SNF/ITO in blank PBS has also been added as an inset of Figure 6a. No gold oxidation peak at ~1.2 V was observed and only a weak reduction peak with very low current near 1.05 V was observed. These results confirm that the deposited gold remains stable during the ECL measurement and does not produce interfering electrochemical signals. The corresponding ECL signals measured on the three types of electrodes are shown in Figure 6b. Compared to bare ITO, the CV oxidation peak of the SNF-modified electrode significantly increases. It can also be observed that SNF/ITO exhibits a 12.4-fold enhancement in ECL intensity compared to the ITO electrode, resulting from the enrichment of negatively charged nanochannels on positively charged $Ru(bpy)_3^{2+}$. Thus, the significant enrichment of the ECL emitter by nanochannels can markedly improve the ECL signal. When AuNPs were further in situ modified, the oxidation-reduction peak current in the CV curve further increased, and the ECL intensity increased by an additional

34% compared to SNF/ITO, demonstrating that AuNPs can act as nanocatalysts to enhance the ECL signal. This may be attributed to the electrocatalytic oxidation of TPrA by AuNPs, generating more TPrA cationic-free radicals ($TPrA^{+}$), promoting the generation of more excited states and enhancing the ECL signal.

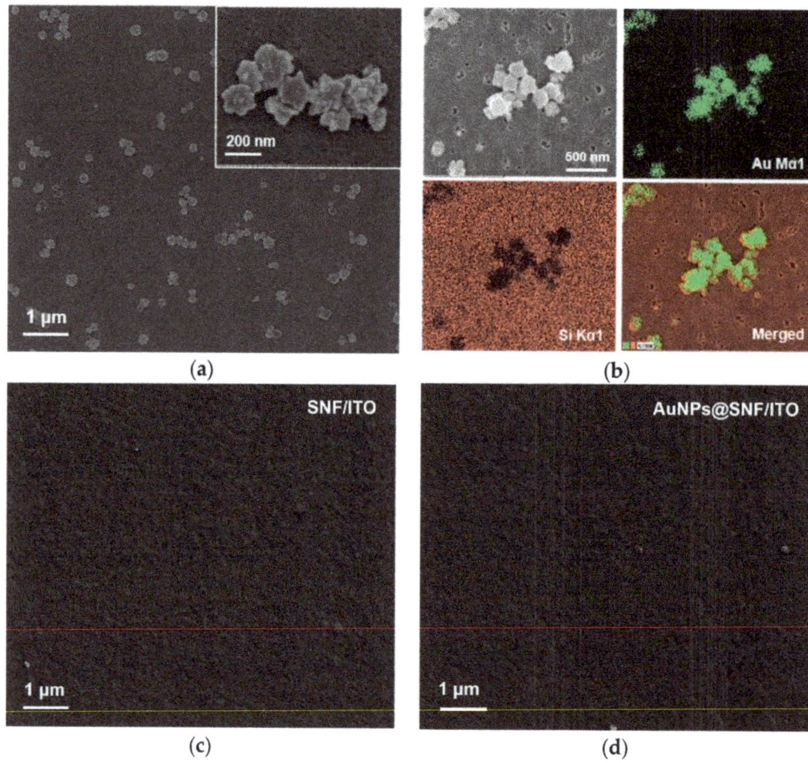

Figure 5. (**a**) SEM image of the electrode surface after removal of SNF. Inset is the corresponding high-resolution SEM image. (**b**) Elemental mapping of the surface of AuNPs@SNF/ITO after dissolving SNF. The image in the top left is the SEM photograph of the region where element mapping was conducted. The top right image is the gold element mapping, the bottom left image is the silicon element mapping. Green and red correspond to signal points of Au or Si elements, respectively. The bottom right image is an overlay of gold and silicon mapping signals. (**c**) SEM image of SNF/ITO. (**d**) SEM overhead image of AuNPs@SNF/ITO.

2.4. Feasibility of the Construction of the Aptasensor

To investigate the feasibility of fabricating the aptasensor, the electrodes obtained in the stepwise modifications were investigated through electrochemical methods. As shown in Figure 6c,d, the signal slightly decreased. Upon covalently attaching the aptamer to the surface of the AuNPs@O-SNF/ITO electrode, the ECL signal of the Apt/AuNPs@O-SNF/ITO electrode decreased. This can be attributed to the potential hindrance of the aptamer attachment, which may obstruct the entry of co-reactants and ECL probes into the nanochannels, resulting in a reduction in the ECL signal.

After blocking non-specific sites with BSA, the electrode surface was covered with a non-conductive, large-sized protein layer, further impeding the transfer of probes and electron transfer, resulting in a further decrease in the ECL signal of BSA/Apt/AuNPs@O-SNF/ITO. In the presence of CRP, the formation of complexes due to CRP binding to the aptamer further hindered probe transfer and electron transfer, causing a significant reduction in the ECL signal. These results confirm the successful construction of the aptasensor and the feasibility of CRP detection.

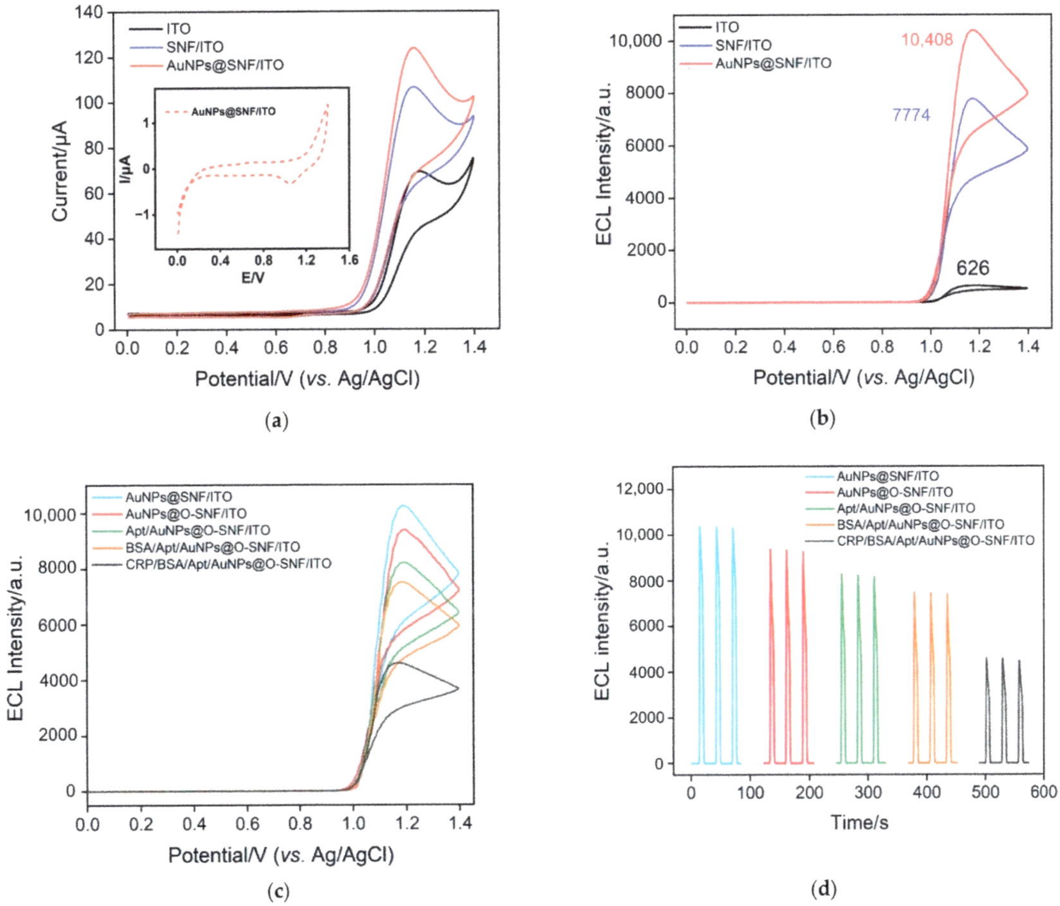

Figure 6. (**a**) CV curves obtained on ITO, SNF/ITO and AuNPs@SNF/ITO electrodes in 0.01 M PBS (pH 7.4) solution containing 10 μM $Ru(bpy)_3^{2+}$ and 3 mM TPrA. Inset is the CV of AuNPs@SNF/ITO in blank PBS. (**b**) ECL signal plots of ITO SNF/ITO and AuNPs@SNF/ITO electrodes in 0.01 M PBS (pH 7.4) solution containing 10 μM $Ru(bpy)_3^{2+}$ and 3 mM TPrA. ECL intensity-potential curves (**c**) and ECL intensity-time curves (**d**) were obtained at different electrodes. Scan rate of CV: 100 mV s^{-1}. Scan potential: 0–1.4 V. Photomultiplier tube (PMT) voltage was 400 V.

2.5. Optimization of Experimental Conditions

To achieve high detection performance of the fabricated aptasensor, certain relevant conditions for the experiment were optimized. The deposition time of AuNPs, the concentration of the aptamer for the fabrication of the recognitive interface, the incubation time for CRP, and the results are shown in Figure 7. As observed in Figure 7a, compared to the electrode without AuNPs (deposition time of 0 s), the ECL signals of the electrode exhibit an initial increase followed by a decrease as the AuNPs deposition time increases. The increase in ECL signal is attributed to the enhanced electrode active surface area due to the presence of AuNPs, which also catalyze the ECL luminescence process. As a prolonged deposition time may lead to blockage within the nanochannels, resulting in reduced enrichment of $Ru(bpy)_3^{2+}$ by the nanochannels and subsequently a decrease in the ECL signal of the sensor, the optimal deposition time for AuNPs was chosen at 3 s (Figure 7a). The modification of the bio-recognitive interface is crucial for the fabrication of the aptasensor, taking into account both time and cost factors. When the incubation time for aptamer immobilization was fixed at 90 min [54], the aptamer concentration was optimized. As shown in Figure 7b, the fixed incubation time was 30 min, and ECL signal decreases with increasing aptamer concentration and stabilizes at around 0.3 μM. Further increasing the aptamer concentration only results in minimal changes in the ECL signal. Thus, the aptamer concentration of 0.3 μM was chosen as the optimal condition. Similarly, in the case of antigen concentration of 1 ng/mL, the optimal incubation time for CRP, as obtained through optimization, was 80 min (Figure 7c).

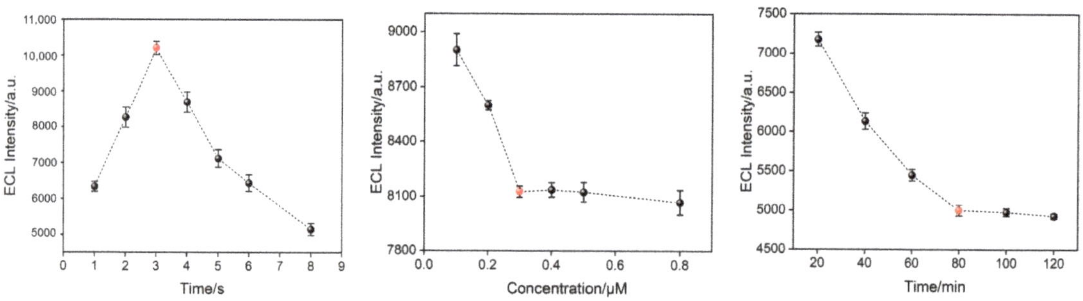

(a) (b) (c)

Figure 7. (**a**) Effect of deposition time of AuNPs on ECL signal of AuNPs@O-SNF/ITO electrode. (**b**) Optimization of the aptamer concentration on the ECL signal. (**c**) Optimization of CRP (1 ng/mL) incubation time on the ECL signal. Electrolyte solution: 0.01 M PBS (pH 7.4) solution containing 10 μM $Ru(bpy)_3^{2+}$ and 3 mM TPrA. Scan rate: 100 mV s^{-1}. Scanning potential: 0~1.4 V. PMT = 400 V. The red dots represent the optimized conditions.

2.6. ECL Determination of CRP

Under optimal conditions, the proposed aptasensor exhibits a negative correlation between ECL intensity and CRP concentration (Figure 8a). Commonly, logarithmic processing is usually performed on the electrochemical data to explore a good coefficient of determination across a wide concentration range [55]. As observed from Figure 8b, within the range from 10 pg/mL to 1 μg/mL, the ECL signal (I_{ECL}) demonstrates a linear relationship with the logarithm of CRP concentration ($logC_{CRP}$). The corresponding linear regression equation ($I_{ECL} = -863.3\ logC_{CRP} + 5717$) exhibits a high linear correlation coefficient of 0.9995. The detection limit (DL) calculated using three signals to noise (S/N = 3) is 7.4 pg/mL. Comparison of CRP detection performance with other sensors is shown in Table S1 (SI) [18,56–61]. The LOD of the fabricated aptasensor is lower than that of electrochemical sensors using square wave voltammetry [56] or differential pulse voltammetry [57], or ECL immunosensors based on platinum nanowire/titania nanotube composites [58] or Ir(III) compound and β-cyclodextrin complex [59]. The LOD is higher than that of the ECL immunosensor based on the resonance energy transfer of $Ru(bpy)_3^{2+}@Cu_3(HHTP)_2$ and graphene oxide-gold nanoparticles (GO-AuNPs) [60], an ECL immunosensor using $Ru(bpy)_3^{2+}$-labeled AuNPs' modified screen-printed electrode (SPE) combined with a flow measurement [61], or an electrochemical aptasensor based on AuNPs anchored in covalent poly deep eutectic solvent functionalized graphene-modified glassy carbon electrode (Apt/AuNPs/GO/PDES/GCE) [18]. However, the fabricated aptasensor does not conducted a complex preparation of nanocomposite materials or labelled bioligands and exhibits a wider detection range.

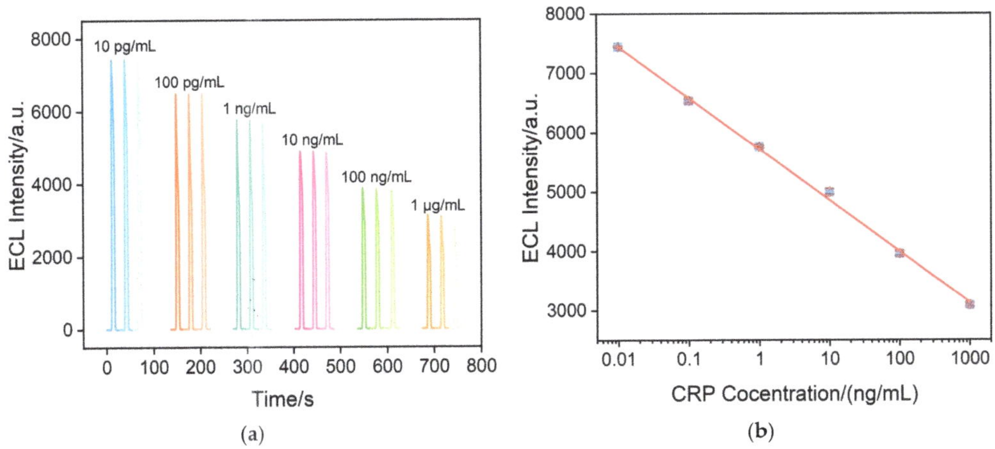

Figure 8. (**a**) ECL intensity of the aptasensor at different CRP concentrations. (**b**) Calibration curves of ECL intensity versus CRP concentration in the range from 0.01 to 1000 ng/mL. The error bars show the standard deviation of measurements taken from three experiments.

2.7. Specificity, Selectivity, Repeatability and Stability of the Fabricated Aptasensors

Other aptamers with an amino label, such as the AFB_1 aptamer, PSA aptamer, and CEA aptamer, were employed to assess the specificity of the sensor. As shown in Figure 9a, only sensors modified with the CRP aptamer exhibit significant signal changes. Selectivity is also a crucial parameter for assessing sensor performance. Figure 9b displayed the change of ECL signal (ΔECL intensity) when the aptasensor incubated with CRP (10 ng/mL), or one (100 ng/mL) of the other tumor biomarkers (AFP or CA15-3) or substances (glucose, L-serine, Na^+, K^+ and Ca^{2+}) found in serum. As observed, even the concentrations of other substances are ten times higher than that of CRP, and the change of ECL signal is very minimal. In contrast, a significant ECL change is observed when the aptasensor incubates with CRP because of the decrease in ECL signal resulting from the specific binding with aptamer, indicating that the aptasensor possesses excellent selectivity. Six aptasensors are individually fabricated to measure CRP, as shown in Figure 9c. The results indicated a relative standard deviation (RSD) of 2.3%, demonstrating excellent repeatability of the sensors (Figure 9c). As shown in the upper image in Figure 9b, when the electrode undergoes continuous electrochemical scanning, the RSD of the continuously measured ECL signals is only 1.1%, demonstrating high signal stability. Furthermore, even after 15 days of storage at 4 °C, the ECL intensity of the sensors remains at 85% of the initial signal, indicating excellent storage stability (the bottom image of Figure 9d).

2.8. Real Sample Analysis

Compared to turn-on sensors, turn-off sensors are more susceptible to matrix effects when analyzing samples with complex matrices, thereby affecting detection sensitivity and accuracy. The potential application of the aptasensor in measuring CRP in real biological samples was assessed through standard addition experiments conducted in serum. The easy detection of CRP levels even when the serum is diluted to 1/50 is achieved. After incubating the sensor with serum containing 0.1–100 ng/mL CRP, ECL signals were measured, and recovery rates were calculated. The results are shown in Table 1. As shown, the recovery ranges between 94.4% and 104%, with a low RSD of less than 2%, indicating high detection reliability. Thus, the impact of complex matrices on the detection is low thanks to the size exclusion and charge-selective permeability of the employed nanochannel array. Although the dilution of the sample diminishes the convenience of the detection process compared with turn-on sensors, the proposed strategy can potentially be extended to the detection of other biomolecules with a simple replacement of the corresponding aptamer.

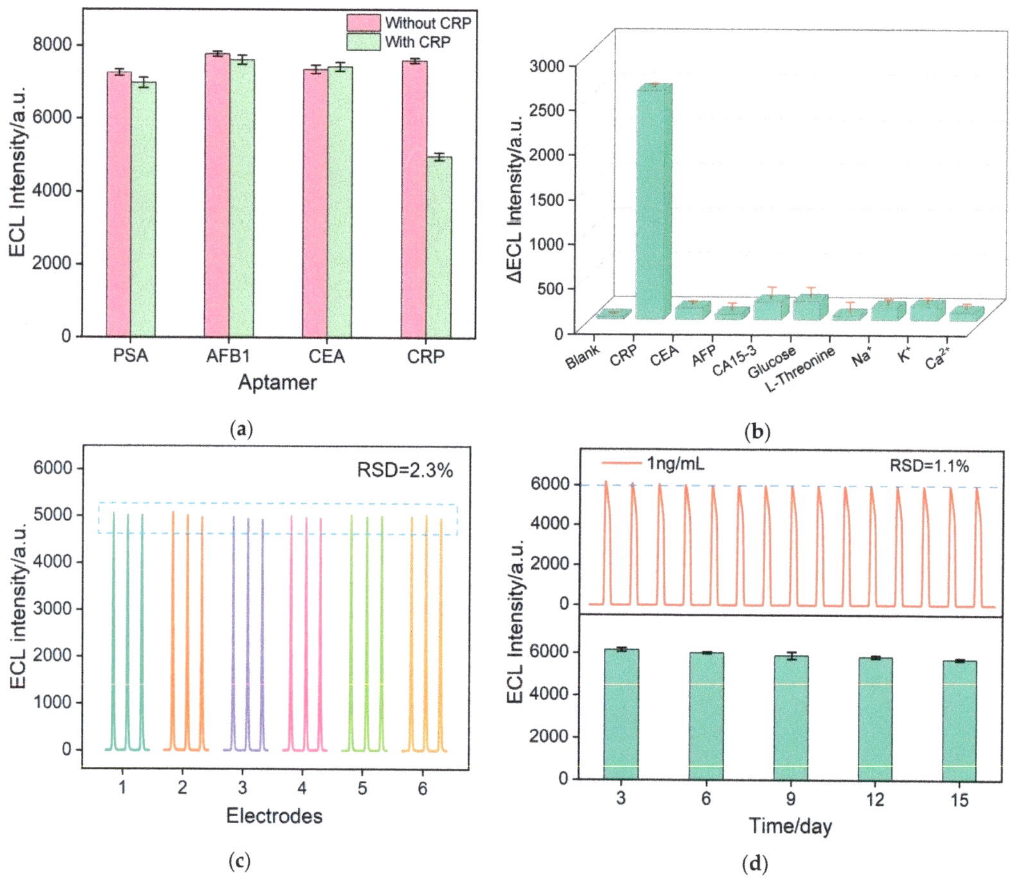

Figure 9. (**a**) ECL signal measured before and after incubation with 10 ng/mL CRP in PBS solution containing 10 μM $Ru(bpy)_3{}^{2+}$ and 3 mM TPrA. The recognitive interface was fabricated using different aptamer PSA, AFB1, and CEA. (**b**) ECL signal change values of BSA/Apt/AuNPs@O-SNF/ITO after incubation with 10 ng/mL CRP or other substances. The concentrations were 100 ng/mL for CEA, 100 ng/mL for AFP, 100 ng/mL for CA15-3, 100 μM for Glucose, 100 μM for L-Threonine, 1 mM for Na^+, 1 mM for k^+ and 1 mM for Ca^{2+}. (**c**) The reproducibility of the aptasensor. ECL intensity after incubation with different batches of six electrodes and 10 ng/mL CRP. (**d**) ECL signal stability of the aptasensor for a single measurement.

Table 1. Determination of CRP in serum sample using the fabricated aptasensor.

Sample	Added [a] (ng/mL)	Found [b] (ng/mL)	Recovery (%)	RSD (%, n = 3)
	0.100	0.104	104	1.8
Serum [c]	1.00	0.947	94.4	1.6
	100	102	102	1.2

[a] The concentration was obtained after serum dilution. [b] The concentration represents the average of measurements on three serum samples. [c] The sample was diluted by a factor of 50.

3. Materials and Methods

3.1. Chemicals and Materials

The recombinant human CRP protein was purchased from Okay Biotechnology Co., Ltd. (Nanjing, China). Carcinoembryonic antigen (CEA), bovine serum albumin (BSA), alpha-fetoprotein (AFP), and cancer antigen 15-3 (CA15-3) were obtained from Key-Bio Biotech Co., Ltd. (Beijing, China). Human blood serum from a healthy male donor was provided by the Hangzhou Institute of Occupational Diseases (Hangzhou, China). The CRP aptamer, PSA aptamer, CEA aptamer, and AFB_1 aptamer with amino labeling at the 5′ end were purchased from Sangon Biotech Co., Ltd. (Shanghai, China). The aptamer sequence is shown below: CRP (5′-NH_2-CGA AGG GGA TTC GAG GGG TGA TTG CGT GCT CCA TTT GGT G-3′); PSA (5′-NH_2-ATT AAA GCT CGC CAT CAA ATA GCT GC-3′); CEA (5′-ATACCAGCTTATTCAATT-NH_2-3′); AFB1 (5′-GTT GGG CAC GTG TTG TCT CTC TGT GTC TCG TGC CCT TCG CTA GGC CCA CA-NH_2-3′). Chemicals, including potassium ferrocyanide ($[K_4Fe(CN)_6]$, 99%), ethanol (EtOH, ≥99.7%), tri-n-propylamine (TPrA, 98.0%), hydrochloric acid, hydroxymethyl ferrocene (FcMeOH, 98%) and potassium ferricyanide ($[K_3Fe(CN)_6]$, 98%), were purchased from Aladdin Reagents (Shanghai, China). Hexaammineruthenium (III) chloride ($[Ru(NH_3)_6]Cl_3$, 98%), (3-glycidyloxypropyl) trimethoxysilane (GPTMS), tetraethyl orthosilicate (TEOS, ≥99%) and hexadecyl trimethylammonium bromide (CTAB, 99%) were sourced from Sigma–Aldrich (Shanghai, China). Tripropylamine and potassium hydrogen phthalate (KHP, 99%) were obtained from Maclin Biochemical Technology Co., Ltd. (Shanghai, China). Indium tin oxide (ITO)-coated glasses were purchased from Kaivo Optoelectronics Technology Co., LTD. (Zhuhai, China). The dilution of biological reagents and the detection of CRP were carried out in phosphate buffer solution (PBS, 0.01 M, pH = 7.4) obtained with KH_2PO_4 and Na_2HPO_4. Ultrapure water (18.2 MΩ·cm) was prepared using a Milli-Q water purification system. All chemicals were used as received without further purification.

3.2. Characteriaztions and Instrumentations

The film's microscopic structure and morphology were examined using a transmission electron microscope (TEM, HT7700, Hitachi, Japan) and field-emission scanning electron microscope (SEM, SU-8010, Hitachi, Japan), respectively. AuNPs were characterized using X-ray photoelectron spectroscopy (XPS) with a PHI5300 instrument, employing a Mg Kα source excitation at 250 W and 14 kV. Elemental mapping was conducted with an Oxford Xplore 50 energy spectrometer. The acceleration voltage was 15 kV. Cyclic voltammetry (CV) measurements were conducted on an Autolab electrochemical workstation (PGSTAT302N, Metrohm, Switzerland). A three-electrode system was employed, consisting of Ag/AgCl as the reference electrode, Pt electrode as the counter electrode, and ITO or modified ITO as the working electrode. ECL measurements were carried out in a quartz cell using an electrochemiluminescence detector (MPI-E, Xi'an Remex Analytical Instrument Co., Ltd., Xi'an, China). ECL signal was measured during continuous CV scanning with a potential range from 0 to 1.4 V at a scanning rate of 100 mV/s. The luminescent body was $Ru(bpy)_3^{2+}$, and the luminescent wavelength is 610 nm [62]. A bias voltage of 400 V was employed. All measurements were performed at room temperature.

3.3. Synthesis of AuNPs' Confined SNF/ITO Electrode

ITO electrodes were pre-treated before use [61]. They were firstly sonicated in a 1 M NaOH aqueous solution to create a negatively charged surface for 10 min, followed by ultrasonic cleaning in acetone, ethanol, and water for 10 min, respectively. Afterward, they were dried at 60 °C before use.

SNF was grown on the pretreated ITO surface using the Stöber-solution growth method using CTAB as the surfactant template [63]. This approach created a dense array of nanochannels that were oriented vertically to the ITO surface. Subsequently, the ITO glasses with SNF on the surface were cut into dimensions of 0.5 cm × 5 cm, and the working electrode area was fixed to 0.5 cm × 1 cm using insulating glue. Given that the nanochannel interiors were filled with surfactant micelle (SM), the resulting electrode was designated as SM@SNF/ITO. SM can be easily removed by immersing the SM@SNF/ITO electrode in HCl (0.1 M in ethanol) for 8 min [63]. The final electrode with an open nanochannel array that was oriented vertically to the surface of the underlying ITO electrode was named as SNF/ITO.

To deposit AuNPs within the nanochannels of SNF, the working electrode was immersed in a 0.5% aqueous solution of $HAuCl_4$, and a constant voltage of −0.5 V was applied for 3 s. The AuNPs-modified electrode was obtained after rinsing with ultrapure water.

3.4. *Fabrication of Aptamosensor*

For the covalent immobilization of recognitive aptamer, the outer surface of SNF is firstly derivatized by epoxy groups. To ensure that the epoxyization reaction occurs primarily on the outer surface of SNF rather than inside the nanochannels, an electrode with SM-sealed nanochannels (SM@SNF/ITO) was employed for epoxy derivatization, followed by the removal of SM. Specifically, SM@SNF/ITO was immersed in a 25 mL ethanol solution containing GPTMS (2.26 mM) for 1 h to introduce reactive epoxy groups onto the external surface of the SNF. Then, the obtained electrode was cleaned with ultrapure water followed with removal of SM. The obtained electrode was designated as O-SNF/ITO.

Subsequently, AuNPs were confined within the nanochannels of O-SNF/ITO, through simple electrodeposition [34]. Briefly, O-SNF/ITO was subjected to a 5% $HAuCl_4$ solution and electrodeposition was performed at −0.5 V for 3 s, yielding AuNPs@O-SNF/ITO.

For the fabrication of the recognitive interface, AuNPs@O-SNF/ITO was immersed in the CRP aptamer (Apt) solution (0.3 μM in 0.01 M PBS) and incubated at 4 °C for 80 min. This process led to the covalent immobilization of the CRP aptamer, resulting in Apt/AuNPs@O-SNF/ITO. Finally, Apt/AuNPs@O-SNF/ITO was incubated with BSA (10 mg/mL) at 4 °C for 30 min to block non-specific active sites, creating the aptasensor, BSA/Apt/AuNPs@O-SNF/ITO. The fabricated aptasensor was directly stored in a refrigerator at 4 °C when not in use.

3.5. *ECL Determination of CRP*

The ECL intensity before and after CRP binding was assessed by incubating the fabricated aptasensor with different concentrations of CRP at 37 °C for 80 min. Specifically, the aptasensor before (BSA/Apt/AuNPs@O-SNF/ITO) or after CRP binding (CRP/BSA/Apt/AuNPs@O-SNF/ITO) were immersed in a PBS solution (0.01 M, pH = 7.4) containing the ECL emitter ($Ru(bpy)_3^{2+}$, 10 μM) and co-reactor (TPrA, 3 mM). The ECL signal of the electrode was then recorded. For the analysis of real samples, CRP in human serum (from a healthy male) was determined using a standard addition method. Briefly, human serum was spiked with different concentrations of CRP and then diluted 50-fold with PBS (0.01 M, pH 7.4) before measurement.

4. Conclusions

In this work, a simple and label-free ECL aptasensor has been successfully developed for the highly sensitive detection of CRP in biological samples. Large-area SNF-modified electrodes were fabricated on cost-effective and readily available ITO glass, allowing the simultaneous preparation of multiple SNF-modified electrodes. Due to the low cost of both ITO and silica-based SNF, the cost of SNF/ITO is very low (commonly 1 USD/cm^2). The modified electrode used can be easily miniaturized and patterned. Subsequently, AuNPs were grown in situ within the nanochannels of SNF through electrodeposition. This achievement established a dual ECL signal amplification system based on the efficient enrichment of ECL emitters within SNF nanochannels and the catalytic enhancement by AuNPs. By functionalizing the outer surface of SNF with epoxy groups when the nanochannels was blocked with a surfactant template, a covalently immobilized recognition interface was obtained for aptamer immobilization. In comparison with the traditional sandwich-type immunoassay, e.g., enzyme-linked immunosorbent assay (ELISA), the fabricated aptasensor employs DNA aptamers as recognitive ligands and the immobilization process of aptamers is similar to the immobilization step of antibodies in traditional immunosensors. However, aptamers are cost-effective and exhibit much higher stability than antibodies [64,65]. For CRP detection, the constructed aptasensor exhibited a wide linear range, low detection limit, high selectivity, and excellent storage stability. Furthermore, the aptasensor also demonstrated advantages such as low cost and rapid analysis, resulting in satisfactory recovery rates in CRP detection in serum samples. The aptasensor developed in this study has great promise for potential applications in disposable, portable detection or point-of-care diagnostics.

Supplementary Materials: The following supporting information can be downloaded at: https://www.mdpi.com/article/10.3390/molecules28227664/s1, Table S1: Comparison of CRP detection performance with other sensors.

Author Contributions: Investigation, N.M. and S.X.; data curation, N.M. and S.X.; writing—original draft preparation, S.X.; conceptualization and supervision, W.W. and J.L.; writing—review and editing, W.W. and J.L. All authors have read and agreed to the published version of the manuscript.

Funding: This research was funded by National Natural Science Foundation of China (22374131).

Institutional Review Board Statement: Not applicable.

Informed Consent Statement: Not applicable.

Data Availability Statement: The data presented in this study are available on request from the corresponding author.

Conflicts of Interest: The authors declare no conflict of interest.

References

1. Boncler, M.; Wu, Y.; Watala, C. The multiple faces of C-reactive protein-physiological and pathophysiological implications in cardiovascular disease. *Molecules* **2019**, *24*, 2062. [CrossRef]
2. Sproston, N.R.; Ashworth, J.J. Role of C-reactive protein at sites of inflammation and infection. *Front. Immunol.* **2018**, *9*, 754. [CrossRef]
3. Pepys, M.B.; Hirschfield, G.M. C-reactive protein: A critical update. *J. Clin. Investig.* **2003**, *111*, 1805–1812. [CrossRef] [PubMed]
4. Yang, Y.; Xie, J.; Guo, F.; Longhini, F.; Gao, Z.; Huang, Y.; Qiu, H. Combination of C-reactive protein, procalcitonin and sepsis-related organ failure score for the diagnosis of sepsis in critical patients. *Ann. Intensive Care* **2016**, *6*, 51. [CrossRef]
5. Prajapati, A.; Verma, N.; Pandya, A. Highly sensitive vertical flow based point-of-care immunokit for rapid and early detection of human CRP as a cardiovascular risk factor. *Biomed. Microdevices* **2020**, *22*, 28. [CrossRef]
6. Macwan, I.; Aphale, A.; Bhagvath, P.; Prasad, S.; Patra, P. Detection of cardiovascular CRP protein biomarker using a novel nanofibrous substrate. *Biosensors* **2020**, *10*, 72. [CrossRef]
7. Rasmussen, L.J.H.; Schultz, M.; Gaardsting, A.; Ladelund, S.; Garred, P.; Iversen, K.; Eugen-Olsen, J.; Helms, M.; David, K.P.; Kjær, A.; et al. Inflammatory biomarkers and cancer: CRP and suPAR as markers of incident cancer in patients with serious nonspecific symptoms and signs of cancer. *Int. J. Cancer* **2017**, *141*, 191–199. [CrossRef]
8. Lv, Y.; Wu, R.; Feng, K.; Li, J.; Mao, Q.; Yuan, H.; Shen, H.; Chai, X.; Li, L.S. Highly sensitive and accurate detection of C-reactive protein by CdSe/ZnS quantum dot-based fluorescence-linked immunosorbent assay. *J. Nanobiotechnol.* **2017**, *15*, 35. [CrossRef] [PubMed]
9. Bravin, C.; Amendola, V. Wide range detection of C-Reactive protein with a homogeneous immunofluorimetric assay based on cooperative fluorescence quenching assisted by gold nanoparticles. *Biosens. Bioelectron.* **2020**, *169*, 112591. [CrossRef] [PubMed]
10. Zhang, L.; Li, H.-Y.; Li, W.; Shen, Z.-Y.; Wang, Y.-D.; Ji, S.-R.; Wu, Y. An ELISA Assay for Quantifying Monomeric C-Reactive Protein in Plasma. *Front. Immunol.* **2018**, *9*, 511. [CrossRef]
11. Zhong, S.; Chen, L.; Shi, X.; Chen, G.; Sun, D.; Zhang, L. Recent advances in electrochemical aptasensors for detecting cardiac biomarkers: A review. *Microchem. J.* **2023**, *193*, 109063. [CrossRef]
12. Tang, M.-Q.; Xie, J.; Rao, L.-M.; Kan, Y.-J.; Luo, P.; Qing, L.-S. Advances in aptamer-based sensing assays for C-reactive protein. *Anal. Bioanal. Chem.* **2022**, *414*, 867–884. [CrossRef] [PubMed]
13. Sun, D.; Lu, J.; Zhang, L.; Chen, Z. Aptamer-based electrochemical cytosensors for tumor cell detection in cancer diagnosis: A review. *Anal. Chim. Acta* **2019**, *1082*, 1–17. [CrossRef] [PubMed]
14. Wang, Y.; Liu, X.; Wu, L.; Ding, L.; Effah, C.Y.; Wu, Y.; Xiong, Y.; He, L. Construction and bioapplications of aptamer-based dual recognition strategy. *Biosens. Bioelectron.* **2022**, *195*, 113661. [CrossRef]

15. Zhang, T.; Yang, L.; Yan, F.; Wang, K. Vertically-ordered mesoporous silica film based electrochemical aptasensor for highly sensitive detection of alpha-fetoprotein in human serum. *Biosensors* **2023**, *13*, 628. [CrossRef] [PubMed]
16. Zhong, W.; Sczepanski, J.T. A mirror image fluorogenic aptamer sensor for live-cell imaging of microRNAs. *ACS Sens.* **2019**, *4*, 566–570. [CrossRef]
17. Valizadeh Shahbazlou, S.; Vandghanooni, S.; Dabirmanesh, B.; Eskandani, M.; Hasannia, S. Biotinylated aptamer-based SPR biosensor for detection of CA125 antigen. *Microchem. J.* **2023**, *194*, 109276. [CrossRef]
18. Mahyari, M.; Hooshmand, S.E.; Sepahvand, H.; Gholami, S.; Rezayan, A.H.; Zarei, M.A. Gold nanoparticles anchored onto covalent poly deep eutectic solvent functionalized graphene: An electrochemical aptasensor for the detection of C-reactive protein. *Mater. Chem. Phys.* **2021**, *269*, 124730. [CrossRef]
19. López, L.; Hernández, N.; Reyes Morales, J.; Cruz, J.; Flores, K.; González-Amoretti, J.; Rivera, V.; Cunci, L. Measurement of neuropeptide Y using aptamer-modified microelectrodes by electrochemical impedance spectroscopy. *Anal. Chem.* **2021**, *93*, 973–980. [CrossRef] [PubMed]
20. Jiang, M.; Wang, M.; Lai, W.; Song, X.; Li, J.; Liu, D.; Wei, Z.; Hong, C. Construction of electrochemical and electrochemiluminescent dual-mode aptamer sensors based on ferrocene dual-functional signal probes for the sensitive detection of Alternariol. *Anal. Chim. Acta* **2023**, *1272*, 341476. [CrossRef] [PubMed]
21. Liu, X.; Luo, L.; Li, L.; Di, Z.; Zhang, J.; You, T. An electrochemiluminescence aptasensor for analysis of bisphenol A based on carbon nanodots composite as co-reaction of Ru(bpy)32+ nanosheets. *Electrochim. Acta* **2019**, *319*, 849–858. [CrossRef]
22. Nikolaou, P.; Valenti, G.; Paolucci, F. Nano-structured materials for the electrochemiluminescence signal enhancement. *Electrochim. Acta* **2021**, *388*, 138586. [CrossRef]
23. Li, J.; Luo, M.; Jin, C.; Zhang, P.; Yang, H.; Cai, R.; Tan, W. Plasmon-enhanced electrochemiluminescence of PTP-decorated Eu MOF-based Pt-tipped Au bimetallic nanorods for the lincomycin assay. *ACS Appl. Mater. Interfaces* **2022**, *14*, 383–389. [CrossRef]
24. Gu, W.; Wang, H.; Jiao, L.; Wu, Y.; Chen, Y.; Hu, L.; Gong, J.; Du, D.; Zhu, C. Single-Atom Iron Boosts Electrochemiluminescence. *Angew. Chem. Int. Ed.* **2020**, *59*, 3534–3538. [CrossRef]
25. Xu, S.; Zhang, S.; Li, Y.; Liu, J. Facile synthesis of iron and nitrogen co-doped carbon dot nanozyme as highly efficient peroxidase mimics for visualized detection of metabolites. *Molecules* **2023**, *28*, 6064. [CrossRef]
26. Chang, Q.; Huang, J.; He, L.; Xi, F. Simple immunosensor for ultrasensitive electrochemical determination of biomarker of the bone metabolism in human serum. *Front. Chem.* **2022**, *10*, 940795. [CrossRef] [PubMed]
27. Zhu, Q.; Liu, H.; Zhang, J.; Wu, K.; Deng, A.; Li, J. Ultrasensitive QDs based electrochemiluminescent immunosensor for detecting ractopamine using AuNPs and Au nanoparticles@PDDA-graphene as amplifier. *Sens. Actuators B Chem.* **2017**, *243*, 121–129. [CrossRef]
28. Wang, L.; Liu, Y.; Yan, J.; Li, H.; Tu, Y. Novel electrochemiluminescent immunosensor using dual amplified signals from a CoFe Prussian blue analogue and Au nanoparticle for the detection of Lp-PLA2. *ACS Sens.* **2023**, *8*, 2859–2868. [CrossRef]
29. Gai, Q.-Q.; Wang, D.-M.; Huang, R.-F.; Liang, X.-X.; Wu, H.-L.; Tao, X.-Y. Distance-dependent quenching and enhancing of electrochemiluminescence from tris(2,2′-bipyridine) ruthenium (II)/tripropylamine system by gold nanoparticles and its sensing applications. *Biosens. Bioelectron.* **2018**, *118*, 80–87. [CrossRef]
30. Zhou, Z.; Yu, F.; Ma, J. Nanoconfinement engineering for enchanced adsorption of carbon materials, metal–organic frameworks, mesoporous silica, MXenes and porous organic polymers: A review. *Environ. Chem. Lett.* **2022**, *20*, 563–595. [CrossRef]
31. Huang, L.; Su, R.; Xi, F. Sensitive detection of noradrenaline in human whole blood based on Au nanoparticles embedded vertically-ordered silica nanochannels modified pre-activated glassy carbon electrodes. *Front. Chem.* **2023**, *11*, 1126213. [CrossRef]
32. Zhang, C.; Zhou, X.; Yan, F.; Lin, J. N-doped graphene quantum dots confined within silica nanochannels for enhanced electrochemical detection of doxorubicin. *Molecules* **2023**, *28*, 6443. [CrossRef]
33. Ding, L.H.; Su, B. A non-enzymatic hydrogen peroxide sensor based on platinum nanoparticle-polyaniline nanocomposites hosted in mesoporous silica film. *J. Electroanal. Chem.* **2015**, *736*, 83–87. [CrossRef]
34. Ding, L.; Li, W.; Sun, Q.; He, Y.; Su, B. Gold nanoparticles confined in vertically aligned silica nanochannels and their electrocatalytic activity toward ascorbic acid. *Chem.-Eur. J.* **2014**, *20*, 12777–12780. [CrossRef]
35. Liang, R.; Dong, J.; Li, J.; Jin, H.; Wei, M.; Bai, T.; Ren, W.; Xu, Y.; He, B.; Suo, Z. DNAzyme-driven bipedal DNA walker and catalytic hairpin assembly multistage signal amplified electrochemical biosensor based on porous AuNPs@Zr-MOF for detection of Pb^{2+}. *Food Chem.* **2024**, *435*, 137503. [CrossRef] [PubMed]
36. Ma, X.; Qian, K.; Ejeromedoghene, O.; Kandawa-Schulz, M.; Wang, Y. Electrochemical detection of microRNA based on SA-PPy/AuNPs nanocomposite with the signal amplification through catalytic hairpin assembly reaction and the spontaneous catalytic reaction of Fe^{3+}/Cu^{2+}. *Electrochim. Acta* **2020**, *362*, 137168. [CrossRef]
37. Zhao, J.; Duan, W.; Liu, X.; Xi, F.; Wu, J. Microneedle patch integrated with porous silicon confined dual nanozymes for synergistic and hyperthermia-enhanced nanocatalytic ferroptosis treatment of melanoma. *Adv. Funct. Mater.* **2023**, 2308183. [CrossRef]
38. Liu, X.; Chen, Z.; Wang, T.; Jiang, X.; Qu, X.; Duan, W.; Xi, F.; He, Z.; Wu, J. Tissue imprinting on 2D nanoflakes-capped silicon nanowires for lipidomic mass spectrometry imaging and cancer diagnosis. *ACS Nano* **2022**, *16*, 6916–6928. [CrossRef]
39. Cui, Y.; Duan, W.; Jin, Y.; Wo, F.; Xi, F.; Wu, J. Ratiometric fluorescent nanohybrid for noninvasive and visual monitoring of sweat glucose. *ACS Sens.* **2020**, *5*, 2096–2105. [CrossRef]

40. Walcarius, A. Electroinduced surfactant self-assembly driven to vertical growth of oriented mesoporous films. *Acc. Chem. Res.* **2021**, *54*, 3563–3575. [CrossRef] [PubMed]
41. Zhou, H.; Ding, Y.; Su, R.; Lu, D.; Tang, H.; Xi, F. Silica nanochannel array film supported by ß-cyclodextrin-functionalized graphene modified gold film electrode for sensitive and direct electroanalysis of acetaminophen. *Front. Chem.* **2022**, *9*, 812086. [CrossRef]
42. Zhu, X.; Xuan, L.; Gong, J.; Liu, J.; Wang, X.; Xi, F.; Chen, J. Three-dimensional macroscopic graphene supported vertically-ordered mesoporous silica-nanochannel film for direct and ultrasensitive detection of uric acid in serum. *Talanta* **2022**, *238*, 123027. [CrossRef]
43. Cui, Y.; Zhang, S.; Zhou, X.; Yan, F.; Hu, W. Silica nanochannel array on co-electrodeposited graphene-carbon nanotubes 3D composite film for antifouling detection of uric acid in human serum and urine samples. *Microchem. J.* **2023**, *190*, 108632. [CrossRef]
44. Yan, F.; Su, B. Tailoring molecular permeability of nanochannel-micelle membranes for electrochemical analysis of antioxidants in fruit juices without sample treatment. *Anal. Chem.* **2016**, *88*, 11001–11006. [CrossRef]
45. Lin, X.; Yang, Q.; Ding, L.; Su, B. Ultrathin silica membranes with highly ordered and perpendicular nanochannels for precise and fast molecular separation. *ACS Nano* **2015**, *9*, 11266–11277. [CrossRef]
46. Zou, Y.; Zhou, X.; Xie, L.; Tang, H.; Yan, F. Vertically-ordered mesoporous silica films grown on boron nitride-graphene composite modified electrodes for rapid and sensitive detection of carbendazim in real samples. *Front. Chem.* **2022**, *10*, 939510. [CrossRef]
47. Su, R.; Tang, H.; Xi, F. Sensitive electrochemical detection of p-nitrophenol by pre-activated glassy carbon electrode integrated with silica nanochannel array film. *Front. Chem.* **2022**, *10*, 954748. [CrossRef]
48. Yang, L.; Zhang, T.; Zhou, H.; Yan, F.; Liu, Y. Silica nanochannels boosting $Ru(bpy)_3^{2+}$-mediated electrochemical sensor for the detection of guanine in beer and pharmaceutical samples. *Front. Nutr.* **2022**, *9*, 987442. [CrossRef] [PubMed]
49. Zhang, M.; Zou, Y.; Zhou, X.; Yan, F.; Ding, Z. Vertically-ordered mesoporous silica films for electrochemical detection of Hg(II) ion in pharmaceuticals and soil samples. *Front. Chem.* **2022**, *10*, 952936. [CrossRef] [PubMed]
50. Zheng, W.; Su, R.; Yu, G.; Liu, L.; Yan, F. Highly sensitive electrochemical detection of paraquat in environmental water samples using a vertically ordered mesoporous silica film and a nanocarbon composite. *Nanomaterials* **2022**, *12*, 3632. [CrossRef]
51. Chen, H.; Huang, J.; Zhang, R.; Yan, F. Dual-mode electrochemiluminescence and electrochemical sensor for alpha-fetoprotein detection in human serum based on vertically ordered mesoporous silica films. *Front. Chem.* **2022**, *10*, 1023998. [CrossRef] [PubMed]
52. Chen, D.; Luo, X.; Xi, F. Probe-integrated electrochemical immunosensor based on electrostatic nanocage array for reagentless and sensitive detection of tumor biomarker. *Front. Chem.* **2023**, *11*, 1121450. [CrossRef] [PubMed]
53. Zhou, P.; Su, B. Enhanced electrochemiluminescence at silica nanochannel membrane studied by scanning electrochemical microscopy. *J. Electroanal. Chem.* **2022**, *904*, 115943. [CrossRef]
54. Villalonga, A.; Vegas, B.; Paniagua, G.; Eguílaz, M.; Mayol, B.; Parrado, C.; Rivas, G.; Díez, P.; Villalonga, R. Amperometric aptasensor for carcinoembryonic antigen based on a reduced graphene oxide/gold nanoparticles modified electrode. *J. Electroanal. Chem.* **2020**, *877*, 114511. [CrossRef]
55. Kong, D.; Zhao, J.; Tang, S.; Shen, W.; Lee, H.K. Logarithmic data processing can be used justifiably in the plotting of a calibration curve. *Anal. Chem.* **2021**, *93*, 12156. [CrossRef] [PubMed]
56. Yang, H.J.; Kim, M.W.; Raju, C.V.; Cho, C.H.; Park, T.J.; Park, J.P. Highly sensitive and label-free electrochemical detection of C-reactive protein on a peptide receptor−gold nanoparticle−black phosphorous nanocomposite modified electrode. *Biosens. Bioelectron.* **2023**, *234*, 115382. [CrossRef] [PubMed]
57. Cheng, Y.-Y.; Feng, X.-Z.; Zhan, T.; An, Q.-Q.; Han, G.-C.; Chen, Z.; Kraatz, H.-B. A facile indole probe for ultrasensitive immunosensor fabrication toward C-reactive protein sensing. *Talanta* **2023**, *262*, 124696. [CrossRef] [PubMed]
58. Rong, Z.; Chen, F.; Jilin, Y.; Yifeng, T. A C-reactive protein immunosensor based on platinum nanowire / titania nanotube composite sensitized electrochemiluminescence. *Talanta* **2019**, *205*, 120135. [CrossRef]
59. Yang, X.; Xu, Y.; Huang, X.; Hang, J.; Guo, W.; Dai, Z. Multicolor iridium(iii) complexes with host–guest recognition motifs for enhanced electrochemiluminescence and modular labeling. *Anal. Chem.* **2023**, *95*, 4543–4549. [CrossRef]
60. Cui, C.; Lin, X.; Lv, J.; Guo, H.; Shen, L.; Xiang, G.; Zhao, W.; Jiang, D. Electrochemiluminescence resonance energy transfer between $Ru(bpy)_3^{2+}$@$Cu_3(HHTP)_2$ and GO-Au composites for C-reactive protein detection. *Talanta* **2023**, *263*, 124709. [CrossRef]
61. Hong, D.; Kim, K.; Jo, E.-J.; Kim, M.-G. Electrochemiluminescence-Incorporated Lateral Flow Immunosensors Using $Ru(bpy)_3^{2+}$-Labeled Gold Nanoparticles for the Full-Range Detection of Physiological C-Reactive Protein Levels. *Anal. Chem.* **2021**, *93*, 7925–7932. [CrossRef] [PubMed]
62. Yan, L.; Zhang, C.; Xi, F. Disposable amperometric label-free immunosensor on chitosan–graphene-modified patterned ITO electrodes for prostate specific antigen. *Molecules* **2022**, *27*, 5895. [CrossRef] [PubMed]
63. Teng, Z.; Zheng, G.; Dou, Y.; Li, W.; Mou, C.-Y.; Zhang, X.; Asiri, A.M.; Zhao, D. Highly ordered mesoporous silica films with perpendicular mesochannels by a simple Stöber-solution growth approach. *Angew. Chem. Int. Ed.* **2012**, *51*, 2173–2177. [CrossRef]

64. Hassan, E.M.; DeRosa, M.C. Recent advances in cancer early detection and diagnosis: Role of nucleic acid based aptasensors. *TrAC Trend Anal. Chem.* **2020**, *124*, 115806. [CrossRef]
65. Negahdary, M. Aptamers in nanostructure-based electrochemical biosensors for cardiac biomarkers and cancer biomarkers: A review. *Biosens. Bioelectron.* **2020**, *152*, 112018. [CrossRef]

molecules

MDPI

Article

Broad-Spectrum Antibody-Based Immunochromatographic Strip Assay for Rapid Screening of Bisphenol A Diglycidyl Ether and Its Derivatives in Canned Foods

Chundi Yu [1], Jinnuo Hu [2], Wei Wu [1,3,4], Yongfei Zhou [2], Can Zhang [2,*] and Qingli Yang [1,*]

1 College of Food Science and Engineering, Qingdao Agricultural University, Qingdao 266109, China; naturalyu@qau.edu.cn (C.Y.); wuweiouc@126.com (W.W.)
2 School of Food and Biological Engineering, Jiangsu University, Zhenjiang 212013, China; qdxdhou@qau.edu.cn (J.H.); tangjuan@qau.edu.cn (Y.Z.)
3 Academy of Dongying Efficient Agricultural Technology and Industry on Saline and Alkaline Land in Collaboration with Qingdao Agricultural University, Qingdao 266109, China
4 Qingdao Institute of Special Food, Qingdao 266109, China
* Correspondence: zhangcan@ujs.edu.cn (C.Z.); yql@qau.edu.cn (Q.Y.)

Abstract: Bisphenol A diglycidyl ether (BADGE) is widely present in the inner coating of metal food cans, from which it can migrate into food and generate harmful derivatives during storage, such as bisphenol A (2,3-dihydroxypropyl) glycidyl ether, bisphenol A (3-chloro-2-hydroxypropyl) glycidyl ether, and bisphenol A (3-chloro-2-hydroxypropyl) (2,3-dihydroxypropyl) glycidyl ether. Here, a gold-nanoparticle-based immunochromatographic strip assay based on a broad-spectrum polyclonal antibody was developed for the simultaneous detection of BADGE and its derivatives, which could be accomplished within 15 min. The quantitative analysis of the visualization results was performed using Adobe Photoshop CC 2021, and the detection limit, defined as the concentration causing 15% inhibition, was 0.97 ng/mL. The recoveries of BADGE and its derivatives at various spiking levels in canned food samples ranged from 79.86% to 93.81%. The detection results of the proposed immunochromatographic strip assay were validated via high-performance liquid chromatography, showing a good correlation coefficient (R^2 = 0.9580).

Keywords: bisphenol A diglycidyl ether; derivatives; broad-spectrum polyclonal antibodies; gold nanoparticles; immunochromatographic strip assay; canned food

Citation: Yu, C.; Hu, J.; Wu, W.; Zhou, Y.; Zhang, C.; Yang, Q. Broad-Spectrum Antibody-Based Immunochromatographic Strip Assay for Rapid Screening of Bisphenol A Diglycidyl Ether and Its Derivatives in Canned Foods. *Molecules* **2024**, *29*, 13. https://doi.org/10.3390/molecules29010013

Academic Editor: Masayasu Kuwahara

Received: 19 September 2023
Revised: 2 December 2023
Accepted: 4 December 2023
Published: 19 December 2023

1. Introduction

Bisphenol A diglycidyl ether (2,2-bis(4-glycidyloxyphenyl) propane, BADGE) is a condensation product of bisphenol A (BPA) and epichlorohydrin. It can be used as an additive in polyester fibers and as a hydrochloric acid scrubber. In particular, BADGE is often utilized to remove the hydrochloric acid in the organosol resins used as inner coatings of metal food cans [1–3], where it can remain if the chemical reaction is not complete during the coating manufacturing process [4,5] and migrate into food during processing and storage [6,7]. Due to the complexity of the food matrix, BADGE can react with acidic or greasy food via hydrolysis or chlorination, generating various derivatives, such as bisphenol A (2,3-dihydroxypropyl) glycidyl ether (BADGE·H_2O), bisphenol A (3-chloro-2-hydroxypropyl) glycidyl ether (BADGE·HCl), and bisphenol A (3-chloro-2-hydroxypropyl) (2,3-dihydroxypropyl) glycidyl ether (BADGE·HCl·H_2O) [8,9], which have been detected along with BADGE in canned foods such as canned seafood, canned meat products, and energy drinks [10,11]. BADGE may lead to abnormalities in the human endocrine, immune, and nervous systems and affect normal reproductive and genetic functions [12–14]. Therefore, European legislation has set specific migration limits for BADGE and its hydrolysis derivative BADGE·H_2O at 9 mg/kg in foodstuffs and food simulants as well as for the hydrochloric derivatives BADGE·HCl and BADGE·HCl·H_2O at 1 mg/kg [15].

Various analytical methods based on high-performance liquid chromatography–fluorescence detection (HPLC–FLD) [16–18], liquid chromatography–tandem mass spectrometry (LC-MS) [19–21], and gas chromatography–mass spectrometry (GC–MS) have been developed for the detection of bisphenol-dihydrate glycerol ether [22]. Although LC-MS and GC–MS have good accuracy and high sensitivity, they are expensive, complex, and require trained operators, which is not conducive to rapid on-site detection. In contrast, immunoassays are suitable for the simultaneous detection of a large quantity of samples and can meet daily monitoring requirements. The immunochromatographic strip assay, also known as lateral-flow immunoassay, is a combination of chromatography and immunoassay. It has gained attention as an alternative to enzyme-linked immunosorbent assay (ELISA). Immunochromatographic strip test is considered to be a useful tool for the rapid screening of food and raw materials. It is in demand for rapid and point-of-care testing in all parts of the food-processing chain "from farm to fork" [23,24].

Gold nanoparticles are commonly used as detector reagents in lateral-flow immunochromatography for the visualization of signals. In particular, gold nanoparticle (AuNP-based immunochromatographic strip assays have been widely used for detecting hazardous macromolecular and small-molecule substances [25–31].

To the best of our knowledge, although the production of monoclonal antibodies that can recognize BADGE and the establishment of an ELISA method for detecting BADGE in lake water have been previously reported [32], broad-spectrum polyclonal antibodies that can simultaneously recognize BADGE and its derivatives using an immunochromatographic strip assay have not yet been described.

Here, we developed an immunochromatographic strip assay using AuNPs and a broad-spectrum polyclonal antibody to simultaneously detect BADGE and its derivatives. The visualized results were quantitatively analyzed using Adobe Photoshop CC software. Canned foods were selected as samples, and HPLC was used as the analysis method to evaluate the proposed immunochromatographic strip assay, which is a convenient tool for the fast and efficient detection of BADGE and its derivatives.

2. Results

2.1. Screening of Broad-Spectrum Antibodies

Standard solutions of BADGE derivatives were freshly prepared before analysis. The concentration causing 50% inhibition (IC_{50}) values were determined, and the cross-reactivities were calculated using the indirect competitive ELISA (ic-ELISA) method. The recognition capability of selected antisera (PAb-1, PAb-2, PAB-3, and PAb-4) for BADGE derivatives was also investigated. As shown in Table 1, the cross-reactivities of antiserum PAb-1 with BADGE·HCl, BADGE·H_2O, and BADGE·HCl·H_2O were 79.6%, 175.8%, and 110.8%, respectively. Compared with other antisera, antiserum PAb-1 exhibited better recognition capability for BADGE and its derivatives; therefore, it was selected as a broad-spectrum recognition antiserum. PAb-1 was purified using protein A-Sepharose 4B affinity chromatography (Figure 1a), and the obtained antibody was characterized using SDS-PAGE (Figure 1b).

Table 1. Cross-reactivities of antisera with BADGE, BADGE·HCl, BADGE·H_2O, and BADGE·HCl·H_2O.

Target	Antisera							
	PAb-1		PAb-2		PAb-3		PAb-4	
	IC_{50} (ng/mL)	CR (%)	IC_{50} (ng/mL)	CR (%)	IC_{50} (ng/mL)	CR (%)	IC_{50} (ng/mL)	CR (%)
BADGE	51	100	65	100	26	100	29	100
BADGE·HCl	64	79.6	114	57.0	57	45.6	60	48.3
BADGE·H_2O	29	175.8	51	127.4	76	34.2	47	61.7
BADGE·HCl·H_2O	46	110.8	59	110.1	82	31.7	70	41.4

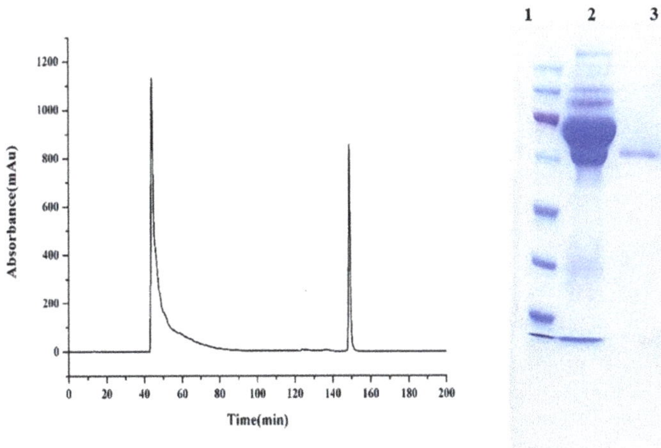

(a) (b)

Figure 1. (**a**) The antiserum purified via protein A-Sepharose 4B affinity chromatography. (**b**) The results of SDS-PAGE: 1: marker; 2: antiserum; 3: purified antibody.

2.2. Optimization of AuNP-Labeled Antibody

The AuNPs and AuNP-labeled antibody (Au NPs@PAb-1) were characterized using their UV–vis spectrum, as shown in Figure 2. The adsorption peak of AuNPs shifted from 521 nm to 532 nm, which indicated that the antibody successfully conjugated to the AuNPs.

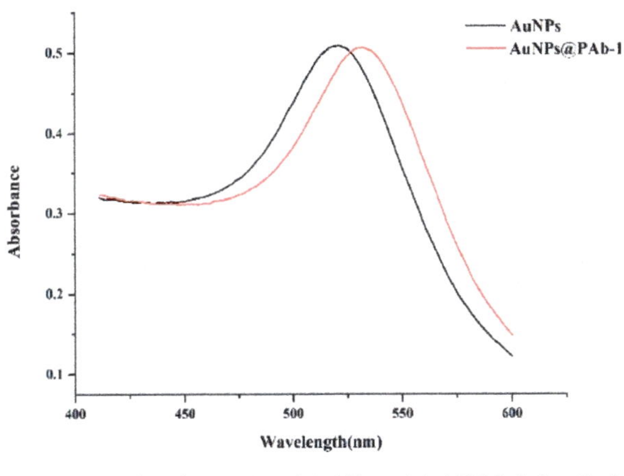

Figure 2. Wavelength scanning of AuNPs and AuNP-labeled antibody.

The combination of AuNPs and antibody mainly depended on the pH and the amount of binding antibody. When the pH of the system was close to the isoelectric point of the antibody protein, the binding between AuNPs and antibody was more stable. For 1 mL of AuNP solution, the pH value was adjusted by adding different amounts of a 0.2 M potassium carbonate (K_2CO_3) solution. As shown in Figure 3, the absorbance first increased and then decreased with an increasing amount of K_2CO_3 solution. When the volume was set to 15 µL, the solution absorbance of the AuNP-labeled antibody reached a maximum. Therefore, 15 µL of K_2CO_3 solution was used for adjusting the pH of 1 mL of AuNP solution.

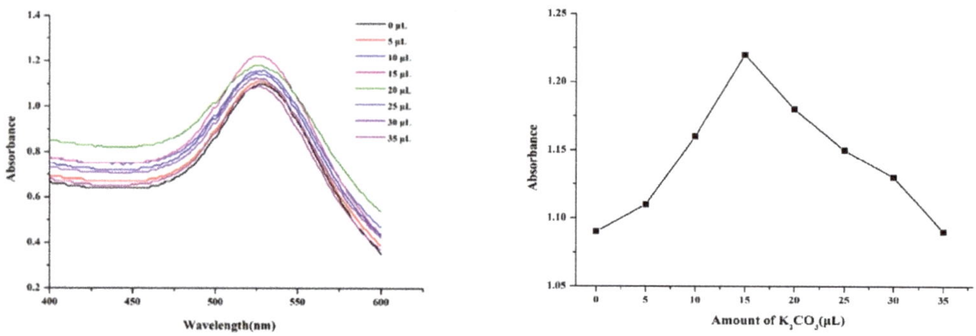

(a) (b)

Figure 3. (**a**) Wavelength scanning of AuNP-labeled antibody with different amounts of K_2CO_3. (**b**) Absorbance of AuNP-labeled antibody with different amounts of K_2CO_3.

The amount of antibody used for labeling was also a key factor; an insufficient amount of antibody would result in the surface instability of AuNPs, while excess antibody would lead to the waste of antibody. According to Figure 4, when the antibody amount was 9.35 µg (experimental group 5 in Figure 4b), the amount of labeling antibody had already reached saturation. Based on these parameters, for 1 mL of AuNP solution, the optimum labeling conditions were set to 15 µL of 0.2 M K_2CO_3 solution for pH adjustment and 9.35 µg of antibody for conjugation.

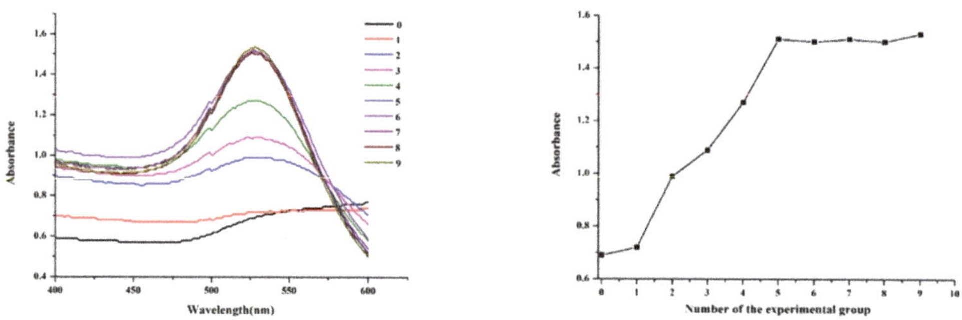

(a) (b)

Figure 4. (**a**) Wavelength scanning of AuNP-labeled antibody with different amounts of antibody. (**b**) The absorbance of AuNP-labeled antibody with different amounts of antibody.

2.3. Establishment of AuNP Lateral-Flow Immunochromatographic Strip Assay

An immunochromatographic strip assay was developed on the basis of an antibody–antigen competitive immunoreaction. The process is depicted in Figure 5. The reaction principle is as follows: In the absence of BADGE in the sample, the AuNP-labeled antibody binds to the coating antigen at the test (T) zone, which turns red. This result is considered negative. Meanwhile, BADGE competes with the coating antigen (T zone) to bind the AuNP-labeled antibody, thereby weakening the color of the T zone. This color weakening is more pronounced as the BADGE concentration increases. The concentrations of the coating antigen and AuNP-labeled antibody in the lateral-flow immunoassay strip assay were optimized. As shown in Figure 6a, the color of the T zone gradually darkened with increasing concentration of the coating antigen up to 20 µg/mL. As shown in Figure 6b, when the volume of AuNP-labeled antibody reached 20 µL, the color of the T zone did not change considerably. Therefore, 20 μL of AuNP-labeled antibody solution was selected for the following test.

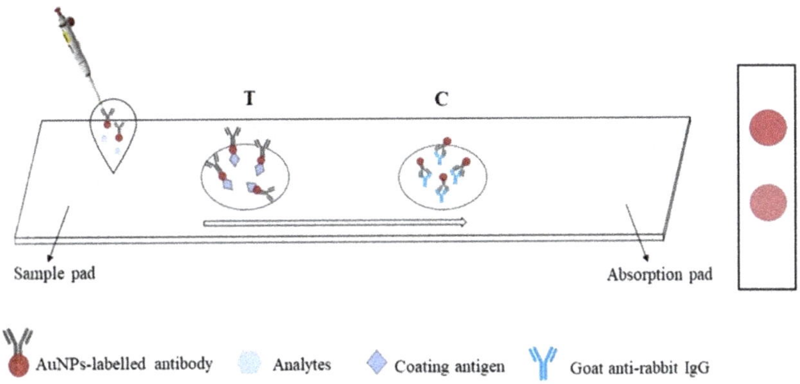

Figure 5. Schematic diagram of AuNPs lateral-flow immunoassay.

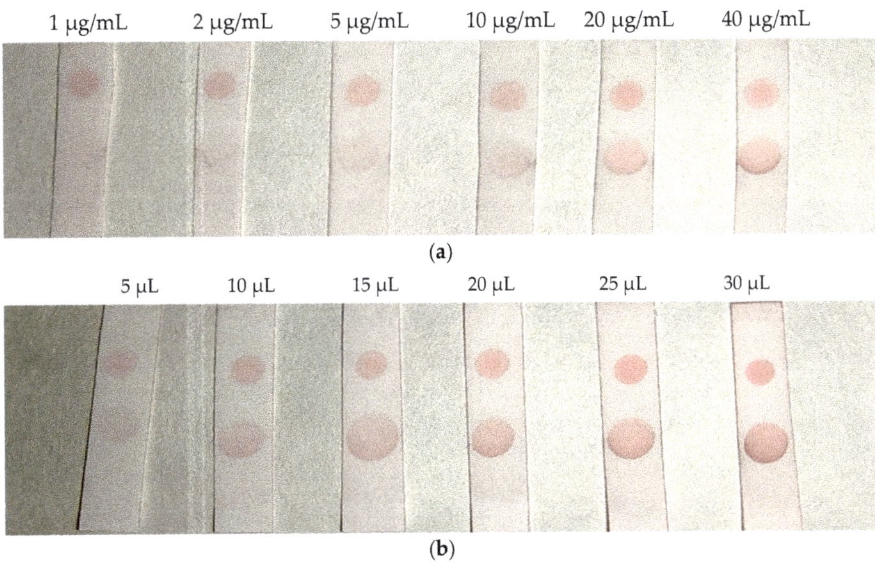

Figure 6. (**a**) Optimization of the coating antigen. (**b**) Optimization of AuNP-labeled antibody.

Under optimal conditions, the AuNP-labeled antibody was mixed with different concentrations of BADGE and added to the strip. As shown in Figure 7a, the intensity of the color of the T zone decreased as the BADGE concentration increased. When the BADGE concentration reached 1 ng/mL, the color of the T zone could be distinguished from that of the negative sample (without BADGE); therefore, the visual detection limit was set to 1 ng/mL.

Color intensities were quantified using Adobe Photoshop CC software. The grayscale variation in the T zone and the standard curve drawn by calculating the inhibition ratio are shown in Figure 7b and Figure 7c, respectively. The calculated limit of detection, which was defined as the concentration causing 15% inhibition (IC15), was 0.97 ng/mL. This result is consistent with the visual result.

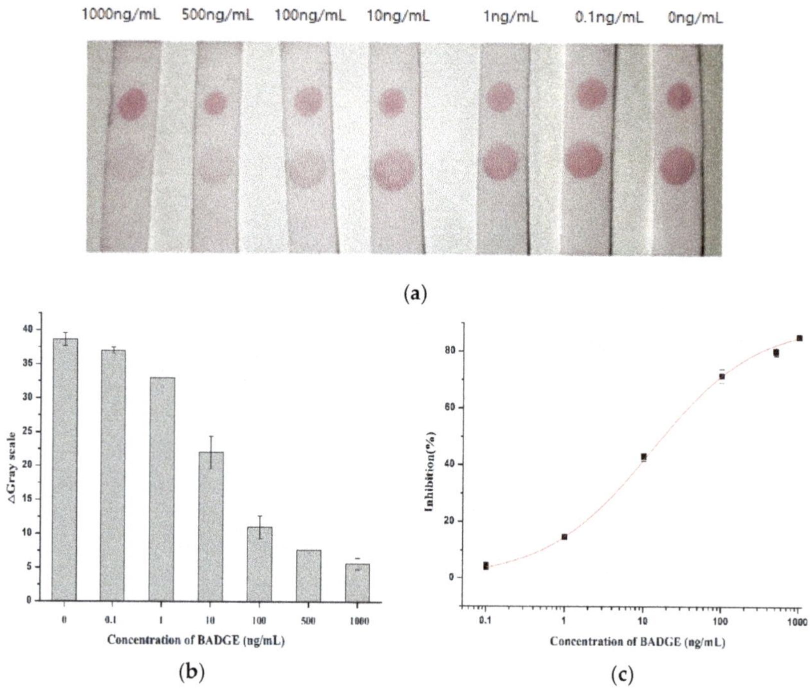

Figure 7. (**a**) Visual results of AuNPs lateral-flow immunoassay strip assay (the concentration of BADGE: 0, 0.1, 1, 10, 100, 500, 1000 ng/mL). (**b**) Grayscale variation in T zones for different concentrations of BADGE. (**c**) Inhibition analysis for BADGE.

2.4. Stability Analysis of AuNPs-PAb Immunochromatographic Strip

The test strips were stored at 4 °C for 1, 3, 5, and 7 days, and the color was observed via the immunochromatographic strip assay using a standard 1 ng/mL BADGE solution. The results are shown in Table 2. No remarkable change was observed in the T zone, indicating the good stability of the test strips for the detection of BADGE.

Table 2. Stability analysis of AuNP-labeled antibody immunochromatographic strip assay.

1 (Day)	3 (Day)	5 (Day)	7 (Day)
	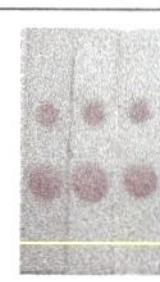		

2.5. Matrix Effect

In the analysis of real samples, the matrix can interfere with the detection accuracy. Therefore, to eliminate the matrix effect, the samples were diluted to different concentrations (5-, 10-, 20-, and 40-fold dilution) and analyzed via ic-ELISA. As shown in Figure 8, the 40-fold dilution could remove the matrix effect almost completely, enabling the analysis using the assay.

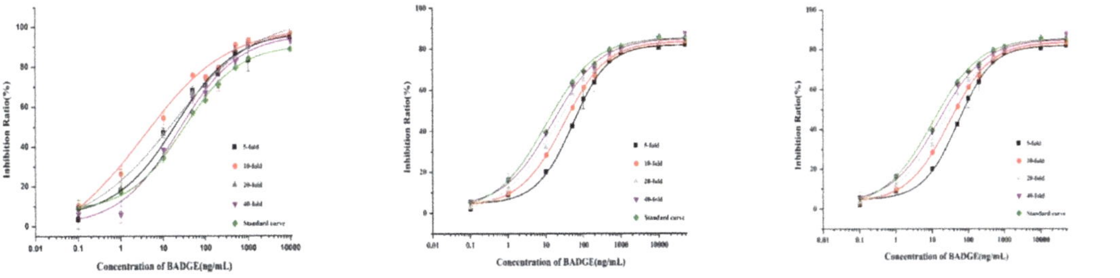

(a) (b) (c)

Figure 8. The optimized dilution of the sample-extracting solution: (**a**) canned luncheon meat; (**b**) canned yellow peach; (**c**) Red Bull drink (Hainan Red Bull Beverage Co., Ltd., Lingao, China).

2.6. Sample Recovery Analysis

Spiking and recovery analyses were performed using the AuNP-labeled antibody immunochromatographic strip assay, which was validated via HPLC analysis (Table 3). The recoveries of BADGE in canned food samples ranged from 79.86% to 93.81%. The correlation coefficient between the AuNP-based lateral-flow immunochromatographic strip assay and HPLC analysis was 0.9580 (Figure 9).

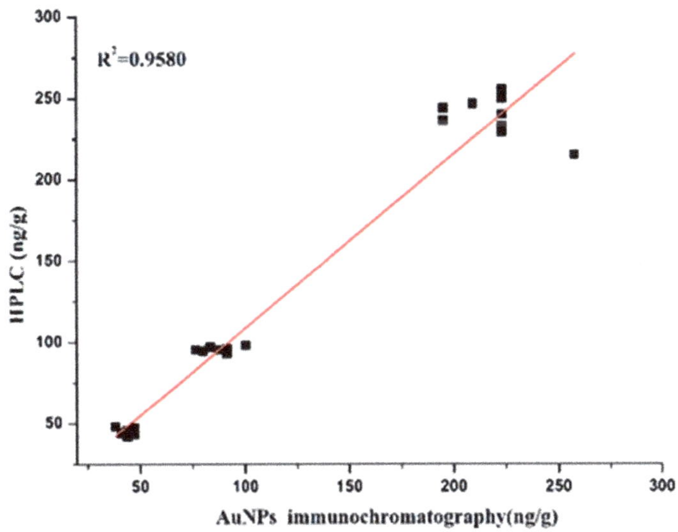

Figure 9. Correlation analysis between AuNPs lateral-flow immunochromatographic strip assay and HPLC.

BADGE and its derivatives may exist simultaneously in real samples, thus the proposed immunochromatographic strip assay was used to detect the total amount of BADGE and its derivatives and validated via HPLC analysis. As shown in Table 4, the total amount of BADGE compounds in the samples determined using the immunochromatographic strip assay was consistent with that obtained via HPLC, demonstrating that the developed AuNP-based lateral-flow immunochromatographic assay could be used as a practical biological monitoring method for the rapid screening of BADGE and its derivatives.

Table 3. Recoveries of BADGE and its derivatives in spiked samples by AuNPs lateral-flow immunochromatographic strip assay and HPLC (n = 3).

Sample	Spiked Conc. (ng/g)	HPLC (n = 3)		AuNPs Lateral-Flow Immunochromatographic Strip Assay (n = 3)		
		Mean ± SD (ng/g)	Recovery (%)		Mean ± SD (ng/g)	Recovery (%)
Canned luncheon meat	50	45.28 ± 0.02	90.56	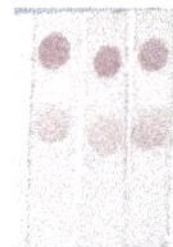	45.67 ± 1.66	91.35
	100	95.71 ± 0.10	95.71	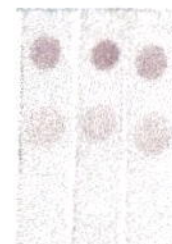	79.86 ± 3.51	79.86
	250	247.56 ± 0.11	99.02		209.10 ± 5.59	83.64
Canned yellow peach	50	43.30 ± 0.03	86.60		44.10 ± 2.60	88.20
	100	96.93 ± 0.09	96.93		91.69 ± 6.94	91.69
	250	243.95 ± 0.74	97.58		213.76 ± 5.27	85.50
Red Bull drink	50	46.97 ± 0.03	93.94		43.17 ± 3.78	86.35
	100	94.82 ± 0.11	94.82		87.35 ± 3.97	87.35
	250	231.41 ± 0.50	92.56	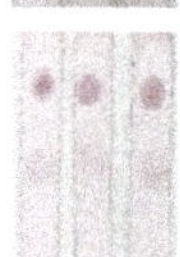	234.54 ± 6.48	93.81

Table 4. Detection of BADGE and its derivatives in samples using AuNPs lateral-flow immunochromatographic strip assay and HPLC (n = 3).

Sample		BADGE and Its Derivatives	HPLC		AuNPs Lateral-Flow Immunochromatographic Strip Assay		
			Mean ± SD (ng/g)	Total Amount	Detection of Concentration Mean ± SD (ng/g)		Actual Concentration Mean ± SD (ng/g)
Canned luncheon meat	1	BADGE BADGE·HCl BADGE·H_2O BADGE·HCl·H_2O	45.03 ± 0.83 85.03 ± 0.36 ND ND	130.06	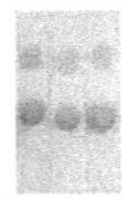	3.14 ± 5.44	125.72
	2	BADGE BADGE·HCl BADGE·H_2O BADGE·HCl·H_2O	39.00 ± 0.80 71.76 ± 0.19 ND ND	110.76		2.57 ± 1.09	103.69
Canned yellow peach	3	BADGE BADGE·HCl BADGE·H_2O BADGE·HCl·H_2O	47.85 ± 0.39 ND 54.25 ± 0.31 ND	102.11	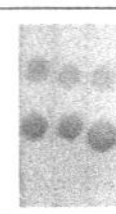	2.48 ± 0.84	99.42
	4	BADGE BADGE·HCl BADGE·H_2O BADGE·HCl·H_2O	61.39 ± 0.25 ND 83.42 ± 0.07 ND	144.78	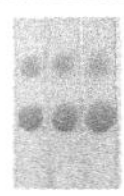	3.41 ± 1.72	136.50
Red Bull drink	5	BADGE BADGE·HCl BADGE·H_2O BADGE·HCl·H_2O	50.54 ± 0.77 ND 126.37 ± 0.82 ND	176.92		4.27 ± 2.64	170.81
	6	BADGE BADGE·HCl BADGE·H_2O BADGE·HCl·H_2O	56.49 ± 2.16 ND 137.18 ± 0.62 ND	193.68		4.67 ± 5.44	187.14

3. Discussion

BADGE is widely used in the preparation of the inner coatings of metal cans. It is often used to avoid direct contact between the food matrix and metals in packaging materials. During processing, storage, and transportation, BADGE is prone to migrating into the food matrix and generating hydrated or chlorinated derivatives, which can accumulate in the human body through the food chain, causing abnormalities in the endocrine and nervous systems. Anna et al. [33] studied the toxicity and potential lipid destruction of BADGE in human placental JEG-3 cells and found that it can interfere with lipid metabolism and alter the cellular lipidome, ultimately causing disease. Instrumental analysis methods have been used to detect BADGE and its derivatives. For instance, Gallo et al. [34] simultaneously determined eight kinds of bisphenol substances, including BADGE, in soft drinks via liquid chromatography–fluorescence.

Due to the wide polarities of BADGE and its derivatives, conventional LC requires complex sample pretreatment to complete the separation, which cannot meet the need for rapid detection. Moreover, traditional instrumental analysis methods cannot realize the simple and rapid analysis of large quantities of samples or the real-time monitoring of migrated residual pollution.

In contrast, immunoassays based on antibodies or other biological molecules enable rapid analysis with simple pretreatments and constitute an effective supplementary method to conventional instrumental analysis methods. In fact, immunoassays have been widely used for the analysis of various pollutants in the field of food safety because of their reliability, efficiency, and cost savings. Accordingly, they have been recommended by institutional analysis departments for the rapid screening of large quantities of samples. Considering the health hazards posed by BADGE compounds, the monitoring of food via long-term, multifrequency, high-throughput sample analysis is required, for which immunoassays are suitable methods. Guan [35] developed a fluorescence polarization assay for the simultaneous monitoring of BPA, BPF, BADGE, and BFDGE in canned tuna with detection limits of 0.35, 0.08, 0.10, and 0.49 mg/L, respectively. The method was not sufficiently sensitive for the detection of BADGE, which may be because the receptor cannot completely replace the biological antibody to achieve high-affinity recognition of the target compound. The broad-spectrum polyclonal antibody used in this study can simultaneously recognize BADGE and its derivatives, enabling the development of a rapid AuNP-based lateral-flow immunochromatographic strip assay for the detection of BADGE and its hydrolyzed and chlorinated derivatives in canned food. The assay can be performed within 15 min. The visual results were processed using Adobe Photoshop CC software, and the detection limit (IC15) was 0.97 ng/mL. Therefore, this method can meet the need for real-time screening and the detection of large quantities of samples. The developed AuNPs-based lateral-flow immunochromatography assay is compared with other methods reported for the detection of BADGE and its derivatives in Table 5.

Table 5. Comparison of analytical methods for the detection of bisphenol derivatives (including BADGE).

Detection Method	Detection Targets	Detection Results	References
HPLC-FLD	Bisphenols, bisphenol diglycidyl ethers and their derivatives	Quantitation limits for the analytes ranged between 0.9 and 3.5 μg kg^{-1}	Ref. [16]
HPLC-FLD	5 bisphenol derivatives including BADGE	Limits of detection (LODs) were between 21 and 28 ng/mL	Ref. [17]
HPLC-FLD	Bisphenol A diglycidyl ether and its derivatives	LODs varied from 0.01 to 0.20 ng/g	Ref. [18]
LC–MS/MS	BPA and its derivatives including BADGE	Quantification limits were in the range of 2–10 μg kg^{-1}	Ref. [19]
UHPLC-ESI-MS/MS	Bisphenol ethers and their derivatives	Limits of quantitation (LOQs) for the analytes ranged from 0.02 to 5 mg/kg	Ref. [21]
GC-MS	Bisphenols and their diglycidyl ethers	Migration of BPA is between 104.67 and 181.46 μg L^{-1}	Ref. [22]
ic-ELISA	BADGE BADGE·HCl BADGE·H_2O BADGE·HCl·H_2O	IC_{15} of BADGE, BADGE·H_2O, BADGE·HCl, BADGE·HCl·H_2O were 0.73, 0.39, 0.78, 1.45 ng/mL	Ref. [36]
AuNPs-based immunochromatographic strip assay	BADGE BADGE·HCl BADGE·H_2O BADGE·HCl·H_2O	Simultaneous detection of BADGE and its derivatives within 15 min. Visual detection limit was 1 ng/mL for BADGE.	This work

4. Materials and Methods

4.1. Chemicals and Reagents

Chloroauric acid ($HAuCl_4 \cdot 4H_2O$), trisodium citrate dihydrate, sodium chloride (NaCl), K_2CO_3, disodium hydrogen phosphate dodecahydrate ($Na_2HPO_4 \cdot 12H_2O$), sodium phosphate dibasic dihydrate ($Na_2HPO_4 \cdot 2H_2O$), sulfuric acid (H_2SO_4), Tween-20, ethyl acetate, hexane, acetonitrile, and methanol were purchased from Sinopharm Chemical Reagent (Shanghai, China). The chemical standards BADGE, BADGE·H_2O, BADGE·HCl, and BADGE·HCl·H_2O were purchased from Macklin (Shanghai, China). Bovine serum albumin (BSA), nonfat milk powder, and goat anti-rabbit IgG-HRP were purchased from Sangon

Biotech (Shanghai, China). Canned luncheon meat, canned yellow peach, and Red Bull drinks were purchased from a local supermarket (Zhenjiang, China).

4.2. Instruments

An HH-A magnetic stirrer was purchased from Zhongda Instrument Factory (Changzhou, China), and a UV-1801 ultraviolet spectrophotometer was purchased from Ruili Company (Beijing, China). A smartphone (Huawei P40 with a rear camera of 8 million pixels, Huawei Technologies Co., Ltd, Shenzhen, China) was used to take photos. HPLC was performed using an LC-20AD module (Shimadzu, Japan) coupled to an ultraviolet detector. A Hypersil GOLD C18 column (4.6 mm × 250 mm, 5 µm) was employed for chromatographic separation in the HPLC analysis. The mobile phase was composed of 40% ultrapure water and 60% acetonitrile. The eluent flow rate was 0.5 mL/min, the injection volume was 10 µL, and the temperature of the column oven was maintained at 30 °C.

4.3. Screening of Broad-Spectrum Antibodies

Four antisera with certain recognition abilities for BADGE and its derivatives were screened. The cross-reactivity of the four antisera, which was investigated to determine their broad-spectrum recognition specificity, was calculated as follows:

$$\text{Cross-reactivity (\%)} = IC_{50}\ (\text{BADGE})/IC_{50}\ (\text{other analogs}) \times 100$$

4.4. Preparation of AuNPs

AuNPs were produced according to the sodium citrate method. Briefly, 100 mL of a 0.01% $HAuCl_4$ solution was heated and kept boiling. Then, 4 mL of 1% sodium citrate was quickly added under stirring, and the resulting solution was heated and stirred for 5–10 min until the color was stable. The obtained AuNP solution was cooled to room temperature and stored at 4 °C.

4.5. Preparation of AuNP-Labeled Antibody

4.5.1. Optimization of pH Value of AuNPs Solution

The pH value of the AuNP solution was adjusted by adding different amounts of K_2CO_3 solution. Excess amounts of antibodies were added and allowed to stand at room temperature for 10 min. Further, 10% NaCl was added and vortex-mixed. After standing for 10 min, the color was observed and scanned via UV–vis spectroscopy.

4.5.2. Optimization of the Amounts of Antibody

As shown in Table 6, different amounts of antibody were added to the pH-adjusted AuNP solution, and the mixture was allowed to stand at room temperature for 10 min. Then, 10% NaCl was added and vortex-mixed. After standing for 10 min, the color was observed and scanned via UV–vis spectroscopy.

Table 6. pH-adjusted AuNPs solution for labeling different amounts of antibody.

Number of Experimental Group	0	1	2	3	4	5	6	7	8	9
Antibody (µg)	0	0.37	1.49	3.36	5.98	9.35	13.51	18.33	23.93	30.32
AuNPs solution (mL)	1	1	1	1	1	1	1	1	1	1

Overall, 10 mL of a AuNP solution was adjusted to the optimal pH value by adding a 0.2 M K_2CO_3 solution. The antibody was added to the resulting AuNP solution, and the mixture was allowed to react at 4 °C for 330 min. Then, 10% BSA was added to block the unbinding sites. The solution was centrifuged at 12,000 rpm for 20 min, and the precipitate was redissolved in 1 mL of PB solution containing 0.05% Tween-20 and 1% BSA. The resulting solution was stored at 4 °C until use.

4.6. Establishment of AuNPs Lateral-Flow Immunochromatographic Assay

A typical immunochromatographic test strip consisted of a sample pad, a conjugate or reagent pad, a capture zone (test line and control line), and an absorbent pad. A nitrocellulose membrane with smooth backing was used to prepare the immunochromatographic strips, which were cut to a certain size. The T zone was coated with the coating antigen, and the control (C) zone was coated with goat anti-rabbit IgG. The strip was then dried at room temperature and stored at 4 °C. A certain volume of AuNP-labeled antibody was mixed with 100 μL of a BADGE standard solution and incubated for 10 min. The mixture was added to the sample zone. After 15 min, the liquid completely flowed through the T and C zones. The visualized results were recorded using a smartphone (Huawei Technologies Co., Ltd., Shenzhen, China) and processed using Adobe Photoshop CC software for quantitative analysis. All photos were recorded during the daytime without a flashlight.

Inhibition was calculated using the following equation:

$$\text{Inhibition } (\%) = (\triangle G - \triangle G1)/(\triangle G) \times 100\%,$$

where G and G1 represent the grayscale variation without and with different concentrations of BADGE (or its derivatives), respectively.

4.7. Sample Analysis

Canned luncheon meat, canned yellow peach, and Red Bull drinks obtained from a local supermarket were selected as samples for analysis. The canned luncheon meat samples were treated as follows: 2.0 g of sample was homogenized with 10 mL hexane and subjected to ultrasonic-assisted extraction for 30 min. The mixture was centrifuged at 4500 rpm for 10 min. The supernatant was extracted twice using 5 mL of acetonitrile. The acetonitrile extracts were evaporated to dryness at 40 °C under nitrogen. Subsequently, the residues were redissolved in 2 mL of methanol. The canned yellow peach samples were treated as follows: 2.0 g of sample was weighed, and 5 mL of ethyl acetate was added as the extraction solvent. The mixture was shaken for 20 min in a shaker and then transferred to an ultrasonic bath for 30 min. The mixture was centrifuged at 4500 rpm for 15 min. The supernatant was evaporated to dryness under a nitrogen stream. Furthermore, the extract was redissolved in 2 mL of methanol. The Red Bull samples (5 mL) were centrifuged at 4500 rpm for 10 min, and the supernatant was filtered through a 0.22 μm membrane before analysis. The samples were separately spiked with BADGE at different concentrations (50, 100, and 250 ng/g) to evaluate the accuracy of the AuNP-based lateral-flow immunochromatographic strip assay, and the results were confirmed using HPLC.

5. Conclusions

A broad-spectrum polyclonal antibody that can recognize BADGE and its BADGE·HCl, BADGE·H_2O, and BADGE·HCl·H_2O derivatives was evaluated and coupled with AuNPs to develop a AuNP-based lateral-flow immunochromatographic strip assay for the simultaneous detection of BADGE and its derivatives in canned food. The visualized results were processed using Adobe Photoshop CC software and validated via HPLC analysis. The developed broad-spectrum lateral-flow immunochromatographic strip assay is reliable and accurate and meets the requirements for the rapid screening of large quantities of samples because of its simplicity, rapid operation, and cost effectiveness.

Author Contributions: Q.Y. and C.Z. constructed the project; C.Y., Q.Y., and C.Z. designed the experiments; C.Y., J.H., and Y.Z. performed the experiments; C.Y., C.Z., and W.W. analyzed the data; C.Y. wrote the paper. All authors have read and agreed to the published version of the manuscript.

Funding: This work was financially supported by the Special Foundation for Distinguished Taishan Scholar of Shandong Province (tstp20230630), National Key Research and Development Program of China (No. 2022YFF1102200), National Natural Science Foundation of China (No. U22A20541), the High-level Talent Start-up Fund from Qingdao Agricultural University, Science & Technology Specific Projects in Agricultural High-tech Industrial Demonstration Area of the Yellow River Delta (2022SZX26).

Institutional Review Board Statement: Not applicable.

Informed Consent Statement: Not applicable.

Data Availability Statement: Data are contained within the article.

Conflicts of Interest: The authors declare no conflict of interest.

References

1. Cabado, A.G.; Aldea, S.; Porro, C.; Ojea, G.; Lago, J.; Sobrado, C.; Vieites, J.M. Migration of BADGE (bisphenol A diglycidyl-ether) and BFDGE (bisphenol F diglycidyl-ether) in canned seafood. *Food Chem. Toxicol.* **2008**, *46*, 1674–1680. [CrossRef] [PubMed]
2. García, R.S.; Losada, P.P.; Lamela, C.P. Determination of Compounds from Epoxy Resins in Food Simulants by HPLC-Fluorescence. *Chromatographia* **2003**, *58*, 337–342. [CrossRef]
3. Jordáková, I.; Dobiáš, J.; Voldřich, M.; Postka, J. Determination of bisphenol A, bisphenol F, bisphenol A diglycidyl ether and bisphenol F diglycidyl ether migrated from food cans using Gas Chromatography-Mass Spectrometry. *Czech J. Food Sci.* **2003**, *21*, 85–90. [CrossRef]
4. Jana, S.K.; Okamoto, T.; Kugita, T.; Namba, S. Selective synthesis of bisphenol F catalyzed by microporous H-beta zeolite. *Appl. Catal. A Gen.* **2005**, *288*, 80–85. [CrossRef]
5. Leepipatpiboon, N.; Sae-Khow, O.; Jayanta, S. Simultaneous determination of bisphenol-A-diglycidyl ether, bisphenol-F-diglycidyl ether, and their derivatives in oil-in-water and aqueous-based canned foods by high-performance liquid chromatography with fluorescence detection. *J. Chromatogr. A* **2005**, *1073*, 331–339. [CrossRef] [PubMed]
6. Bolt, H.M.; Stewart, J.D. Highlight report: The bisphenol A controversy. *Arch. Toxicol.* **2011**, *85*, 1491–1492. [CrossRef] [PubMed]
7. Kang, J.H.; Kondo, F. Bisphenol A in the Surface Water and Freshwater Snail Collected from Rivers Around a Secure Landfill. *Bull. Environ. Contam. Toxicol.* **2006**, *76*, 113–118. [CrossRef]
8. García, R.S.; Losada, P.P. Determination of bisphenol A diglycidyl ether and its hydrolysis and chlorohydroxy derivatives by liquid chromatography-mass spectrometry. *J. Chromatogr. A* **2004**, *1032*, 37–43. [CrossRef]
9. Le, H.H.; Carlson, E.M.; Chua, J.P.; Belcher, S.M. Bisphenol A is released from polycarbonate drinking bottles and mimics the neurotoxic actions of estrogen in developing cerebellar neurons. *Toxicol. Lett.* **2008**, *176*, 149–156. [CrossRef]
10. Gallart-Ayala, H.; Moyano, E.; Galceran, M.T. Fast liquid chromatography-tandem mass spectrometry for the analysis of bisphenol A-diglycidyl ether, bisphenol F-diglycidyl ether and their derivatives in canned food and beverages. *J. Chromatogr. A* **2011**, *1218*, 1603–1610. [CrossRef]
11. Fattore, M.; Russo, G.; Barbato, F.; Grumetto, L.; Albrizio, S. Monitoring of bisphenols in canned tuna from Italian markets. *Food Chem. Toxicol.* **2015**, *83*, 68–75. [CrossRef] [PubMed]
12. Poole, A.; van Herwijnen, P.; Weideli, H.; Thomas, M.C.; Ransbotyn, G.; Vance, C. Review of the toxicology, human exposure and safety assessment for bisphenol A diglycidylether (BADGE). *Food Addit. Contam.* **2004**, *21*, 905–919. [CrossRef] [PubMed]
13. Satoh, K.; Ohyama, K.; Aoki, N.; Iida, M.; Nagai, F. Study on anti-androgenic effects of bisphenol a diglycidyl ether (BADGE), bisphenol F diglycidyl ether (BFDGE) and their derivatives using cells stably transfected with human androgen receptor, AR-EcoScreen. *Food Chem. Toxicol.* **2004**, *42*, 983–993. [CrossRef] [PubMed]
14. Ramilo, G.; Valverde, I.; Lago, J.; Vieites, J.M.; Cabado, A.G. Cytotoxic effects of BADGE (bisphenol A diglycidyl ether) and BFDGE (bisphenol F diglycidyl ether) on Caco-2 cells in vitro. *Arch. Toxicol.* **2006**, *80*, 748–755. [CrossRef] [PubMed]
15. European Commission Regulation (EC) No. 1895/2005. The restriction of use of certain epoxy derivatives in materials and articles intended to come into contact with food. *Off. J. Eur. Union* **2005**, *L 302/28*, 28–32.
16. Alabi, A.; Caballero-Casero, N.; Rubio, S. Quick and simple sample treatment for multiresidue analysis of bisphenols, bisphenol diglycidyl ethers and their derivatives in canned food prior to liquid chromatography and fluorescence detection. *J. Chromatogr. A* **2014**, *1336*, 23–33. [CrossRef] [PubMed]
17. Fischnaller, M.; Bakry, R.; Bonn, G.K. A simple method for the enrichment of bisphenols using boron nitride. *Food Chem.* **2016**, *194*, 149–155. [CrossRef]
18. Guo, M.; He, M.; Zhong, J.; He, Q.; Ismail, B.B.; Chen, G.; Liu, D. High-performance liquid chromatography (HPLC)-fluorescence method for determination of bisphenol A diglycidyl ether (BADGE) and its derivatives in canned foods. *Sci. Total. Environ.* **2020**, *710*, 134975. [CrossRef]
19. Míguez, J.; Herrero, C.; Quintás, I.; Rodríguez, C.; Gigosos, P.; Mariz, O. A LC–MS/MS method for the determination of BADGE-related and BFDGE-related compounds in canned fish food samples based on the formation of $[M^{+}NH4]^{+}$ aducts. *Food Chem.* **2012**, *135*, 1310–1315. [CrossRef]

20. Lane, R.; Adams, C.; Randtke, S.; Carter, R. Bisphenol diglycidyl ethers and bisphenol A and their hydrolysis in drinking water. *Water Res.* **2015**, *72*, 331–339. [CrossRef]
21. Cheng, Y.; Nie, X.-M.; Wu, H.-Q.; Hong, Y.-H.; Yang, B.-C.; Liu, T.; Zhao, D.; Wang, J.-F.; Yao, G.-H.; Zhang, F. A high-throughput screening method of bisphenols, bisphenols digycidyl ethers and their derivatives in dairy products by ultra-high performance liquid chromatography-tandem mass spectrometry. *Anal. Chim. Acta* **2017**, *950*, 98–107. [CrossRef] [PubMed]
22. Oca, M.L.; Ortiz, M.C.; Herrero, A.; Sarabia, L.A. Optimization of a GC/MS procedure that uses parallel factor analysis for the determination of bisphenols and their diglycidyl ethers after migration from polycarbonate tableware. *Talanta* **2013**, *106*, 266–280. [CrossRef] [PubMed]
23. Bahadır, E.B.; Sezgintürk, M.K. Lateral flow assays: Principles, designs and labels. *Trends Anal. Chem.* **2016**, *82*, 286–306. [CrossRef]
24. Kim, J.; Shin, M.-S.; Shin, J.; Kim, H.-M.; Pham, X.-H.; Park, S.-M.; Kim, D.-E.; Kim, Y.J.; Jun, B.-H. Recent Trends in Lateral Flow Immunoassays with Optical Nanoparticles. *Int. J. Mol. Sci.* **2023**, *24*, 9600. [CrossRef] [PubMed]
25. Huang, X.; Aguilar, Z.P.; Xu, H.; Lai, W.; Xiong, Y. Membrane-based lateral flow immunochromatographic strip with nanoparticles as reporters for detection: A review. *Biosens. Bioelectron.* **2016**, *75*, 166–180. [CrossRef] [PubMed]
26. Ji, Y.; Ren, M.; Li, Y.; Huang, Z.; Shu, M.; Yang, H.; Xiong, Y.; Xu, Y. Detection of aflatoxin B1 with immunochromatographic test strips: Enhanced signal sensitivity using gold nanoflowers. *Talanta* **2015**, *142*, 206–212. [CrossRef]
27. Jin, Y.; Liu, R.; Zhu, L.; Liu, Z. Development of high sensitivity immunochromatographic test card for zearalenone. *J. Hyg. Res.* **2019**, *48*, 651–658.
28. Kong, D.; Liu, L.; Song, S.; Suryoprabowo, S.; Li, A.; Kuang, H.; Wang, L.; Xu, C. A gold nanoparticle-based semi-quantitative and quantitative ultrasensitive paper sensor for the detection of twenty mycotoxins. *Nanoscale* **2016**, *8*, 5245–5253. [CrossRef]
29. Lin, L.; Wu, X.; Luo, P.; Song, S.; Zheng, Q.; Kuang, H. IC-ELISA and immunochromatographic strip assay based monoclonal antibody for the rapid detection of bisphenol S. *Food Agric. Immunol.* **2019**, *30*, 633–646. [CrossRef]
30. He, F.; Yang, J.; Zou, T.; Xu, Z.; Tian, Y.; Sun, W.; Wang, H.; Sun, Y.; Lei, H.; Chen, Z.; et al. A gold nanoparticle-based immunochromatographic assay for simultaneous detection of multiplex sildenafil adulterants in health food by only one antibody. *Anal. Chim. Acta* **2021**, *1141*, 1–12. [CrossRef]
31. Yang, X.-D.; Wang, F.-Y.; Song, C.-M.; Wu, S.-Y.; Zhang, G.-P.; Zeng, X.-Y. Establishment of a lateral flow colloidal gold immunoassay strip for the rapid detection of estradiol in milk samples. *LWT* **2015**, *64*, 88–94. [CrossRef]
32. Ge, W.-L.; Lin, L.; Xu, C.-L. Study on immunoassay of bisphenol A diglycidyl ether. *Packag. Eng.* **2019**, *40*, 59–63.
33. Marqueño, A.; Pérez-Albaladejo, E.; Flores, C.; Moyano, E.; Porte, C. Toxic effects of bisphenol A diglycidyl ether and derivatives in human placental cells. *Environ. Pollut.* **2018**, *244*, 513–521. [CrossRef] [PubMed]
34. Gallo, P.; Di Marco Pisciottano, I.; Esposito, F.; Fasano, E.; Scognamiglio, G.; Mita, G.D.; Cirillo, T. Determination of BPA, BPB, BPF, BADGE and BFDGE in canned energy drinks by molecularly imprinted polymer cleaning up and UPLC with fluorescence detection. *Food Chem.* **2017**, *220*, 406–412. [CrossRef]
35. Guan, T.; Li, T.; Zhang, T.; Li, Z.; Wang, Y.; Yu, H.; Ruan, P.; Zhang, J.; Wang, Y. Fluorescence polarization assay for the simultaneous determination of bisphenol A, bisphenol F and their diglycidyl ethers in canned tuna. *Int. J. Food Prop.* **2017**, *20*, S1920–S1929. [CrossRef]
36. Zhang, C.; Hu, J.; Song, J.; Wu, M.; Zhang, Z. Development of ic-ELISAs for the Detection of Bisphenol A Diglycidyl Ether and Its Derivatives in Canned Luncheon Meats. *ACS Food Sci. Technol.* **2022**, *2*, 160–168. [CrossRef]

molecules

Communication

Systematic Analysis of 2′-*O*-Alkyl Modified Analogs for Enzymatic Synthesis and Their Oligonucleotide Properties

Kenta Ishida [1,2], Yuuya Kasahara [1,2,*], Hidekazu Hoshino [1], Takumi Okuda [1,2] and Satoshi Obika [1,2,*]

[1] National Institutes of Biomedical Innovation, Health and Nutrition (NIBIOHN), 7-6-8 Saito-Asagi, Ibaraki 567-0085, Osaka, Japan; kenta-ishida@nibiohn.go.jp (K.I.); h-hoshino@nibiohn.go.jp (H.H.)

[2] Graduate School of Pharmaceutical Sciences, Osaka University, 1-6 Yamadaoka, Suita 565-0871, Osaka, Japan

* Correspondence: y-kasahara@nibiohn.go.jp (Y.K.); obika@phs.osaka-u.ac.jp (S.O.); Tel.: +81-72-641-9882 (Y.K.); +81-6-6879-8200 (S.O.)

Abstract: Enzymatic oligonucleotide synthesis is used for the development of functional oligonucleotides selected by in vitro selection. Expanding available sugar modifications for in vitro selection helps the functional oligonucleotides to be used as therapeutics reagents. We previously developed a KOD DNA polymerase mutant, KOD DGLNK, that enzymatically synthesized fully-LNA- or 2′-*O*-methyl-modified oligonucleotides. Here, we report a further expansion of the available 2′-*O*-alkyl-modified nucleotide for enzymatic synthesis by KOD DGLNK. We chemically synthesized five 2′-*O*-alkyl-5-methyluridine triphosphates and incorporated them into the oligonucleotides. We also enzymatically synthesized a 2′-*O*-alkyl-modified oligonucleotide with a random region (oligonucleotide libraries). The 2′-*O*-alkyl-modified oligonucleotide libraries showed high nuclease resistance and a wide range of hydrophobicity. Our synthesized 2′-*O*-alkyl-modified oligonucleotide libraries provide novel possibilities that can promote the development of functional molecules for therapeutic use.

Keywords: enzymatic oligonucleotide synthesis; aptamers; 2′-*O*-alkyl modifications; polymerase mutants

check for updates

Citation: Ishida, K.; Kasahara, Y.; Hoshino, H.; Okuda, T.; Obika, S. Systematic Analysis of 2′-*O*-Alkyl Modified Analogs for Enzymatic Synthesis and Their Oligonucleotide Properties. *Molecules* **2023**, *28*, 7911. https://doi.org/10.3390/molecules28237911

Academic Editors: Sai Wang, Wei Wu and Long Wu

Received: 30 October 2023
Revised: 29 November 2023
Accepted: 30 November 2023
Published: 2 December 2023

1. Introduction

Enzymatic oligonucleotide synthesis is applied for the development of functional oligonucleotides so that their structure can contribute to their activities such as aptamers and nucleic acid enzymes. Aptamers are single-stranded oligonucleotides that specifically bind target molecules [1]. Nucleic acid enzymes are single-stranded oligonucleotides that have catalytic activities such as the site-specific cleavage of RNA [2]. These functional molecules are selected from $10^{12\text{–}15}$ unique oligonucleotide sequences (oligonucleotide library) using in vitro selection [3,4]. Since these functional oligonucleotides mainly consist of DNA or RNA, they are easily digested by nuclease [1,2], which diminishes the potential of using them as therapeutic agents.

The chemical modification of functional oligonucleotides enhances their properties such as their nuclease stability and binding affinity to target molecules. Therefore, expanding the available modified nucleotides for in vitro selection helps to develop functional oligonucleotides. Many nucleobase analogs have been developed and incorporated into aptamers to enhance their binding affinity. For example, incorporating hydrophobic nucleobase analogs like 5-modified uridine analogs into aptamers has been shown to improve their binding affinity [5–8]. On the other hand, since nucleobase-modified analogs contribute less to improving nuclease stability [9], the expansion of available sugar modifications for in vitro selection is necessary to promote the development of the functional molecules for therapeutic use. However, wild-type polymerases have a low tolerance for sugar modifications.

To expand the available sugar modifications for in vitro selection, we focused on 2′-*O*-alkyl modifications. Since the 2′-*O*-methyl (2′-OMe)-modified oligonucleotide exhibited a higher resistance against nucleases and higher thermal duplex stability for RNA than unmodified DNA, 2′-OMe modification has been used for oligonucleotide therapeutics [10,11]. Moreover, many 2′-*O*-alkyl-modified analogs have also been developed and investigated for their impact on oligonucleotides [12–17]. For instance, 2′-*O*-linear-carbon-chain-modified analogs, such as 2′-*O*-ethyl (2′-OEt), 2′-*O*-propyl (2′-OPr), and 2′-*O*-butyl (2′-OBu) modifications, tended to show higher nuclease stability depending on their substituted carbon chain length [13]. On the other hand, the 2′-*O*-substitution of a longer linear carbon chain tended to decrease the duplex stability of RNA [14,15]. Furthermore, the substitution of single carbon to oxygen in their alkyl chain, such as 2′-*O*-methoxyethyl (2′-MOE) and 2′-*O*-hydroxyethyl (2′-HE) modifications, enhance the duplex stability of RNA [14,16]. Moreover, 2′-MOE showed higher nuclease stability than the 2′-OMe modification [17]. The 2′-*O*-branched-carbon-chain-modified analog, 2′-*O*-isopropyl modification (2′-O^{i}Pr), decreased the duplex stability of RNA [12,14].

Although several wild-type polymerases have the potential to polymerize sugar-modified nucleotides [18–20], sugar modification is still challenging to incorporate by wild-type DNA polymerase. A combination of multiple mutations in thermostable DNA polymerases can improve the polymerization of 2′-*O*-substituted nucleotides. Several polymerase mutants have been developed to synthesize 2′-OMe-modified oligonucleotides [21–23]. Chen et al. developed Taq DNA polymerase mutants which could synthesize fully-2′-OMe-modified oligonucleotides [21]. Moreover, Freund et al. recently developed Tgo polymerase mutants, which enzymatically synthesized not only fully-2′-OMe-modified oligonucleotides but also 2′-MOE-modified oligonucleotides [22]. We also developed a KOD polymerase mutant, KOD DGLNK, for LNA-modified oligonucleotide synthesis, which can also efficiently synthesize fully-2′-OMe-modified oligonucleotides [23]. KOD DGLNK had multiple mutations (N210D, Y409G, A485L, D614N, and E664K). The N210D substitution reduces exonuclease activity [24]. The Y409 residue is predicted to sterically clash with the 2′-OH of ribonucleotide triphosphates [25]. The Y409G substitution reduces this steric clash. The A485L substitution promotes the sending of triphosphates to the active site [26]. The D614N and E664K substitutions enhance the binding affinity between the polymerase and primer-template duplex [23,26,27]. These multiple mutations significantly promote the synthesis of fully-LNA- and 2′-OMe-modified oligonucleotides. Additionally, we developed a 2′-OMe-and-LNA-mix aptamer [23]. However, there has been no systematic analysis of the tolerance range for the enzymatic synthesis of the 2′-*O*-alkyl modifications. And the properties of synthesized oligonucleotide will also be of interest.

Here, we synthesized five 2′-*O*-alkyl-5-methyluridine triphosphates (2′-*O*-ethyl 5-methyuridine triphosphate [2′-OEt-^{5m}UTP], 2′-*O*-propyl 5-methyuridine triphosphate [2′-OPr-^{5m}UTP], 2′-*O*-butyl 5-methyuridine triphosphate [2′-OBu-^{5m}UTP], 2′-*O*-isopropyl 5-methyuridine triphosphate [2′-O^{i}Pr-^{5m}UTP], and 2′-*O*-hydroxyethyl 5-methyuridine triphosphate [2′-HE-^{5m}UTP]) and incorporated them into oligonucleotides by KOD DGLNK (Figure 1). Moreover, we evaluated the nuclease stability and hydrophobicity of the resulting 2′-*O*-alkyl-modified oligonucleotides.

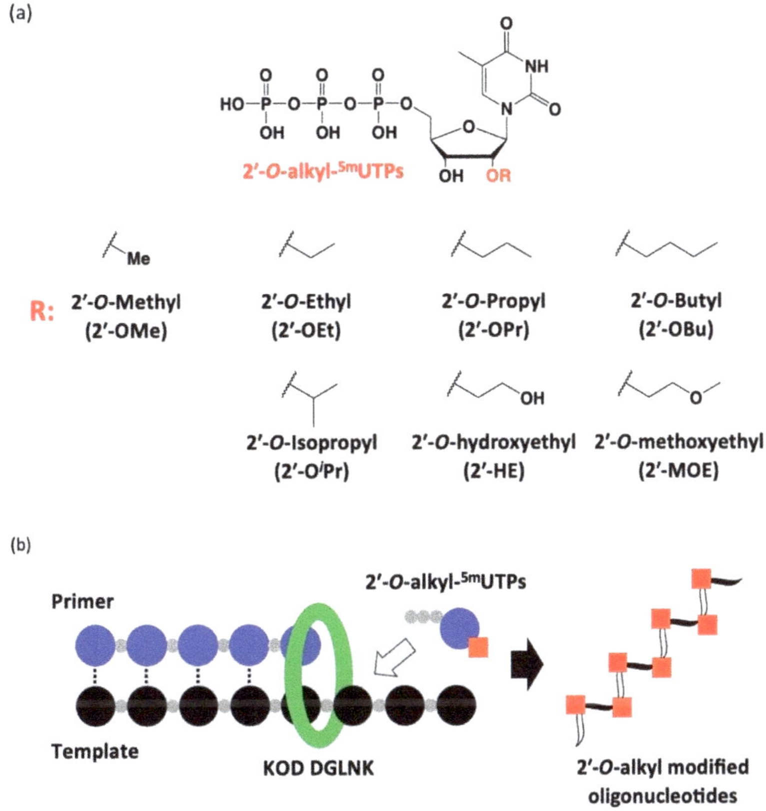

Figure 1. (**a**) Chemical structures of the 2′-O-alkyl-5-methyluridine triphosphates (2′-O-alkyl-^{5m}UTPs) that we used in this study. (**b**) Schematic illustration of the incorporation of 2′-O-alkyl-^{5m}UTPs into an oligonucleotide.

2. Results and Discussion

2.1. Synthesis of 2′-O-Alkyl-ribonucleotide Triphosphates

KOD DGLNK has been modified based on DNA polymerase, and it lacks mutations that enhance its tolerance for uridine. Consequently, we opted for 5-methyluridine (thymidine), a component of DNA, and not uridine, a component of RNA, as the substrate for KOD DGLNK. Subsequently, we synthesized 2′-OEt-^{5m}UTP, 2′-OPr-^{5m}UTP, 2′-OBu-^{5m}UTP, 2′-O^{i}Pr-^{5m}UTP, and 2′-HE-^{5m}UTP. The 2′-O-alkylations were systematically performed with reference to the methods previously used to synthesize other 2′-O-alkyl-U or ^{5m}Us [16,28]. The synthesis of 2′-OEt-^{5m}UTP, 2′-OPr-^{5m}UTP, 2′-OBu-^{5m}UTP and 2′-O^{i}Pr-^{5m}UTP were started from 3′-,5′-silyl- and N^3-benzoyl-protected 5-methyluridine [29]. The 2′-hydroxy group was alkylated via a nucleophilic attack of the 2′-hydroxy group to the corresponding alkyl iodide. After switching the protection group, the resulting 3′-acetyl- and N^3-benzoyl-protected 2′-O-alkyl-5-methyluridines were substituted to triphosphates using the Ludwig and Eckstein method [30] (Scheme S1). The synthesis of 2′-HE-^{5m}UTP was started from 2,2′-anhydro-5-methyluridine. After TBS protection of the 5′-hydroxy group, the alkylation of the 2′-hydroxy group was performed by hydroxyethylation using trihydroxyethyl borate esters. This was followed by switching the protection group, resulting in 3′- and 2′-O-hydroxyethyl-acetyl-protected 5-methyuridine being substituted to triphosphate in the same manner (Scheme S2). All the synthesized triphosphates were purified by the ion pair RP-HPLC.

2.2. Polymerase Incorporation of 2′-O-Alkyl 5-methyluridine Triphosphates into Oligonucleotide

First, we evaluated the incorporation efficiency of 2′-*O*-alkyl-^{5m}UTPs using a primer extension assay against a DNA template (Template**1**) which had a 5nt consecutive adenosine. An extended FAM-labeled DNA primer (Primer**1**) was analyzed by denaturing polyacrylamide gel electrophoresis (PAGE) (Figure 2). Following an incubation period of 30 min at 74 °C with KOD DGLNK (50 ng/μL), we obtained a fully extended product with dTTP, and all of the 2′-*O*-alkyl-^{5m}UTPs showed at least single nucleotide incorporation (Figure 2a). This result suggests that 2′-*O*-alkyl-^{5m}UTPs potentially works as substrates for oligonucleotide polymerization. During the comparison of the relationship between the length of alkyl chains and incorporation efficiency, 2′-OBu-^{5m}UTP, which is the longest alkyl chain, surprisingly showed almost the same incorporation efficiency as 2′-OMe-^{5m}UTP. The efficiency of incorporation improved according to the length of the alkyl chain (incorporation efficiency: 2′-OBu-^{5m}UTP > 2′-OPr-^{5m}UTP > 2′-OEt-^{5m}UTP) (Figure S1). On the other hand, the substitution of a carbon atom for an oxygen atom resulted in decreased incorporation efficiency at any carbon length (incorporation efficiency: 2′-OBu-^{5m}UTP > 2′-MOE-^{5m}UTP, 2′-OPr-^{5m}UTP > 2′-HE-^{5m}UTP) (Figure S1). We saw that 2′-MOE-^{5m}UTP demonstrated higher incorporation efficacy into oligonucleotides than 2′-HE-^{5m}UTP (Figure S1), and 2′-O^{i}Pr-^{5m}UTP showed the lowest incorporation efficiency (Figure S1). Although the steric hindrance of the 2′-*O*-alkyl modification is known to reduce incorporation efficiency [22], 2′-OBu-^{5m}UTP was more effectively incorporated into DNA than 2′-OEt-^{5m}UTP. Our results indicated that factors other than steric hindrance also impact the incorporation efficiency of 2′-*O*-alkyl modifications. Next, to enhance the extension efficiency of 2′-*O*-alkyl-^{5m}UTPs, we attempted to optimize reaction conditions. The addition of Mn^{2+} accelerates the incorporation of modified nucleotide triphosphates [21,27,31]. Under stronger conditions such as a higher amount of KOD DGLNK (150 ng/μL), the addition of Mn^{2+} (1.0 mM), and a longer incubation time (1 h), fully extended products were observed, except for 2′-O^{i}Pr, in the incorporation of 2′-*O*-alkyl-^{5m}UTPs (Figure 2b). On the other hand, during the incorporation of dTTP, we observed ladder-like bands. Moreover, we observed bands under the primer bands. These bands were probably caused by competition between the polymerization activity and exonuclease activity of KOD DGLNK. Extended products were also observed when Primer**1** was replaced with Primer**2**, which was modified with 2′-OMe modifications (Figure S3).

2.3. Enzymatic Synthesis of 2′-O-alkyl-Modified Oligonucleotide Libraries

After the characterization of 2′-*O*-alkyl-^{5m}UTPs for a short-length template, a fully modified 2′-*O*-alkyl-modified oligonucleotide synthesis for a longer template which had a 70 nt DNA sequence with a random region in the center (2′-*O*-alkyl-modified oligonucleotide libraries) was tested. For the incorporation of 50 nucleotide triphosphates, stronger reaction conditions than the reaction conditions for the incorporation of 5 nucleotide triphosphates are assumed to be required. However, such drastic conditions induce digestion on the part of DNA as we observed in Figure 2b. Therefore, primer extension was performed with 2′-OMe-modified primers with 2′-OMe-ATP, 2′-OMe-GTP, 2′-OMe-CTP, and synthesized 2′-*O*-alkyl-^{5m}UTPs for giving exonuclease resistance to the extended oligonucleotide. In the random region, averagely, a 25% nucleotide would contain a target alkyl modification at the end. As we expected, extension efficiency was shown with the same tendency as the result of the short-length template (Figure 3). Except for the 2′-O^{i}Pr modification, fully extended products were observed. Furthermore, by increasing the Mn^{2+} concentration and reaction time, the 2′-O^{i}Pr and 2′-HE modifications were also nicely incorporated and full-length products were observed under 3.0 mM of Mn^{2+} following 4 h of incubation (Figure S4). We noticed that, when Mg^{2+} and Mn^{2+} coexist, 3.0 mM or more of Mn^{2+} promotes the enzymatic synthesis of 2′-*O*-alkyl-modified oligonucleotides

(Figure S4). The enzymatically synthesized 2′-*O*-alkyl-modified oligonucleotide library that include 2′-OEt-5-methyluridine was named ON_Et. Moreover, other 2′-*O*-alkyl-modified oligonucleotide libraries also named as the same manner (ON_Me, ON_Pr, ON_Bu, ON_^{i}Pr, ON_HE, and ON_MOE).

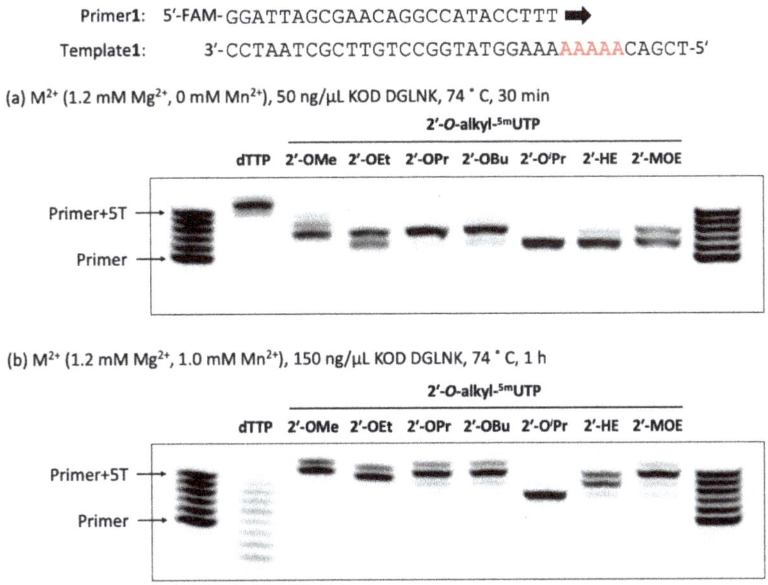

Figure 2. Polymerase incorporation of 2′-*O*-alkyl-^{5m}UTPs with a DNA primer. (**a**) Primer extension was performed with a 0.4 μM 5′-FAM-labeled DNA primer, 0.5 μM DNA template, 1× KOD Dash® buffer, and 0.2 mM 2′-*O*-alkyl-^{5m}UTP by incubation with 50 ng/μL KOD DGLNK at 74 °C for 30 min. (**b**) Primer extension was performed with a 0.4 μM 5′-FAM-labeled DNA primer, 0.5 μM DNA template, 1 × KOD Dash® buffer, 1.0 mM $MnSO_4$, and 0.2 mM 2′-*O*-alkyl-^{5m}UTPs by incubation with 150 ng/μL KOD DGLNK at 74 °C for 1 h.

2.4. *Properties of the 2′-O-Alkyl-Modified Oligonucleotide Libraries*

2.4.1. Hydrophobicity of the 2′-O-Alkyl-Modified Oligonucleotide Libraries

The hydrophobicity of the oligonucleotide libraries was evaluated by the retention time on the ion pair RP-HPLC (Figure 4). The longer alkyl chain modification increases the interaction between the oligonucleotide and C18 modification on the reverse phase column, resulting in slow elution. The substitution of carbon to oxygen in the alkyl chain induced fast elution in the column. The hydrophobic tendency of nucleotide triphosphates is reflected in the hydrophobic tendency of the oligonucleotide libraries (Figure S5). ON_Pr was shown to be hydrophobic as the fully-phosphorothioate-modified DNA library (PS_DNA). With only the substitution of 5-methyluridine, ON_Bu was shown to be more hydrophobic than PS_DNA. This property is highly beneficial in terms of expanding the hydrophobic range.

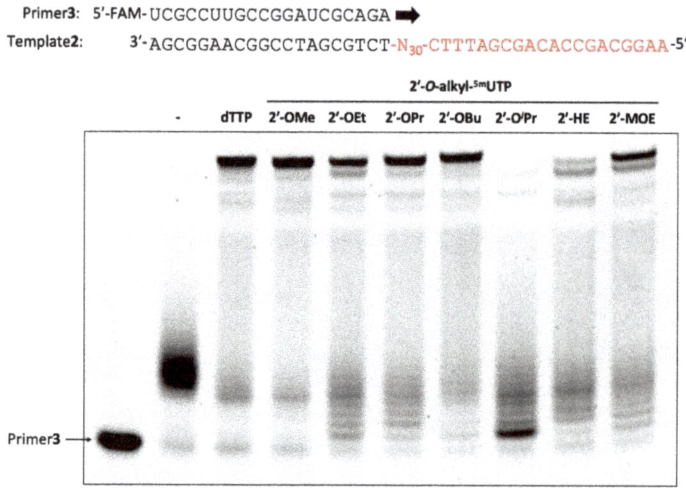

Figure 3. Enzymatic synthesis of 2′-*O*-alkyl-modified oligonucleotide libraries. Primer extension was performed with a 0.4 μM 5′-FAM-labeled 2′-OMe-modified primer, 0.5 μM DNA template, 1× KOD Dash® buffer, 1.0 mM $MnSO_4$, 0.2 mM 2′-OMe-ATP, 0.2 mM 2′-OMe-GTP, 0.2 mM 2′-OMe-CTP, and 2′-*O*-alkyl-^{5m}UTP by incubation with 300 ng/μL KOD DGLNK at 60 °C for 1 h, 60 °C to 74 °C for 4 h, and 74 °C for 1 h. Following the extension, the reaction mixture was treated with 95 mU/μL DNase1 at 37 °C for 30 min.

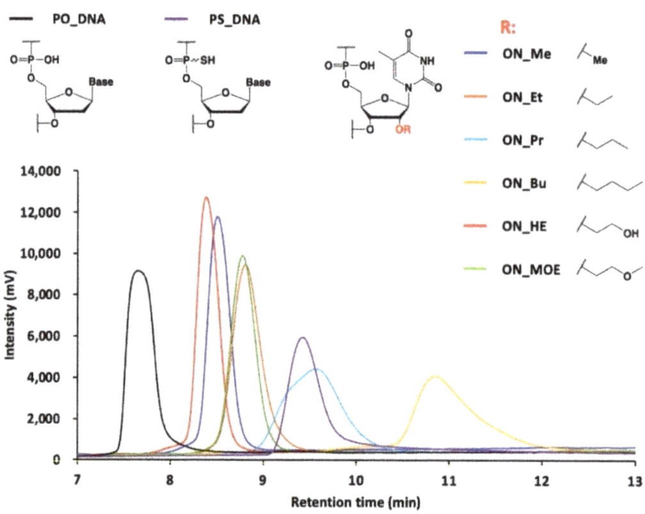

Figure 4. Hydrophobicity analysis of 2′-*O*-alkyl-modified oligonucleotide libraries. Ion pair RP-HPLC was performed by the injection of 5.0 pmol of oligonucleotide libraries onto an XBridge Oligonucleotide BEH C18 column. The column temperature was 50 °C, with a flow rate of 1.0 mL/min, and detection at 260 nm. Mobile phase A was a 100 mM triethylamine acetic acid solution (pH 6.9), and mobile phase B was acetonitrile. The column was initially maintained at 10% mobile phase B, and then at a gradient of 10–40% over 15 min.

2.4.2. Nuclease Stability of the 2′-O-Alkyl-Modified Oligonucleotide Libraries

The nuclease stability of the 2′-*O*-alkyl-modified oligonucleotide libraries in 50% FBS was evaluated by denaturing PAGE (Figure 5). Following 4 h of incubation, the intact and digested products were observed. All the 2′-*O*-alkyl-modified oligonucleotides were more stable than PO_DNA. ON_Bu was as stable as PS_DNA. The substitution of the longer 2′-*O*-alkyl 5-methyluridine of the oligonucleotides tended to show a higher stability in 50% FBS (stability: ON_Bu > ON_Pr > ON_Et > ON_Me). The replacement of a carbon atom for an oxygen atom reduced stability in 50% FBS (stability: ON_Bu > ON_MOE, ON_Pr > ON_HE). On the other hand, after 24 h of incubation, the ratio of the intact product was a little different. This was probably caused by the digestion of the 2′-OMe-modified primer region.

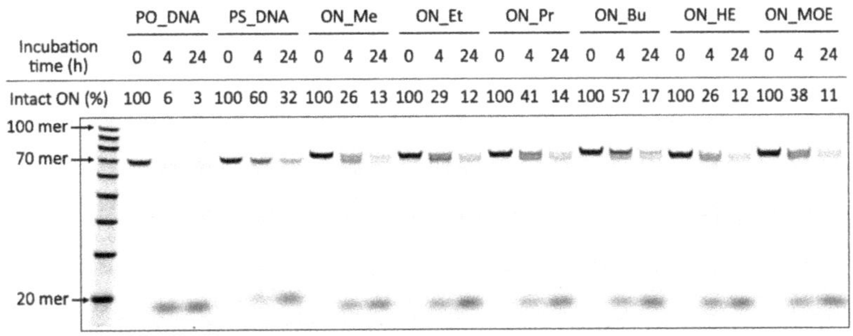

Figure 5. Nuclease stability of the 2′-*O*-alkyl-modified oligonucleotide libraries. FBS digestion of the oligonucleotides was performed with 0.1 μM 5′-6-FAM-labeled oligonucleotides, by incubation with 50% FBS at 37 °C for 0, 4, and 24 h followed by inactivation steps: 95 °C for 5 min. Proteinase K digestion was performed with an FBS digested reaction mixture by incubation with 36 mU/μL Proteinase K at 37 °C for 30 min.

Expanding the available chemical modifications to enzymatic oligonucleotide synthesis accelerates the development of aptamers and nucleic acid enzymes. By mixing 2′-OMe-adenosine, guanosine, and cytidine, oligonucleotide libraries containing 2′-OEt, 2′-OPr-, 2′-OBu or 2′-HE-5-methyluridine were successfully synthesized by KOD DGLNK (Figure 3). This indicated that KOD DGLNK is tolerant of 2′-*O*-linear alkyl modifications in polymerization. On the other hand, KOD DGLNK showed a lower tolerance for 2′-*O*-branched alkyl modification, such as 2′-O^{i}Pr modification, than 2′-*O*-linear alkyl modifications. It would be interesting to explore the kinetics of the incorporation and identify the novel mutations associated with a tolerance for 2′-*O*-branched alkyl modification. Additionally, we investigated the impact of a partial substitution of 2′-*O*-alkyl modification on the nuclease stability and hydrophobicity of the oligonucleotide libraries. Despite only incorporating a 5-methyluridine base into the oligonucleotides, ON_Et, ON_Pr, ON_Bu, ON_HE, and ON_MOE demonstrated a different hydrophobicity (Figure 4). This variety of hydrophobicity difference can be a toolbox which offers the option for aptamer selection for individual targets. Enzymatic stability is also an important factor in the efficacy of aptamers and nucleic acid enzymes in vitro and in vivo. Indeed, approved antisense oligonucleotides and siRNA contain chemical modifications to improve nuclease stability [11]. Moreover, the approved aptamers, Macugen (pegaptanib) [32] and IZERVAY (avacincaptad pegol) [33,34], have sugar modifications (2′-OMe and 2′-fluoro modifications) and 40 kDa PEG modifications by post-modification. However, following post-modification, even partial modification sometimes reduces their binding affinity to the target molecules. On the other hand, the 2′-*O*-alkyl-modified oligonucleotide libraries showed higher nuclease stability than DNA (Figure 5). Moreover, ON_Bu showed the equivalent stability to PS-modified DNA. These results indicate that our study can help develop high nuclease-stable functional oligonucleotides without post-modification. To develop such functional oligonucleotides through in vitro selection, it is important to explore the conversion of 2′-O-alkyl-modified oligonucleotide libraries to DNA. Furthermore, it is important to investigate the error rate of incorporation because the addition of Mn^{2+} for the synthesis of 2′-O-alkyl-modified oligonucleotide libraries reduces fidelity [35]. The 2′-O-alkyl-modified oligonucleotide libraries exhibited a wide range of hydrophobicity and improvement of nuclease stability by only the replacement of 5-methyluridines. The synthesis of 2′-O-alkyl-^{5m}UTPs was systematic and easy; hence, our strategy can be applied for the synthesis of other 2′-O-alkyl ^{5m}UTPs. Moreover, by applying the 2′-O-alkyl-modified oligonucleotides that are synthesized in this study to not only 5-methyluridine but also other nucleobases, 2′-O-alkyl-modified oligonucleotide libraries can show a further wide range of hydrophobicity and enhancement nuclease stability.

3. Materials and Methods

3.1. General Information

All reagents and solvents were purchased from commercial suppliers and used without purification unless otherwise specialized. All reactions were carried out under Ar atmosphere. PAGE images were recorded on Chemi-Doc (Bio-Rad, Hercules, CA, USA) and analyzed by Image Lab™ (Bio-Rad). ^{1}H, ^{13}C and ^{31}P NMR spectra were recorded on an AVHD 400 NB (Bruker Daltonics, Billerica, MA, USA) using $CDCl_3$, DMSO-d_6, and D_2O as the solvents. Mass spectra of all new compounds were measured on a JMS-700 instrument (JEOL, Tokyo, Japan) (for fast atom bombardment, FAB). For HPLC, Shimadzu SLC-20A3R, LC-20AD, CTO-20AC, SPD-20A, and FRC-10A were utilized.

3.2. Synthesis of N^{3}-Benzoyl-2′-O-ethyl-3′,5′-O-(1,1,3,3-tetraisopropyldisiloxane-1,3-diyl)-5-methyluridine (2a)

Compound **1** (1.5 g, 2.5 mmol) was dissolved in dry toluene (7.5 mL). Ag_2O (1460 mg, 6.3 mmol) and iodoethane (5.0 mL, 63 mmol) were added and the reaction mixture was stirred at 60 °C for 13 h. The resulting reaction mixture was filtered by celite pad and the filtrate was diluted with $CHCl_3$. The combined organic layers were washed with water and brine, dried over anhydrous Na_2SO_4, and concentrated in vacuo. The residue was purified by silica gel column chromatography (Hexane/Ethyl acetate = 85/15) to afford compound **2a** (456 mg, 0.72 mmol, 29%) as a white form. ^{1}H NMR (400 MHz, $CDCl_3$): δ 7.93 (dd, J = 1.2 Hz, 8.4 Hz, 2H), 7.73 (d, J = 1.2 Hz, 1H), 7.67–7.63 (m, 1H), 7.49 (t, J = 7.5 Hz, 2H), 5.70 (s, 1H), 4.29–4.21 (m, 2 H), 4.12 (dd, J = 2.0 Hz, 9.6 Hz, 1H), 3.87–3.77 (d, 3H), 1.95 (d, J = 1.1 Hz, 3H), 1.19 (t, J = 7.0 Hz, 3H), 1.21–1.05 (m, 28H). ^{13}C NMR (100 MHz, $CDCl_3$): δ 169.12, 162.95, 149.01, 135.11, 134.85, 131.52, 130.55, 129.17, 110.10, 89.79, 82.00, 81.88, 68.18, 66.97, 59.38, 17.48, 17.43, 17.31, 17.27, 17.21, 17.09, 17.04, 16.92, 15.34, 13.57, 12.91, 12.81, 12.77, 12.60. HRMS (MALDI): calcd for $C_{31}H_{48}N_2O_8NaSi_2$ $[M + Na]^+$ 655.2841, found 655.2844.

3.3. Synthesis of N^{3}-Benzoyl-2′-O-ethyl-5-methyluridine (3a)

Compound **2a** (426 mg, 0.67 mmol) was dissolved in dry THF (6.7 mL). Then, 1 M TBAF in THF (2.0 mL, 2.0 mmol) was added and the reaction mixture was stirred at room temperature for 30 min. The resulting reaction mixture was concentrated in vacuo. The residue was purified by silica gel column chromatography ($CHCl_3$/MeOH = 99/1 to 97/3) to afford compound **3a** (238 mg, 0.61 mmol, 91%) as a white form. ^{1}H NMR (400 MHz, $CDCl_3$): δ 7.92 (dd, J = 1.2 Hz, 8.4 Hz, 2H), 7.75 (s, 1H), 7.68–7.63 (m, 1H), 7.49 (t, J = 7.6 Hz, 2H), 5.78 (d, J = 2.8 Hz, 1H), 4.32 (d, J = 5.3 Hz, 1H), 4.09 (dd, J = 3.0 Hz, 5.2 Hz, 1H), 4.03 (dd, J = 2.2 Hz, 8.3 Hz, 2H), 3.90–3.83 (m, 2H), 3.67–3.59 (m, 1H), 2.73 (d, J = 6.2 Hz, 1H), 2.50 (br, 1H), 1.95 (d, J = 1.2 Hz, 3H), 1.23 (t, J = 7.0 Hz, 3H). ^{13}C NMR (100 MHz, $CDCl_3$): δ 168.77, 162.79, 149.31, 136.74, 135.18, 131.43, 130.49, 129.20, 110.88, 89.85, 84.80, 81.23, 68.35, 66.79, 61.10, 15.24, 12.61. HRMS (MALDI): calcd for $C_{19}H_{22}N_2O_7Na$ $[M + Na]^+$ 413.1319, found 413.1316.

3.4. Synthesis of 3′-O-Acetyl-N^3-benzoyl-5′-O-dimethoxytrityl-2′-O-ethyl-5-methyluridine (4a)

Compound **3a** (220 mg, 0.56 mmol) was dissolved in dry pyridine (5.6 mL). DMTrCl (209 mg, 0.62 mmol) was added and the reaction mixture was stirred at room temperature for 6 h. DMAP (7 mg, 0.06 mmol) and Ac_2O (262 μL, 2.8 mmol) were added and the reaction mixture was stirred at room temperature for 18 h. The resulting reaction mixture was concentrated in vacuo and diluted with ethyl acetate. The combined organic layers were washed with saturated $NaHCO_3$ aq. and brine, dried over anhydrous Na_2SO_4, andconcentrated in vacuo. The residue was purified by silica gel column chromatography (Hexane/Ethyl acetate = 3/1 to 2/1) to afford compound **4a** (291 mg, 0.40 mmol, 71%) as a white form. ^{1}H NMR (400 MHz, $CDCl_3$): δ 7.94 (d, *J* = 7.2 Hz, 2H), 7.72 (d, *J* = 1.0 Hz, 1H), 7.64 (t, *J* = 7.4 Hz, 1H), 7.49 (t, *J* = 7.8 Hz, 2H), 7.41 (d, *J* = 7.2 Hz, 2H), 7.34–7.26 (m, 7H), 6.86 (dd, *J* = 1.9 Hz, 8.9 Hz, 4H), 6.03 (d, *J* = 4.4 Hz, 1H), 5.36 (t, *J* = 5.3 Hz, 1H), 4.32–4.28 (m, 2H), 3.81 (s, 6H), 3.76–3.69 (m, 1H), 3.63–3.56 (m, 2H), 3.39 (dd, *J* = 2.3 Hz, 11.0 Hz, 1H), 2.10 (s, 3H), 1.40 (s, 3H), 1.16 (t, *J* = 7.0 Hz, 3H). ^{13}C NMR (100 MHz, $CDCl_3$): δ 170.13, 168.85, 162.76, 158.85, 158.83, 149.41, 144.05, 135.12, 135.00, 131.59, 130.52, 130.14, 129.11, 128.18, 128.09, 127.31, 113.36, 113.35, 114.41, 87.59, 87.24, 81.16, 80.18, 70.58, 66.93, 62.09, 55.29, 20.68, 15.16, 11.82. HRMS (MALDI): calcd for $C_{42}H_{42}N_2O_{10}Na$ $[M + Na]^+$ 757.2732, found 757.2738.

3.5. Synthesis of 3′-O-Acetyl-N^3-benzoyl-2′-O-ethyl-5-methyluridine (5a)

Compound **4a** (276 mg, 0.38 mmol) was dissolved in CH_2Cl_2 (3.8 mL). TFA (1.0 mL) was added and the reaction mixture was stirred at room temperature for 30 min. The resulting reaction mixture was concentrated in vacuo and diluted with $CHCl_3$. The combined organic layers were washed with saturated $NaHCO_3$ aq. and brine, dried over anhydrous Na_2SO_4, and concentrated in vacuo. The residue was purified by silica gel column chromatography (Hexane/Ethyl acetate = 1/1 to 1/2) to afford compound **5a** (125 mg, 0.29 mmol, 76%) as a white form. ^{1}H NMR (400 MHz, $CDCl_3$): δ 7.92 (dd, *J* = 1.1 Hz, 8.3 Hz, 2H), 7.66 (t, *J* = 7.4 Hz, 1H), 7.62 (br, 1H), 7.50 (t, *J* = 7.9 Hz, 2H), 5.69 (d, *J* = 4.8 Hz, 1H), 5.27 (t, *J* = 5.0 Hz, 1H), 4.37 (t, *J* = 5.0 Hz, 1H), 4.23–4.21 (m, 1H), 3.97 (dd, *J* = 1.8 Hz, 12.6 Hz, 1H), 3.76 (dd, *J* = 2.0 Hz, 12.6 Hz, 1H), 3.70–3.53 (m, 2H), 2.15 (s, 3H), 1.96 (d, *J* = 1.0 Hz, 3H), 1.15 (t, *J* = 7.0 Hz, 3H). ^{13}C NMR (100 MHz, $CDCl_3$): δ 170.50, 168.66, 162.72, 149.43, 137.33, 135.20, 131.41, 130.53, 129.21, 111.19, 91.09, 83.01, 79.14, 70.43, 66.94, 61.55, 20.83, 15.22, 12.66. HRMS (MALDI): calcd for $C_{21}H_{24}N_2O_8Na$ $[M + Na]^+$ 455.1425, found 455.1427.

3.6. Synthesis of 2′-O-Ethyl-5-methyluridine-5′-triphospahte (6a)

Compound **5a** was dissolved in dry pyridine and the solution was evaporated to dryness in vacuo. Compound **5a** (23 mg, 0.065 mmol) was dissolved in dry pyridine (0.25 mL) and 1,4-dioxane (0.25 mL). Then, 2-Chloro-4*H*-1,3,2-benzodioxaphosphorin-4-one (15 mg, 0.072 mmol) was added and the reaction mixture was stirred for 30 min at room temperature. Tris(tetrabutylammonium) hydrogen pyrophosphate (43 mg, 0.078 mmol) in DMF (0.5 mL) and tributylamine (52 μL, 0.22 mmol) were added and the reaction mixture was stirred for 1 h at room temperature. Next, 0.1 M I_2 in pyridine/H_2O (98:2, *v*/*v*) was added and the reaction mixture was stirred for 30 min at 0 °C. Then, 10% $Na_2S_2O_3$ aq. was added and the reaction mixture was stirred for 10 min at room temperature. The reaction mixture was purified by ethanol precipitation. The precipitate was dissolved in 7% NH_3 aq. and the reaction mixture was stirred for 2 h at room temperature. The solvent was removed in vacuo and the residue was purified by HPLC to give compound **6a** (14 μmol, 22%). ^{1}H NMR (400 MHz, D_2O): δ7.78 (s, 1H), 6.02 (d, *J* = 5.0 Hz, 1H), 4.53 (t, *J* = 4.5 Hz, 1H), 4.25 (d, *J* = 4.9 Hz, 3H), 4.19 (t, *J* = 5.2 Hz, 1H), 3.71 (dd, *J* = 7.0 Hz, 14.1 Hz, 2H), 1.92 (s, 3H), 1.18 (t, *J* = 7.0 Hz, 3H). ^{31}P NMR (120 MHz, D_2O): δ −9.60 (d, *J* = 14.5 Hz), −12.41 (d, *J* = 14.5 Hz), −23.38 (t, *J* = 14.7 Hz). HRMS (MALDI): calcd for $C_{12}H_{20}N_2O_{15}P_3$ $[M - H]^-$ 525.0082, found 525.0089.

3.7. Synthesis of N^3-Benzoyl-2′-O-propyl-3′,5′-O-(1,1,3,3-tetraisopropyldisiloxane-1,3-diyl)-5-methyluridine (2b)

Compound **1** (1.5 g, 2.5 mmol) was dissolved in dry toluene (7.5 mL). Ag_2O (1460 mg 6.3 mmol) and 1-iodopropane (6.0 mL, 63 mmol) were added and the reaction mixture was stirred at 80 °C for 15 h. The resulting reaction mixture was filtered by celite pad and the filtrate was diluted with $CHCl_3$. The combined organic layers were washed with water and brine, dried over anhydrous Na_2SO_4, and concentrated in vacuo. The residue was purified by silica gel column chromatography (Hexane/Ethyl acetate = 85/15) to afford compound **2b** (369 mg, 0.57 mmol, 23%) as a white form. ^{1}H NMR (400 MHz, $CDCl_3$): δ 7.93 (dd, *J* = 1.2 Hz, 8.4 Hz, 2H), 7.73 (d, *J* = 1.2 Hz, 1H), 7.67–7.63 (m, 1H), 7.49 (t, *J* = 8.0 Hz, 2H), 5.71 (s, 1H), 4.29–4.21 (m, 2 H), 4.16 (dd, *J* = 2.1 Hz, 9.6 Hz, 1H), 3.84 (d, *J* = 4.3 Hz, 1H), 3.71 (t, *J* = 6.4 Hz, 2H), 1.95 (d, *J* = 1.1 Hz, 3H), 1.62–1.53 (m, 2H), 1.13–1.04 (m, 28H), 0.90 (t, *J* = 7.4 Hz, 3H). ^{13}C NMR (100 MHz, $CDCl_3$): δ 169.14, 162.98, 148.98, 135.11, 134.87, 131.53, 130.56, 129.17, 110.06, 89.65, 82.16, 81.90, 73.08, 68.24, 59.39, 23.01, 17.49, 17.43, 17.31, 17.28, 17.21, 17.09, 17.03, 16.93, 13.58, 12.91, 12.81, 12.77, 12.65. HRMS (MALDI): calcd for $C_{32}H_{50}N_2O_8NaSi_2$ $[M + Na]^+$ 699.2998, found 699.2986.

3.8. Synthesis of N^3-Benzoyl-2′-O-propyl-5-methyluridine (3b)

Compound **2b** (185 mg, 0.29 mmol) was dissolved in dry THF (3.0 mL). 1 M TBAF in THF (0.86 mL, 0.86 mmol) was added and the reaction mixture was stirred at room temperature for 30 min. The resulting reaction mixture was concentrated in vacuo. The residue was purified by silica gel column chromatography ($CHCl_3$/MeOH = 100/0 to 97/3) to afford compound **3b** (90 mg, 0.22 mmol, 77%) as a white form. ^{1}H NMR (400 MHz, $CDCl_3$): δ 7.92 (dd, *J* = 1.2 Hz, 8.4 Hz, 2H), 7.73 (s, 1H), 7.68–7.63 (m, 1H), 7.50 (t, *J* = 8.0 Hz, 2H), 5.78 (d, *J* = 2.9 Hz, 1H), 4.33 (dd, *J* = 5.8 Hz, 12.9 Hz, 1H), 4.10 (dd, *J* = 3.0 Hz, 5.2 Hz, 1H), 4.04 (dd, *J* = 2.3 Hz, 8.6 Hz, 2H), 3.90–3.84 (m, 1H), 3.79–3.73 (m, 1H), 3.57–3.51 (m, 1H), 2.70 (d, *J* = 7.3 Hz, 1H), 2.40 (br, 1H), 1.95 (d, *J* = 1.1 Hz, 3H), 1.66–1.57 (m, 2H), 0.92 (t, *J* = 7.4 Hz, 3H). ^{13}C NMR (100 MHz, $CDCl_3$): δ 168.75, 162.77, 149.30, 136.74, 135.16, 131.45, 130.50, 129.19, 110.91, 89.91, 84.82, 81.27, 72.82, 68.46, 61.18, 22.91, 12.62, 10.42. HRMS (MALDI): calcd for $C_{20}H_{24}N_2O_7Na$ $[M + Na]^+$ 427.1476, found 427.1473.

3.9. Synthesis of 3′-O-Acetyl-N^3-benzoyl-5′-O-dimethoxytrityl-2′-O-propyl-5-methyluridine (4b)

Compound **3b** (80 mg, 0.20 mmol) was dissolved in dry pyridine (2.0 mL). DMTrCl (74 mg, 0.22 mmol) was added and the reaction mixture was stirred at room temperature for 4.5 h. DMAP (2 mg, 0.02 mmol) and Ac_2O (93 μL, 1.0 mmol) were added and the reaction mixture was stirred at room temperature for 13 h. The resulting reaction mixture was concentrated in vacuo and diluted with ethyl acetate. The combined organic layers were washed with saturated $NaHCO_3$ aq. and brine, dried over anhydrous Na_2SO_4, and concentrated in vacuo. The residue was purified by silica gel column chromatography (Hexane/Ethyl acetate = 3/1 to 2/1) to afford compound **4b** (104 mg, 0.14 mmol, 69%) as a white form. ^{1}H NMR (400 MHz, $CDCl_3$): δ 7.94 (dd, *J* = 1.1 Hz, 8.4 Hz, 2H), 7.72 (d, *J* = 1.2 Hz, 1H), 7.66–7.62 (m, 1H), 7.49 (t, *J* = 7.8 Hz, 2H), 7.41 (d, *J* = 7.1 Hz, 2H), 7.34–7.26 (m, 7H), 6.86 (dd, *J* = 2.0 Hz, 8.9 Hz, 4H), 6.04 (d, *J* = 4.4 Hz, 1H), 5.36 (t, *J* = 5.3 Hz, 1H), 4.31–4.28 (m, 2H), 3.81 (s, 6H), 3.65–3.59 (m, 2H), 3.51–3.46 (m, 1H), 3.39 (dd, *J* = 2.4 Hz, 11.0 Hz, 1H), 2.10 (s, 3H), 1.59–1.50 (m, 2H), 1.40 (d, *J* = 0.9 Hz, 3H), 0.89 (t, *J* = 7.4 Hz, 3H). ^{13}C NMR (100 MHz, $CDCl_3$): δ 170.13, 168.85, 162.78, 158.86, 158.84, 149.40, 144.07, 135.13, 135.00, 131.60, 130.54, 130.15, 129.11, 128.20, 128.10, 127.32, 113.37, 113.36, 111.40, 87.51, 87.24, 81.21, 80.37, 73.01, 70.62, 62.14, 55.29, 22.86, 20.70, 11.82, 10.41. HRMS (MALDI): calcd for $C_{43}H_{44}N_2O_{10}Na$ $[M + Na]^+$ 771.2888, found 771.2887.

3.10. Synthesis of 3′-O-Acetyl-N^3-benzoyl-2′-O-propyl-5-methyluridine (5b)

Compound **4b** (80 mg, 0.11 mmol) was dissolved in CH_2Cl_2 (1.1 mL). TFA (0.3 mL) was added and the reaction mixture was stirred at room temperature for 30 min. The resulting reaction mixture was concentrated in vacuo and diluted with $CHCl_3$. The combined organic layers were washed with saturated $NaHCO_3$ aq. and brine, dried over anhydrous Na_2SO_4, and concentrated in vacuo. The residue was purified by silica gel column chromatography (Hexane/Ethyl acetate = 1/1 to 1/2) to afford compound **5b** (35 mg, 0.08 mmol, 72%) as a white form. ^{1}H NMR (400 MHz, $CDCl_3$): δ 7.92 (dd, *J* = 1.1 Hz, 8.4 Hz, 2H), 7.68–7.63 (m, 2H), 7.49 (t, *J* = 7.8 Hz, 2H), 5.71 (d, *J* = 4.8 Hz, 1H), 5.27 (t, *J* = 5.0 Hz, 1H), 4.35 (t, *J* = 5.0 Hz, 1H), 4.24–4.21 (m, 1H), 3.97 (dd, *J* = 1.8 Hz, 12.5 Hz, 1H), 3.76 (dd, *J* = 2.0 Hz, 12.5 Hz, 1H), 3.58–3.53 (m, 1H), 3.49–3.43 (m, 1H), 2.14 (s, 3H), 1.96 (d, *J* = 1.0 Hz, 3H), 1.58–1.49 (m, 2H), 0.88 (t, *J* = 7.4 Hz, 3H). ^{13}C NMR (100 MHz, $CDCl_3$): δ 170.49, 168.67, 162.74, 149.40, 137.24, 135.19, 131.42, 130.53, 129.20, 111.16, 90.89, 83.01, 79.36, 73.01, 70.43, 61.53, 22.88, 20.83, 12.66, 10.40. HRMS (MALDI): calcd for $C_{22}H_{26}N_2O_8Na$ $[M + Na]^+$ 469.1581, found 469.1580.

3.11. *Synthesis of 2′-O-Propyl-5-methyluridine-5′-triphospahte* (**6b**)

Compound **5b** was dissolved in dry pyridine and the solution was evaporated to dryness in vacuo. Compound **5b** (47 mg, 0.10 mmol) was dissolved in dry pyridine (0.5 mL) and 1,4-dioxane (0.5 mL). Then, 2-Chloro-4*H*-1,3,2-benzodioxaphosphorin-4-one (24 mg, 0.12 mmol) was added and the reaction mixture was stirred for 30 min at room temperature. Tris(tetrabutylammonium) hydrogen pyrophosphate (66 mg, 0.12 mmol) in DMF (1.0 mL) and tributylamine (86 μL, 0.36 mmol) were added and the reaction mixture was stirred for 1 h at room temperature. Next, 0.1 M I_2 in pyridine/H_2O (98:2, *v*/*v*) was added and the reaction mixture was stirred for 30 min at room temperature. Then, 10% $Na_2S_2O_3$ aq. was added and the reaction mixture was stirred for 10 min at room temperature. The reaction mixture was purified by ethanol precipitation. The precipitate was dissolved in 7% NH_3 aq. and the reaction mixture was stirred for 2 h at room temperature. The solvent was removed in vacuo and the residue was purified by HPLC to give compound **6b** (46 μmol, 46%). ^{1}H NMR (400 MHz, D_2O): δ7.77 (s, 1H), 6.01 (d, *J* = 5.2 Hz, 1H), 4.53–4.50 (m, 1H), 4.24 (d, *J* = 4.2 Hz, 3H), 4.17 (t, *J* = 5.3 Hz, 1H), 3.66–3.55 (m, 2H), 1.91 (s, 3H), 1.59–1.50 (m, 2H), 0.84 (t, *J* = 7.4 Hz, 3H). ^{31}P NMR (120 MHz, D_2O): δ −9.57 (d, *J* = 13.4 Hz), −12.66 (d, *J* = 14.6 Hz), −23.64 (t, *J* = 14.7 Hz). HRMS (MALDI): calcd for $C_{13}H_{22}N_2O_{15}P_3$ $[M - H]^-$ 539.0239, found 539.0243.

3.12. *Synthesis of N^3-Benzoyl-2′-O-butyl-3′,5′-O-(1,1,3,3-tetraisopropyldisiloxane-1,3-diyl)-5-methyluridine* (**2c**)

Compound **1** (1.5 g, 2.5 mmol) was dissolved in dry toluene (7.5 mL). Ag_2O (1460 mg, 6.3 mmol) and 1-iodobutane (7.1 mL, 63 mmol) were added and the reaction mixture was stirred at 80 °C for 15 h. The resulting reaction mixture was filtered by celite pad and the filtrate was diluted with $CHCl_3$. The combined organic layers were washed with water and brine, dried over anhydrous Na_2SO_4, and concentrated in vacuo. The residue was purified by silica gel column chromatography (Hexane/Ethyl acetate = 85/15) to afford compound **2c** (303 mg, 0.46 mmol, 18%) as a white form. ^{1}H NMR (400 MHz, $CDCl_3$): δ 7.93 (dd, *J* = 1.2 Hz, 8.4 Hz, 2H), 7.73 (d, *J* = 1.2 Hz, 1H), 7.67–7.62 (m, 1H), 7.49 (t, *J* = 8.0 Hz, 2H), 5.70 (s, 1H), 4.29–4.21 (m, 2 H), 4.15 (dd, *J* = 2.1 Hz, 9.6 Hz, 1H), 4.00 (dd, *J* = 2.5 Hz, 13.6 Hz, 1H), 3.84 (d, *J* = 4.3 Hz, 1H), 3.75 (t, *J* = 6.3 Hz, 2H), 1.95 (d, *J* = 1.1 Hz, 3H), 1.58–1.50 (m, 2H), 1.42–1.33 (m, 2H), 1.13–1.03 (m, 28H), 0.87 (t, *J* = 7.4 Hz, 3H). ^{13}C NMR (100 MHz, $CDCl_3$): δ 169.14, 162.97, 148.96, 135.09, 134.87, 131.53, 130.55, 129.16, 110.05, 89.64, 82.19, 81.88, 71.24, 68.25, 59.38, 31.84, 19.15, 17.49, 17.42, 17.30, 17.28, 17.21, 17.09, 17.01, 16.92, 13.83, 13.58, 12.91, 12.81, 12.76, 12.64. HRMS (MALDI): calcd for $C_{33}H_{52}N_2O_8NaSi_2$ $[M + Na]^+$ 683.3154, found 683.3158.

3.13. *Synthesis of N^3-Benzoyl-2′-O-butyl-5-methyluridine* (**3c**)

Compound **2c** (192 mg, 0.29 mmol) was dissolved in dry THF (3.0 mL). 1 M TBAF in THF (0.87 mL, 0.87 mmol) was added and the reaction mixture was stirred at room temperature for 30 min. The resulting reaction mixture was concentrated in vacuo. The residue was purified by silica gel column chromatography ($CHCl_3$/MeOH = 100/0 to 97/3) to afford compound **3c** (95 mg, 0.23 mmol, 78%) as a white form. ^{1}H NMR (400 MHz, $CDCl_3$): δ 7.92 (dd, *J* = 1.2 Hz, 8.4 Hz, 2H), 7.73 (s, 1H), 7.68–7.63 (m, 1H), 7.50 (t, *J* = 7.8 Hz, 2H), 5.78 (d, *J* = 2.8 Hz, 1H), 4.33 (dd, *J* = 5.8 Hz, 12.8 Hz, 1H), 4.09 (dd, *J* = 3.0 Hz, 5.1 Hz, 1H), 4.04–4.02 (m, 2H), 3.89–3.84 (m, 1H), 3.83–3.77 (m, 1H), 3.60–3.55 (m, 1H), 2.69 (d, *J* = 7.3 Hz, 1H), 2.43 (br, 1H), 1.95 (d, *J* = 1.1 Hz, 3H), 1.61–1.54 (m, 2H), 1.41–1.32 (m, 2H), 0.91 (t, *J* = 7.4 Hz, 3H). ^{13}C NMR (100 MHz, $CDCl_3$): δ 168.74, 162.77, 149.27, 136.73, 135.15, 131.43, 130.49, 129.17, 110.87, 89.81, 84.80, 81.31, 71.02, 68.43, 61.14, 31.69, 19.16, 13.81, 12.60. HRMS (MALDI): calcd for $C_{21}H_{26}N_2O_7Na$ [M + Na]$^+$ 441.1632, found 441.1633.

3.14. *Synthesis of 3′-O-Acetyl-N^3-benzoyl-2′-O-butyl-5′-O-dimethoxytrityl-5-methyluridine* (**4c**)

Compound **3c** (85 mg, 0.20 mmol) was dissolved in dry pyridine (2.0 mL). DMTrCl (74 mg, 0.22 mmol) was added and the reaction mixture was stirred at room temperature for 4.5 h. DMAP (2 mg, 0.02 mmol) and Ac_2O (93 μL, 1.0 mmol) were added and the reaction mixture was stirred at room temperature for 13 h. The resulting reaction mixture was concentrated in vacuo and diluted with ethyl acetate. The combined organic layers were washed with saturated $NaHCO_3$ aq. and brine, dried over anhydrous Na_2SO_4, and concentrated in vacuo. The residue was purified by silica gel column chromatography (Hexane/Ethyl acetate = 3/1 to 2/1) to afford compound **4c** (113 mg, 0.15 mmol, 75%) as a white form. ^{1}H NMR (400 MHz, $CDCl_3$): δ 7.94 (d, *J* = 7.2 Hz, 2H), 7.71 (d, *J* = 1.0 Hz, 1H), 7.64 (t, *J* = 7.4 Hz, 1H), 7.49 (t, *J* = 7.8 Hz, 2H), 7.41 (d, *J* = 7.2 Hz, 2H), 7.32–7.26 (m, 7H), 6.86 (dd, *J* = 2.0 Hz, 8.8 Hz, 4H), 6.04 (d, *J* = 4.5 Hz, 1H), 5.36 (t, *J* = 5.2 Hz, 1H), 4.30–4.27 (m, 2H), 3.80 (s, 6H), 3.69–3.60 (m, 2H), 3.56–3.47 (m, 1H), 3.39 (dd, *J* = 2.2 Hz, 11.0 Hz, 1H), 2.09 (s, 3H), 1.57–1.49 (m, 2H), 1.40 (s, 3H), 1.38–1.31 (m, 2H), 0.89 (t, *J* = 7.4 Hz, 3H). ^{13}C NMR (100 MHz, $CDCl_3$): δ 170.13, 168.86, 162.79, 158.86, 158.85, 149.40, 144.08, 135.14, 135.03, 135.01, 131.63, 130.54, 130.16, 129.12, 128.20, 128.10, 127.33, 113.38, 113.37, 114.40, 87.46, 87.25, 81.22, 80.35, 73.02, 71.15, 70.63, 62.16, 55.30, 31.65, 22.87, 20.70, 19.10, 13.83, 11.82, 10.42. HRMS (MALDI): calcd for $C_{44}H_{46}N_2O_{10}Na$ [M + Na]$^+$ 785.3045, found 785.3035.

3.15. *Synthesis of 3′-O-Acetyl-N^3-benzoyl-2′-O-butyl-5-methyluridine* (**5c**)

Compound **4c** (120 mg, 0.16 mmol) was dissolved in CH_2Cl_2 (1.6 mL). TFA (0.4 mL) was added and the reaction mixture was stirred at room temperature for 30 min. The resulting reaction mixture was concentrated in vacuo and diluted with $CHCl_3$. The combined organic layers were washed with saturated $NaHCO_3$ aq. and brine, dried over anhydrous Na_2SO_4, and concentrated in vacuo. The residue was purified by silica gel column chromatography (Hexane/Ethyl acetate = 1/1 to 1/2) to afford compound **5c** (57 mg, 0.12 mmol, 75%) as a white form. ^{1}H NMR (400 MHz, $CDCl_3$): δ 7.92 (dd, *J* = 1.2 Hz, 8.4 Hz, 2H), 7.68–7.63 (m, 2H), 7.49 (t, *J* = 7.8 Hz, 2H), 5.71 (d, *J* = 4.8 Hz, 1H), 5.26 (t, *J* = 5.0 Hz, 1H), 4.33 (t, *J* = 5.0 Hz, 1H), 4.23–4.21 (m, 1H), 3.96 (dd, *J* = 1.8 Hz, 12.5 Hz, 1H), 3.76 (dd, *J* = 2.0 Hz, 12.5 Hz, 1H), 3.62–3.56 (m, 1H), 3.53–3.47 (m, 1H), 2.14 (s, 3H), 1.96 (d, *J* = 1.1 Hz, 3H), 1.53–1.46 (m, 2H), 1.37–1.29 (m, 2H), 0.88 (t, *J* = 7.4 Hz, 3H). ^{13}C NMR (100 MHz, $CDCl_3$): δ 170.47, 168.67, 162.74, 149.39, 137.19, 135.17, 131.43, 130.51, 129.19, 111.14, 90.79, 82.99, 79.44, 71.20, 70.42, 61.50, 31.68, 20.81, 19.10, 13.84, 12.64. HRMS (MALDI): calcd for $C_{23}H_{28}N_2O_8Na$ [M + Na]$^+$ 483.1738, found 483.0741.

3.16. *Synthesis of 2′-O-Butyl-5-methyluridine-5′-triphospahte* (**6c**)

Compound **5c** was dissolved in dry pyridine and the solution was evaporated to dryness in vacuo. Compound **5c** (30 mg, 0.065 mmol) was dissolved in dry pyridine (0.25 mL) and 1,4-dioxane (0.25 mL). Then, 2-Chloro-4*H*-1,3,2-benzodioxaphosphorin-4-one (15 mg, 0.072 mmol) was added and the reaction mixture was stirred for 30 min at room temperature. Tris(tetrabutylammonium) hydrogen pyrophosphate (43 mg, 0.078 mmol) in DMF (0.5 mL) and tributylamine (52 μL, 0.22 mmol) were added and the reaction mixture was stirred for 1 h at room temperature. Next, 0.1 M I_2 in pyridine/H_2O (98:2, *v*/*v*) was added and the reaction mixture was stirred for 30 min at 0 °C. Then, 10% $Na_2S_2O_3$ aq. was added and the reaction mixture was stirred for 10 min at room temperature. The reaction mixture was purified by ethanol precipitation. The precipitate was dissolved in 7% NH_3 aq. and the reaction mixture was stirred for 2 h at room temperature. The solvent was removed in vacuo and the residue was purified by HPLC to give compound **6c** (6 µmol, 10%). ^{1}H NMR (400 MHz, D_2O): δ7.79 (s, 1H), 6.03 (d, *J* = 5.4 Hz, 1H), 4.52 (t, *J* = 4.5 Hz, 1H), 4.27–4.24 (m, 3H), 4.18 (t, *J* = 5.4 Hz, 1H), 3.73–3.61 (m, 2H), 1.93 (s, 3H), 1.56–1.49 (m, 2H), 1.34–1.25 (m, 2H), 0.84 (t, *J* = 7.4 Hz, 3H). ^{31}P NMR (120 MHz, D_2O): δ −9.80 (d, *J* = 12.7 Hz), −12.65 (d, *J* = 14.5 Hz), −23.69 (t, *J* = 14.7 Hz). HRMS (MALDI): calcd for $C_{14}H_{24}N_2O_{15}P_3$ $[M-H]^-$ 553.0395, found 553.0424.

3.17. Synthesis of N^3*-Benzoyl-2′-O-isopropyl-3′,5′-O-(1,1,3,3-tetraisopropyldisiloxane-1,3-diyl)-5-methyluridine* (**2d**)

Compound **1** (1.5 g, 2.5 mmol) was dissolved in dry toluene (7.5 mL). Ag_2O (1460 mg, 6.3 mmol) and 2-iodopropane (6.2 mL, 63 mmol) were added and the reaction mixture was stirred at 80 °C for 13 h. The resulting reaction mixture was filtered by celite pad and the filtrate was diluted with $CHCl_3$. The combined organic layers were washed with water and brine, dried over anhydrous Na_2SO_4, and concentrated in vacuo. The residue was purified by silica gel column chromatography (Hexane/Ethyl acetate = 85/15) to afford compound **2d** (372 mg, 0.58 mmol, 23%) as a white form. ^{1}H NMR (400 MHz, $CDCl_3$): δ 7.93 (dd, *J* = 1.2 Hz, 8.4 Hz, 2H), 7.76 (d, *J* = 1.2 Hz, 1H), 7.67–7.62 (m, 1H), 7.49 (t, *J* = 7.8 Hz, 2H), 5.64 (s, 1H), 4.27 (d, *J* = 13.7 Hz, 1H), 4.21–4.14 (m, 2 H), 4.01–3.95 (m, 2 H), 3.93 (d, *J* = 4.1 Hz, 1H), 1.95 (d, *J* = 1.2 Hz, 3H), 1.19–1.03 (m, 34H). ^{13}C NMR (100 MHz, $CDCl_3$): δ 169.19, 163.03, 149.01, 135.11, 134.93, 131.51, 130.59, 129.16, 109.91, 90.44, 82.01, 80.35, 72.83, 67.88, 59.41, 22.59, 22.35, 17.50, 17.42, 17.29, 17.21, 17.12, 17.05, 16.95, 13.58, 12.91, 12.83, 12.77, 12.73. HRMS (MALDI): calcd for $C_{32}H_{50}N_2O_8NaSi_2$ $[M+Na]^+$ 669.2998, found 669.2994.

3.18. Synthesis of N^3*-Benzoyl-2′-O-isopropyl-5-methyluridine* (**3d**)

Compound **2d** (277 mg, 0.43 mmol) was dissolved in dry THF (4.3 mL). 1 M TBAF in THF (1.3 mL, 1.3 mmol) was added and the reaction mixture was stirred at room temperature for 30 min. The resulting reaction mixture was concentrated in vacuo. The residue was purified by silica gel column chromatography ($CHCl_3$/MeOH = 99/1 to 97/3) to afford compound **3d** (150 mg, 0.37 mmol, 86%) as a white form. ^{1}H NMR (400 MHz, $CDCl_3$): δ 7.91 (dd, *J* = 1.1 Hz, 8.4 Hz, 2H), 7.67–7.64 (m, 2H), 7.49 (t, *J* = 7.8 Hz, 2H), 5.68 (d, *J* = 2.9 Hz, 1H), 4.29–4.24 (m, 2H), 4.06–3.91 (m, 3H), 3.85–3.81 (m, 1H), 2.72 (d, *J* = 6.0 Hz, 1H), 2.59 (br, 1H), 1.95 (d, *J* = 1.1 Hz, 3H), 1.19 (dd, *J* = 3.9 Hz, 6.1 Hz, 6H). ^{13}C NMR (100 MHz, $CDCl_3$): δ 168.67, 162.75, 149.38, 137.23, 135.16, 131.42, 130.47, 129.17, 110.91, 91.17, 85.14, 78.75, 72.80, 68.66, 61.43, 22.89, 22.07, 12.58. HRMS (MALDI): calcd for $C_{20}H_{24}N_2O_7Na$ $[M+Na]^+$ 427.1476, found 427.1472.

3.19. Synthesis of 3′-O-Acetyl-N^3*-benzoyl-5′-O-dimethoxytrityl-2′-O-isopropyl-5-methyluridine* (**4d**)

Compound **3d** (150 mg. 0.37 mmol) was dissolved in dry pyridine (4.0 mL). DMTrCl (139 mg, 0.41 mmol) was added and the reaction mixture was stirred at room temperature for 6 h. DMAP (5 mg, 0.04 mmol) and Ac_2O (173 µL, 1.9 mmol) were added and the reaction mixture was stirred at room temperature for 18 h. The resulting reaction mixture was concentrated in vacuo and diluted with ethyl acetate. The combined organic layers were washed with saturated $NaHCO_3$ aq. and brine, dried over anhydrous Na_2SO_4, and concentrated in vacuo. The residue was purified by silica gel column chromatography (Hexane/Ethyl acetate = 3/1 to 2/1) to afford compound **4d** (204 mg, 0.27 mmol, 74%) as a white form. ^{1}H NMR (400 MHz, $CDCl_3$): δ 7.93 (dd, *J* = 1.2 Hz, 8.4 Hz, 2H), 7.74 (d, *J* = 1.2 Hz, 1H), 7.66–7.62 (m, 1H), 7.49 (t, *J* = 7.8 Hz, 2H), 7.41 (dd, *J* = 1.3 Hz, 8.3 Hz, 2H), 7.34–7.26 (m, 7H), 6.86 (dd, *J* = 2.1 Hz, 8.9 Hz, 4H), 6.00 (d, *J* = 5.0 Hz, 1H), 5.37 (t, *J* = 5.1 Hz, 1H), 4.38 (t, *J* = 5.1 Hz, 1H), 4.29–4.27 (m, 1H), 3.80 (d, *J* = 0.8 Hz, 6H), 3.60 (dd, *J* = 2.3 Hz, 11.0 Hz, 1H), 3.38 (dd, *J* = 2.3 Hz, 10.9 Hz, 1H), 2.10 (s, 3H), 1.38 (d, *J* = 1.0 Hz, 3H), 1.15 (d, *J* = 6.0 Hz, 3H), 1.09 (d, *J* = 6.0 Hz, 3H). ^{13}C NMR (100 MHz, $CDCl_3$): δ 170.12, 168.80, 162.77, 158.87, 158.85, 149.46, 144.07, 135.10, 135.08, 134.97, 131.61, 130.50, 130.15, 129.10, 128.18,

128.10, 127.34, 113.37, 113.36, 111.41, 87.74, 87.27, 81.27, 78.07, 72.44, 70.87, 62.42, 55.29, 22.33, 22.06, 20.73, 11.73. HRMS (MALDI): calcd for $C_{43}H_{44}N_2O_{10}Na$ $[M + Na]^+$ 771.2888, found 771.2894.

*3.20. Synthesis of 3′-O-Acetyl-N³-benzoyl-2′-O-isopropyl-5-methyluridine (**5d**)*

Compound **4d** (167 mg, 0.22 mmol) was dissolved in CH_2Cl_2 (2.2 mL). TFA (0.6 mL) was added and the reaction mixture was stirred at room temperature for 30 min. The resulting reaction mixture was concentrated in vacuo and diluted with $CHCl_3$. The combined organic layers were washed with saturated $NaHCO_3$ aq. and brine, dried over anhydrous Na_2SO_4, and concentrated in vacuo. The residue was purified by silica gel column chromatography (Hexane/Ethyl acetate = 1/1 to 1/2) to afford compound **5d** (67 mg, 0.15 mmol, 68%) as a white form. ^1H NMR (400 MHz, $CDCl_3$): δ 7.91 (dd, *J* = 1.1 Hz, 8.4 Hz, 2H), 7.67–7.63 (m, 1H), 7.56 (d, *J* = 1.0 Hz, 1H), 7.49 (t, *J* = 7.8 Hz, 2H), 5.57 (d, *J* = 5.4 Hz, 1H), 5.27 (t, *J* = 4.8 Hz, 1H), 4.49 (t, *J* = 5.3 Hz, 1H), 4.23–4.21 (m, 1H), 3.94 (dd, *J* = 1.9 Hz, 12.5 Hz, 1H), 3.79–3.70 (m, 2H), 2.14 (s, 3H), 1.96 (d, *J* = 1.0 Hz, 3H), 1.10 (dd, *J* = 6.0 Hz, 11.0 Hz, 6H). ^{13}C NMR (100 MHz, $CDCl_3$): δ 170.36, 168.55, 162.71, 149.45, 137.83, 135.19, 131.38, 130.50, 129.18, 111.11, 92.20, 83.21, 72.60, 70.70, 61.75, 22.24, 20.84, 12.60. HRMS (MALDI): calcd for $C_{22}H26_6N_2O_8Na$ $[M + Na]^+$ 469.1581, found 469.1583.

*3.21. Synthesis of 2′-O-Isopropyl-5-methyluridine-5′-triphospahte (**6d**)*

Compound **5d** was dissolved in dry pyridine and the solution was evaporated to dryness in vacuo. Compound **5d** (47 mg, 0.10 mmol) was dissolved in dry pyridine (0.5 mL) and 1,4-dioxane (0.5 mL). Then, 2-Chloro-4*H*-1,3,2-benzodioxaphosphorin-4-one (22 mg, 0.11 mmol) was added and the reaction mixture was stirred for 30 min at room temperature. Tris(tetrabutylammonium) hydrogen pyrophosphate (66 mg, 0.12 mmol) in DMF (1.0 mL) and tributylamine (79 µL, 0.33 mmol) were added and the reaction mixture was stirred for 1 h at room temperature. Next, 0.1 M I_2 in pyridine/H_2O (98:2, *v*/*v*) was added and the reaction mixture was stirred for 30 min at 0 °C. Then, 10% $Na_2S_2O_3$ aq. was added and the reaction mixture was stirred for 10 min at room temperature. The reaction mixture was purified by ethanol precipitation. The precipitate was dissolved in 7% NH_3 aq. and the reaction mixture was stirred for 2 h at room temperature. The solvent was removed in vacuo and the residue was purified by HPLC to give compound **6d** (16 µmol, 16%). ^1H NMR (400 MHz, D_2O): δ7.79 (s, 1H), 6.01 (d, *J* = 5.4 Hz, 1H), 4.57 (t, *J* = 4.5 Hz, 1H), 4.30–4.27 (m, 4H), 3.96–3.90 (m, 1H), 1.96 (s, 3H), 1.20 (d, *J* = 6.2 Hz, 3H), 1.15 (d, *J* = 6.2 Hz, 3H). ^{31}P NMR (120 MHz, D_2O): δ − 8.71 (d, *J* = 14.4 Hz), −12.22 (d, *J* = 13.8 Hz), −22.79 (t, *J* = 14.1 Hz). HRMS (MALDI): calcd for $C_{13}H_{22}N_2O_{15}P_3$ $[M - H]^-$ 539.0239, found 539.0236.

*3.22. Synthesis of 2,2′-Anhydro-5′-O-(tert-butyldimethylsilyl)-5-methyluridine (**8**)*

Compound **7** (500 mg, 2.1 mmol) was dissolved in dry DMF (2.0 mL). Imidazole (357 mg, 5.3 mmol) and TBSCl (362 mg, 2.4 mmol) were added and the reaction mixture was stirred at room temperature for 5 h. The resulting reaction mixture was concentrated in vacuo and diluted with ethyl acetate. The combined organic layers were washed with saturated $NaHCO_3$ aq. and brine, dried over anhydrous Na_2SO_4, and concentrated in vacuo. The residue was purified by silica gel column chromatography ($CHCl_3$/MeOH = 99/1 to 95/5) to afford compound **8** (556 mg, 1.6 mmol, 76%) as a white form. ^1H NMR (400 MHz, DMSO-d_6): δ7.77 (d, *J* = 1.2 Hz, 1H), 6.28 (d, *J* = 5.7 Hz, 1H), 5.94 (d, *J* = 4.4 Hz, 1H), 5.19 (d, *J* = 7.7Hz, 1H), 4.32 (br, 1H), 4.09–4.05 (m, 1H), 3.47–3.37 (m, 2H), 1.78 (d, *J* = 1.1Hz, 3H), 0.78 (s, 9H), −0.06 (s, 3H), −0.08 (s, 3H). ^{13}C NMR (100 MHz, DMSO-d_6): δ 171.52, 159.14, 132.25, 116.90, 89.97, 88.37, 88.05, 74.22, 62.29, 25.69, 18.00, 13.50, −5.52. HRMS (MALDI): calcd for $C_{16}H_{26}N_2O_5NaSi$ $[M + Na]^+$ 377.1503, found 377.1506.

*3.23. Synthesis of 2′-O-Acetoxylethyl-3′-O-acetyl-5′-O-(tert-butyldimethylsilyl)-5-methyluridine (**9**)*

Ethylene glycol (5.0 mL) was added to 0.96 M Boran in THF (2.6 mL, 2.5 mmol). Compound **8** (400 mg. 1.0 mmol) and $NaHCO_3$ (1 mg) were added and the reaction mixture was stirred at 150 °C for 24 h. The resulting reaction mixture was diluted with $CHCl_3$. The combined organic layers were washed with water and brine, dried over anhydrous Na_2SO_4, and concentrated in vacuo. The residue was dissolved in dry pyridine (10 mL). DMAP (12 mg, 0.1 mmol) and Ac_2O (930 µL, 10 mmol) were added and the reaction mixture was stirred at room temperature for 16 h. The resulting reaction mixture was concentrated in vacuo and diluted with ethyl acetate. The combined organic layers were washed with saturated $NaHCO_3$ aq. and brine, dried over anhydrous Na_2SO_4, and concentrated in vacuo. The residue was purified by silica gel column chromatography (Hexane/Ethyl acetate = 1/1) to afford compound **9** (250 mg, 0.50 mmol, 50%) as a white form. ^{1}H NMR (400 MHz, $CDCl_3$): δ 8.85 (s, 1H), 7.48 (d, *J* = 1.0 Hz, 1H), 6.06 (d, *J* = 5.2 Hz, 1H), 5.13 (t, *J* = 4.8 Hz, 1H), 4.22–4.21 (m, 1H), 4.14 (t, *J* = 4.3 Hz, 2H), 4.08 (t, *J* = 5.4 Hz, 1H), 3.98 (dd, *J* = 1.7 Hz, 11.8 Hz, 1H), 3.82–3.70 (m, 3H), 2.14 (s, 3H), 2.03 (s, 3H),1.92 (d, *J* = 0.9 Hz, 3H), 0.94 (s, 9H), 0.13 (s, 6H). ^{13}C NMR (100 MHz, $CDCl_3$): δ 170.88, 170.25, 163.59, 150.37, 134.84, 111.30, 86.64, 82.40, 80.96, 70.22, 69.15, 63.15, 62.34, 25.98, 20.85, 20.77, 18.47, 12.60, −5.32, −5.40. HRMS (MALDI): calcd for $C_{22}H_{36}N_2O_9NaSi$ $[M + Na]^+$ 523.2082, found 523.2082.

3.24. Synthesis of 2′-O-Acetoxylethyl-3′-O-acetyl-5-methyluridine (10)

Compound **9** (250 mg, 0.50 mmol) was dissolved in dry THF (5.0 mL). 1M TBAF in THF (1.5 mL, 1.5 mmol) was added and the reaction mixture was stirred at room temperature for 30 min. The resulting reaction mixture was concentrated in vacuo. The residue was purified by silica gel column chromatography (Hexane/Ethyl acetate = 1/9) to afford compound **10** (170 mg, 0.44 mmol, 88%) as a white form. ^{1}H NMR (400 MHz, $CDCl_3$): δ 9.12 (s, 1H), 7.41 (d, *J* = 1.2 Hz, 1H), 5.62 (d, *J* = 5.5 Hz, 1H), 5.32 (dd, *J* = 4.1 Hz, 5.2 Hz, 1H), 4.49 (t, *J* = 5.4 Hz, 1H), 4.24–4.22 (m, 1H), 4.20–4.11 (m, 2H), 3.95 (dd, *J* = 1.9 Hz, 12.5 Hz, 1H), 3.81–3.73 (m, 3H), 3.25 (br, 1H), 2.15 (s, 3H), 2.03 (s, 3H), 1.92 (d, *J* = 1.1 Hz, 3H). ^{13}C NMR (100 MHz, $CDCl_3$): δ 170.90, 170.35, 163.74, 150.45, 137.94, 111.27, 91.21, 83.23, 79.28, 70.67, 69.19, 63.24, 61.78, 20.86, 20.82, 12.51. HRMS (MALDI): calcd for $C_{16}H_{22}N_2O_9Na$ $[M + Na]^+$ 409.1218, found 409.1219.

3.25. Synthesis of 2′-O-Hydroxylethyl-5-methyluridine-5′-triphospahte (11)

Compound **10** was dissolved in dry pyridine and the solution was evaporated to dryness in vacuo. Compound **10** (39 mg, 0.10 mmol) was dissolved in dry pyridine (0.5 mL) and 1,4-dioxane (0.5 mL). Then, 2-Chloro-4*H*-1,3,2-benzodioxaphosphorin-4-one (22 mg, 0.11 mmol) was added and the reaction mixture was stirred for 30 min at room temperature. Tris(tetrabutylammonium) hydrogen pyrophosphate (66 mg, 0.12 mmol) in DMF (1.0 mL) and tributylamine (133 µL, 0.56 mmol) were added and the reaction mixture was stirred for 1 h at room temperature. Next, 0.1 M I_2 in pyridine/H_2O (98:2, *v*/*v*) was added and the reaction mixture was stirred for 30 min at 0 °C. Then, 10% $Na_2S_2O_3$ aq. was added and the reaction mixture was stirred for 10 min at room temperature. The reaction mixture was purified by ethanol precipitation. The precipitate was dissolved in 7% NH_3 aq. And the reaction mixture was stirred for 2 h at room temperature. The solvent was removed in vacuo and the residue was purified by HPLC to give compound **11** (16 µmol, 16%). ^{1}H NMR (400 MHz, D_2O): δ7.80 (s, 1H), 6.08 (d, *J* = 4.7 Hz, 1H), 4.56 (d, *J* = 4.9 Hz, 1H), 4.31–4.24 (m, 4H), 3.85–3.75 (m, 4H), 1.96 (s, 3H). ^{31}P NMR (120 MHz, D_2O): δ −9.49 (d, *J* = 11.7 Hz), −11.84 (d, *J* = 14.2 Hz), −22.72 (s). HRMS (MALDI): calcd for $C_{12}H_{20}N_2O_{16}P_3$ $[M - H]^-$ 541.0031, found 541.0044.

3.26. Synthesis of 2′-O-Methoxyethyl-5-methyluridine-5′-triphospahte (13)

Compound **12** [36] was dissolved in dry pyridine and the solution was evaporated to dryness in vacuo. Compound **12** (100 mg, 0.28 mmol) was dissolved in dry pyridine (1 mL) and 1,4-dioxane (2.0 mL). Then 2-Chloro-4*H*-1,3,2-benzodioxaphosphorin-4-one (69 mg, 0.34 mmol) was added and the reaction mixture was stirred for 1 h at room temperature. Tris(tetrabutylammonium) hydrogen pyrophosphate (307 mg, 0.56 mmol) in DMF (1.0 mL) and tributylamine (133 μL, 0.56 mmol) were added and the reaction mixture was stirred for 1.5 h at room temperature. Next, 0.1 M I_2 in pyridine/H_2O (98:2, v/v) was added and the reaction mixture was stirred for 1 h at room temperature. Then, 10% $Na_2S_2O_3$ aq. was added and the reaction mixture was stirred for 10 min at room temperature. The reaction mixture was purified by ethanol precipitation. The precipitate was dissolved in 7% NH_3 aq. and the reaction mixture was stirred for 1 h at room temperature. The solvent was removed in vacuo and the residue was purified by HPLC to give compound **13** (44 μmol, 16%). ^{1}H NMR (400 MHz, D_2O): δ7.79 (s, 1H), 6.06 (d, J = 5.0 Hz, 1H), 4.55 (t, J = 4.8 Hz, 1H), 4.32–4.23 (m, 4H), 3.87–3.85 (m, 2H), 3.66–3.64 (m, 2H), 3.37 (s, 3H), 1.96 (s, 3H). ^{31}P NMR (120 MHz, D_2O): δ −9.12(d, J = 13.7 Hz), −11.96 (d, J = 14.5 Hz), −22.69 (t, J = 14.0 Hz). HRMS (MALDI): calcd for $C_{13}H_{22}N_2O_{15}P_3$ $[M + Na]^+$ 579.0153, found 579.0144.

3.27. Expression and Purification of KOD DGLNK

KOD DGLNK was expressed and purified according to a previous report [23].

3.28. Polymerase Incorporation of 2′-O-Alkyl-5-methyluridine Triphosphates into Oligonucleotide

The reaction mixture [1 × KOD Dash® buffer (TOYOBO, Osaka, Japan), DNA or 2′-OMe primer (Primer**1** or Primer**2**: 0.4 μM), DNA template (Template**1**: 0.5 μM), dTTP or 2′-*O*-alkyl-^{5m}UTPs (0.2 mM), and $MnSO_4$ (0 or 0.1 or 0.5 or 1.0 mM)] was prepared. Then, the reaction mixture was denatured at 94 °C for 1 min and annealed at 25 °C for 1 h (1.2 °C/min), KOD DGLNK (50 or 150 ng/μL) was added and incubated at 74 °C for 30 or 60 min. The reaction was stopped by stop buffer (aqueous 3 mM EDTA containing 0.1% bromophenol blue and aqueous 7 M Urea) and heated for 5 min at 95 °C. The reaction was analyzed by using a 20% denaturing urea-PAGE (70 min, 55 °C).

3.29. Enzymatic Synthesis of 2′-O-Alkyl-modified Oligonucleotide Libraries

The reaction mixture [1× KOD Dash® buffer (TOYOBO, Osaka, Japan), 2′-OMe primer (Primer**3**: 0.4 μM), DNA template (Template**2**: 0.5 μM), dTTP or 2′-*O*-alkyl-^{5m}UTPs (0.2 mM), 2′-OMe-ATP (0.2 mM), 2′-OMe-GTP (0.2 mM), 2′-OMe-CTP (0.2 mM), $MnSO_4$ (1.0 or 3.0 or 5.0 mM) and KOD DGLNK (300 ng/μL)] was prepared. Then, the reaction mixture was denatured at 94 °C for 1 min and annealed at 25 °C for 1 h (1.2 °C/min), KOD DGLNK was added and incubated from 60 °C for 1 h, 60 °C to 74 °C for 4 or 8 h, and 74 °C for 1 h. After polymerization, DNase1 (95 mU/μL) was added and incubated at 37 °C for 30 min. The reaction was stopped by stop buffer (aqueous 3 mM EDTA containing 0.1% bromophenol blue and aqueous 7 M Urea) and heated for 5 min at 95 °C. The reaction was analyzed by using a 10% denaturing urea-PAGE (32 min, 55 °C).

3.30. Preparation of 2′-O-ALKYL-modified Oligonucleotide Libraries for Analysis of Properties

The reaction mixture [1× KOD Dash® buffer (TOYOBO, Osaka, Japan), 2′-OMe primer (Primer**3**: 0.4 μM), DNA template (Template**2**: 0.5 μM), 2′-*O*-alkyl-^{5m}UTPs (0.2 mM), 2′-OMe-ATP (0.2 mM), 2′-OMe-GTP (0.2 mM), 2′-OMe-CTP (0.2 mM), $MnSO_4$ (1.0 or 3.0 mM) and KOD DGLNK (300 ng/μL)] was prepared. Then, the reaction mixture was denatured at 94 °C for 1 min and annealed at 25 °C for 1 h (1.2 °C/min), KOD DGLNK (300 ng/μL) was added and incubated from 60 °C for 1 h, 60 °C to 74 °C for 4 or 8 h, and 74 °C for 1 h. After polymerization, DNase1 (95 mU/μL) was added and incubated at 37 °C for 30 min. Then, Proteinase K (36 mU/μL) was added and incubated at 37 °C for 30 min. The reaction was stopped by stop buffer (0.1% bromophenol blue and aqueous 7 M Urea) and heated for 5 min at 95 °C. The oligonucleotides were purified by using a 10% denaturing urea-PAGE (120 min, 55 °C).

3.31. Enzymatic Stability of Oligonucleotide Libraries

The reaction mixture [oligonucleotide libraries (0.1 μM) and 50% FBS] was prepared. Then, the reaction mixture was incubated at 37 °C. After incubation, the reaction mixture was heated at 94 °C for 5 min. Then, Proteinase K (36 mU/μL) was added and incubated at

37 °C for 30 min. The reaction was stopped by stop buffer (aqueous 3 mM EDTA containing 0.1% bromophenol blue and aqueous 7 M Urea). After adding Cy5-labeled DNA (83 nM) as an internal standard to the reaction mixtures, they were heated for 5 min at 95 °C. The reaction was analyzed by using a 10% denaturing urea-PAGE (30 min, 55 °C).

3.32. *Ion Pair RP-HPLC Analysis of Oligonucleotide Libraries*

Oligonucleotide libraries (5.0 pmol) were injected onto a XBridge Oligonucleotide BEH C18 column. The column temperature was 50 °C, with a flow rate of 1.0 mL/min, and detection at 260 nm. Mobile phase A consisted of 100 mM triethylamine acetic acid solution, pH 6.9, and mobile phase B was acetonitrile. The column was initially maintained at 10% mobile phase B, and then at a gradient of 10% to 40% over 15 min.

3.33. *Ion Pair RP-HPLC Analysis of Nucleotide Triphosphates*

Nucleotide triphosphates (700 pmol) were injected onto a XBridge Oligonucleotide BEH C18 column. The column temperature was 50 °C, with a flow rate of 1.0 mL/min, and detection at 260 nm. Mobile phase A consisted of 100 mM triethylamine acetic acid solution (pH 6.9), and mobile phase B was acetonitrile. The column was initially maintained at 5% mobile phase B, and then at a gradient of 5% to 20% over 15 min.

4. Conclusions

In conclusion, 2′-OEt, 2′-OPr, 2′-OBu, 2′-O^{i}Pr, and 2′-HE-^{5m}UTPs were newly synthesized and incorporated them into oligonucleotides by KOD DGLNK. By mixing the 2′-OMe modification, oligonucleotide libraries were enzymatically synthesized. These 2′-*O*-alkyl-modified oligonucleotides showed various hydrophobicities. Furthermore, these 2′-*O*-alkyl-modified oligonucleotide libraries were more stable than unmodified DNA. Our results can accelerate the development of aptamers or nucleic acid enzymes for therapeutic use.

Supplementary Materials: The following supporting information can be downloaded at: https://www.mdpi.com/article/10.3390/molecules28237911/s1, Table S1: Sequences of the primers and templates that were used in this study; Scheme S1: Synthesis of 2′-OEt-^{5m}UTP (**6a**), 2′-OPr-^{5m}UTP (**6b**), 2′-OBu-^{5m}UTP (**6c**), and 2′-O^{i}Pr-^{5m}UTP (**6d**); Scheme S2: Synthesis of 2′-HE-^{5m}UTP (**11**); Scheme S3: Synthesis of 2′-MOE-^{5m}UTP (**13**); Figure S1: Polymerase incorporation of 2′-*O*-alkyl-^{5m}UTPs with a DNA primer; Figure S2: Polymerase incorporation of 2′-*O*-alkyl-^{5m}UTPs with DNA primer at different Mn^{2+} concentrations; Figure S3: Polymerase incorporation of 2′-*O*-alkyl-^{5m}UTPs with a 2′-OMe-modified primer at different Mn^{2+} concentrations; Figure S4: Enzymatic synthesis of ON_^{i}Pr, ON_HE, and ON_MOE at different Mn^{2+} concentrations and incubation times; Figure S5: Ion pair RP-HPLC analysis of dTTP and 2′-*O*-alkyl-^{5m}UTPs; Figures S5–S29: NMR spectra of the all prepared compounds.

Author Contributions: Conceptualization, Y.K. and S.O.; investigation, K.I.; resources, H.H.; data curation, K.I.; writing—original draft preparation, K.I.; writing—review and editing, K.I., Y.K., H.H., T.O. and S.O.; visualization, K.I.; supervision, Y.K. and S.O.; project administration, Y.K. and S.O.; funding acquisition, K.I., Y.K. and H.H. All authors have read and agreed to the published version of the manuscript.

Funding: This research was funded by AMED under Grant Number JP18ak0101102, JP22gm1610010, JP223fa727001, and JP23ak0101207; JSPS KAKENHI under Grant Number 19K15702, 21K14743; and JST SPRING under Grant Number JPMJSP2138.

Institutional Review Board Statement: Not applicable.

Informed Consent Statement: Not applicable.

Data Availability Statement: Data are contained within the article and supplementary materials.

Conflicts of Interest: The authors declare no conflict of interest.

References

1. Zhou, J.; Rossi, J. Aptamers as targeted therapeutics: Current potential and challenges. *Nat. Rev. Drug Discov.* **2017**, *16*, 181–202. [CrossRef]
2. Larcher, L.M.; Pitout, I.L.; Keegan, N.P.; Veedu, R.N.; Fletcher, S. DNAzymes: Expanding the Potential of Nucleic Acid Therapeutics. *Nucleic Acid Ther.* **2023**, *33*, 178–192. [CrossRef] [PubMed]
3. Tuerk, C.; Gold, L. Systematic evolution of ligands by exponential enrichment: RNA ligands to bacteriophage T4 DNA polymerase. *Science* **1990**, *249*, 505–510. [CrossRef] [PubMed]
4. Ellington, A.D.; Szostak, J.W. In vitro selection of rna molecules that bind specific ligands. *Nature* **1990**, *346*, 818–822. [CrossRef] [PubMed]
5. Vaught, J.D.; Bock, C.; Carter, J.; Fitzwater, T.; Otis, M.; Schneider, D.; Ronaldo, J.; Waugh, S.; Wilcox, S.K.; Eaton, B.E. Expanding the chemistry of DNA for in vitro selection. *J. Am. Chem. Soc.* **2010**, *132*, 4141–4151. [CrossRef] [PubMed]
6. Gold, L.; Ayers, D.; Bertino, J.; Bock, C.; Bock, A.; Brody, E.N.; Carter, J.; Dalby, A.B.; Eaton, B.E.; Fitzwater, T.; et al. Aptamer-based multiplexed proteomic technology for biomarker discovery. *PLoS ONE* **2010**, *5*, e15004. [CrossRef] [PubMed]
7. Davis, D.R.; Gelinas, A.D.; Zhang, C.; Rohloff, J.C.; Carter, J.D.; O'Connell, D.; Waugh, S.M.; Wolk, S.K.; Mayfield, W.S.; Burgin, A.B.; et al. Unique motifs and hydrophobic interactions shape the binding of modified DNA ligands to protein targets. *Proc. Natl. Acad. Sci. USA* **2012**, *109*, 19971–19976. [CrossRef]
8. Uemachi, H.; Kasahara, Y.; Tanaka, K.; Okuda, T.; Yoneda, Y.; Obika, S. Hybrid-Type SELEX for the selection of artificial nucleic acid aptamers exhibiting cell internalization activity. *Pharmaceutics* **2021**, *13*, 888. [CrossRef]
9. Kasahara, Y.; Irisawa, Y.; Fujita, H.; Yahara, A.; Ozaka, H.; Obika, S.; Kuwahara, M. Capillary electrophoresis–systematic evolution of ligands by exponential enrichment selection of base- and sugar-modified DNA aptamers: Target binding dominated by 2′-*O*,4′-*C*-methylene-bridged/locked nucleic acid primer. *Anal. Chem.* **2013**, *85*, 4961–4967. [CrossRef]
10. Inoue, H.; Hayase, Y.; Imura, A.; Iwai, S.; Miura, K.; Ohtsuka, E. Synthesis and hybridization studies on two complementary nona(2′-*O*-methyl)ribonucleotides. *Nucleic Acids Res.* **1987**, *15*, 6131–6148. [CrossRef]
11. Egli, M.; Mahoharan, M. Chemistry, structure and function of approved oligonucleotide therapeutics. *Nucleic Acids Res.* **2023**, *51*, 2529–2573. [CrossRef]
12. Martin, P. Ein neuer Zugang zu 2′-O-Alkylribonucleosiden und Eigenschaften deren Oligonucleotide. *Helv. Chim. Acta* **1995**, *78*, 486–504. [CrossRef]
13. Cummins, L.L.; Owens, S.R.; Risen, L.M.; Lesnik, E.A.; Freier, S.M.; McGee, D.; Guinosso, C.J.; Cook, P.D. Characterization of fully 2′-modified oligoribonucleotide hetero- and homoduplex hybridization and nuclease sensitivity. *Nucleic Acids Res.* **1995**, *23*, 2019–2024. [CrossRef] [PubMed]
14. Freier, S.M.; Altmann, K.H. The ups and downs of nucleic acid duplex stability: Structure-stability studies on chemically-modified DNA:RNA duplexes. *Nucleic Acids Res.* **1997**, *25*, 4429–4443. [CrossRef] [PubMed]
15. Lesnik, E.A.; Guinosso, C.J.; Kawasaki, A.M.; Sasmor, H.; Zounes, M.; Cummins, L.L.; Ecker, D.J.; Cook, P.D.; Freier, S.M. Oligodeoxynucleotides containing 2′-*O*-modified adenosine: Synthesis and effects on stability of DNA:RNA duplexes. *Biochemistry* **1993**, *32*, 7832–7838. [CrossRef]
16. Egli, M.; Minasov, G.; Tereshko, V.; Pallan, P.S.; Teplova, M.; Inamati, G.B.; Lesnik, E.A.; Owens, S.R.; Ross, B.S.; Prakash, T.P.; et al. Probing the influence of stereoelectronic effects on the biophysical properties of oligonucleotides: Comprehensive analysis of the RNA affinity, nuclease resistance, and crystal structure of ten 2′-*O*-ribonucleic acid modifications. *Biochemistry* **2005**, *44*, 9045–9057. [CrossRef] [PubMed]
17. Altmann, K.H.; Dean, N.M.; Fabbro, D.; Freier, S.M.; Geiger, T.; Häner, R.; Hüsken, D.; Martina, P.; Monia, B.P.; Müller, M.; et al. Second Generation of Antisense Oligonucleotides: From Nuclease Resistance to Biological Efficacy in Animals. *Chimia* **1996**, *50*, 168–176. [CrossRef]
18. Veedu, R.N.; Vester, B.; Wengel, J. Enzymatic incorporation of LNA nucleotides into DNA strands. *Chembiochem* **2007**, *8*, 490–492. [CrossRef]
19. Masaki, Y.; Ito, H.; Oda, Y.; Yamazaki, K.; Tago, N.; Ohno, K.; Ishii, N.; Tsunoda, H.; Kanamori, T.; Ohkubo, A.; et al. Enzymatic synthesis and reverse transcription of RNAs incorporating 2′-*O*-carbamoyl uridine triphosphate. *Chem. Commun.* **2016**, *52*, 12889–12892. [CrossRef]
20. Diafa, S.; Evéquoz, D.; Leumann, C.J.; Hollenstein, M. Enzymatic Synthesis of 7′,5′-Bicyclo-DNA Oligonucleotides. *Chem Asian J.* **2017**, *12*, 1347–1352. [CrossRef]
21. Chen, T.; Hongdilokkul, N.; Liu, Z.; Adhikary, R.; Tsuen, S.S.; Romesberg, F.E. Evolution of thermophilic DNA polymerases for the recognition and amplification of C2′-modified DNA. *Nat. Chem.* **2016**, *8*, 556–562. [CrossRef] [PubMed]
22. Freund, N.; Taylor, A.I.; Arangundy-Franklin, S.; Subramanian, N.; Peak-Chew, S.Y.; Whitaker, A.M.; Freudenthal, B.D.; Abramov, M.; Herdewijn, P.; Holliger, P. A two-residue nascent-strand steric gate controls synthesis of 2′-*O*-methyl- and 2′-*O*-methoxyethyl)-RNA. *Nat. Chem.* **2023**, *15*, 91–100. [CrossRef] [PubMed]
23. Hoshino, H.; Kasahara, Y.; Kuwahara, M.; Obika, S. DNA polymerase variants with high processivity and accuracy for encording and decording locked nucleic acid sequences. *J. Am. Chem. Soc.* **2020**, *142*, 21530–21537. [CrossRef] [PubMed]
24. Nishioka, M.; Mizuguchi, H.; Fujiwara, S.; Komatsubara, S.; Kitabayashi, M.; Uemura, H.; Takagi, M.; Imanaka, T. Long and accurate PCR with a mixture of KOD DNA polymerase and its exonuclease deficient mutant enzyme. *J. Biotechnol.* **2001**, *88*, 141–149. [CrossRef]

25. Cozens, C.; Pinheiro, V.B.; Vaisman, A.; Woodgate, R.; Holliger, P. A short adaptive path from DNA to RNA polymerases. *Proc. Natl. Acad. Sci. USA* **2012**, *109*, 8067–8072. [CrossRef] [PubMed]
26. Gardner, A.F.; Jack, W.E. Acyclic and dideoxy terminator preferences denote divergent sugar recognition by archaeon and Taq DNA polymerases. *Nucleic Acids Res.* **2002**, *30*, 605–613. [CrossRef] [PubMed]
27. Pinheiro, V.B.; Taylor, A.I.; Cozens, C.; Abramov, M.; Renders, M.; Zhang, S.; Chaput, J.C.; Wengel, J.; Peak-Chew, S.Y.; McLaughlin, S.H.; et al. Synthetic genetic polymers capable of heredity and evolution. *Science* **2012**, *336*, 341–344. [CrossRef]
28. Roy, S.K.; Tang, J. Efficient Large Scale Synthesis of 2′-*O*-Alkyl Pyrimidine Ribonucleosides. *Org. Proc. Res. Dev.* **2000**, *4*, 170–171. [CrossRef]
29. Matsuda, A.; Yasuoka, J.; Sasaki, T.; Ueda, T. Nucleosides and Nucleotides. 95. Improved synthesis of 1-(2-azido-2-deoxy-β-*D*-arabinofuranosyl)cytosine(Cytarazid) and -thymine. Inhibitory spectrum of Cytarazid of the growth of various human tumor cells in vitro. *J. Med. Chem.* **1991**, *34*, 999–1002. [CrossRef]
30. Ludwig, J.; Eckstein, F. Rapid and efficient synthesis of nucleotide 5′-*O*-(1-thiotriphosphates), 5′-triphosphates and 2′,3′-cyclophosphorothioates using 2-chloro-4*H*-1,3,2-benzodioxaphosphorin-4-one. *J. Org. Chem.* **1989**, *54*, 631–635. [CrossRef]
31. Dunn, M.R.; Otto, C.; Fenton, K.E.; Chaput, J.C. Improving polymerase activity with unnatural substrates by sampling mutations in homologous protein architectures. *ACS Chem. Biol.* **2016**, *11*, 1210–1219. [CrossRef] [PubMed]
32. Ng, E.W.M.; Shima, D.T.; Calias, P.; Cunningham, E.T., Jr.; Guyer, D.R.; Adamis, A.P. Pegaptanib, atarget anti-VEGF aptamer for ocular vascular disease. *Nat. Rev. Drug Discov.* **2006**, *5*, 123–132. [CrossRef]
33. Biesecker, G.; Dihel, L.; Enney, K.; Bendele, R.A. Derivation of RNA aptamer inhibitors of human complement C5. *Immunopharmacology* **1999**, *42*, 219–230. [CrossRef] [PubMed]
34. Jaffe, G.J.; Westby, K.; Csaky, K.G.; Mones, J.; Pearlman, J.A.; Patel, S.S.; Joondeph, B.C.; Randolph, J.; Masonson, H.; Rezaei, K.A. C5 inhibitor avacincaptad pegol for geographic atrophy due to age-related macular degeneration: A randomized pivotal phase 2/3 trial. *Ophthalmology* **2021**, *128*, 576–586. [CrossRef] [PubMed]
35. Beckman, R.A.; Mildvan, A.S.; Loeb, L.A. On the fidelity of DNA replication: Manganese mutagenesis in vitro. *Biochemistry* **1985**, *24*, 5810–5817. [CrossRef] [PubMed]
36. Rajeev, K.G.; Prakash, T.P.; Manoharan, M. 2′-Modified-2-thiothymidine oligonucleotides. *Org. Lett.* **2003**, *5*, 3005–3008. [CrossRef]

MDPI
St. Alban-Anlage 66
4052 Basel
Switzerland
www.mdpi.com

Molecules Editorial Office
E-mail: molecules@mdpi.com
www.mdpi.com/journal/molecules

www.ingramcontent.com/pod-product-compliance
Lightning Source LLC
Chambersburg PA
CBHW062035220926
51850CB00010B/2556
* 9 7 8 3 7 2 5 8 1 3 7 6 6 *